PATHWAYS TO INDUSTRIALIZATION AND REGIONAL DEVELOPMENT

PATHWAYS TO INDUSTRIALIZATION AND REGIONAL DEVELOPMENT

Edited by Michael Storper and Allen J. Scott

London and New York

First published 1992
by Routledge
11 New Fetter Lane, London EC4P 4EE

Simultaneously published in the USA and Canada
by Routledge
a division of Routledge, Chapman and Hall, Inc.
29 West 35th Street, New York, NY 10001

Typeset in Garamond by
J&L Composition Ltd, Filey, North Yorkshire
Printed and bound in Great Britain by
Mackays of Chatham plc, Chatham, Kent

British Library Cataloguing in Publication Data

A catalogue record of this book is available from the British Library.
ISBN 0-415-08752-X

Library of Congress Cataloging in Publication Data

Pathways to industrialization and regional development / edited by
Michael Storper and Allen J. Scott.
p. cm.
Includes bibliographical references and index.
ISBN 0-415-08752-X
1. Industrialization—Congresses. 2. Regional planning—Congresses. 3. Industry and state—Congresses. I. Storper, Michael. II. Scott, Allen John.
HD2329.P38 1992
338.9—dc20
92-10397
CIP

CONTENTS

FIGURES

TABLES

CONTRIBUTORS

Elmar Altvater is Professor of Political Economy, Department of Political Science, Free University of Berlin.

Patrizio Bianchi is Professor of Applied Economics at the Universities of Udine and Bologna, and Director of the Laboratorio di Politica Industriale, NOMISMA, Bologna.

Benjamin Coriat is Professor of Economics and Director of the Centre de Recherche en Economie Industrielle (CREI) at the University of Paris XIII.

Bruno Courault is a researcher at the Centre d'Etudes de l'Emploi, Paris, and teaches at the University of Paris X – Nanterre.

Giovanni Dosi is Professor of Applied Economics at the University of Rome, La Sapienza.

Bernard Ganne is a researcher at the Groupe Lyonnais de Sociologie Industrielle, Maison Rhône-Alpes des Sciences de l'Homme, Lyon.

Paul Hirst is Professor of Social Theory at Birkbeck College University of London.

Makoto Itoh is Professor of Economics, University of Tokyo.

Bob Jessop is Professor of Sociology, University of Lancaster, United Kingdom.

Danièle Leborgne is a researcher at the Centre d'Etudes Prospectives d'Economie Appliquée (CEPREMAP), Paris.

Alain Lipietz is Director of Research at the Centre d'Etudes Prospectives d'Economie Appliquée (CEPREMAP), Paris.

Edward H. Lorenz is in the Department of Economics and is a Fellow of the Kellogg Institute for International Studies, University of Notre Dame.

CONTRIBUTORS

Michael Piore is Professor of Economics at the Massachusetts Institute of Technology.

Claudine Romani is Deputy Officer for European Affairs at the Centre d'Etudes et de Recherches sur les Qualifications (CEREQ), Paris.

Jean Saglio is a researcher with the Groupe Lyonnais de Sociologie Industrielle, Maison Rhône-Alpes des Sciences de L'Homme, Lyon.

Robert Salais is Director of the Research Group Institutions, Emploi et Politique Economique (IEPE) at the CNRS, and Administrator at INSEE, Paris.

Roberta Salavatore is a researcher at Bonifica s.p.a., Rome.

AnnaLee Saxenian is Assistant Professor of City and Regional Planning at the University of California, Berkeley.

Allen J. Scott is Professor of Geography and Director of the Lewis Center for Regional Policy Studies at the University of California, Los Angeles.

Guy Standing is labor market research coordinator at the International Labour Office, Geneva.

Michael Storper is Professor of Planning, University of California, Los Angeles.

Jonathan Zeitlin is Associate Professor of History and Industrial Relations at the University of Wisconsin – Madison.

PREFACE

This book represents the main proceedings of a conference that was held at Lake Arrowhead, California, in March 1990. The conference brought together some eighty participants from all over the world to discuss theoretical, empirical, and policy problems generated by the current perplexing turn in actual patterns and processes of industrialization and regional development. This turn is often identified – somewhat ambiguously – as a shift in socioeconomic relations to a "post-Fordist" configuration. The present book is an attempt to construct a theoretical synthesis around this socioeconomic configuration (in its various national and regional variations) involving a combination of concepts dealing with social regulation of the economic order, flexible production organization, and industrial district formation. We believe that this synthesis is coherent and sustainable and is likely to set the terms of the debate on issues of industrialization and regional development for a number of years to come. We offer the book in the hope that it will not only stimulate further creative scientific work on these issues, but that it will also provide a much-needed antidote to the neoconservative point of view, which dominates current policymaking in many of the major capitalist countries at the present time, and which, we believe, cannot substitute for a reasoned and theoretically informed analysis of social and economic realities.

The efforts of many individuals went into the organization of the Lake Arrowhead conference and into the production of this book. We should like to thank Dean Richard Weinstein and Associate Dean Edward Soja, of the Graduate School of Architecture and Urban Planning at UCLA for their support for the project from its inception, as well as Dan Mitchell, Director of the Institute of Industrial Relations at the same university. Mr. Alain d'Iribarne of the PIRTTEM program at the CNRS in Paris also graciously accorded us support, and Mr. Robert Salais of CNRS/IEPE took over responsibility for administration of the French side of the endeavor. At UCLA, the tireless efforts of Vanessa Dingley in organizing the conference, and the diligence of Jane Pollard and Jan Lawrence were indispensable. The staffs of the Urban Planning Program and the Department of Geography at UCLA were always ready to help, and we thank them all as well.

Michael Storper and Allen J. Scott

ACKNOWLEDGMENTS

The editors, authors and publisher wish to acknowledge the generous financial support of: The Lewis Center for Regional Policy Studies, UCLA, Los Angeles; Le Centre National de la Recherche Scientifique, Paris; and The Institute of Industrial Relations, UCLA, Los Angeles.

Part I
INTRODUCTION

1

INDUSTRIALIZATION AND REGIONAL DEVELOPMENT[1]

Allen J. Scott and Michael Storper

THEORIES OF INDUSTRIALIZATION AND REGIONAL DEVELOPMENT: REGULATION, INSTITUTIONS, AND EVOLUTION

From the 1920s down to the 1960s, across the advanced capitalist world, many leading industrial sectors were converted to mass production methods. Over this same period, various institutions and practices were constructed to regulate the social and economic effects of mass production. These regulatory apparatuses consisted of rather similar basic elements throughout the advanced capitalist world. One such element was the strongly oligopolistic structure of industry, with its large-scale markets and rising barriers to entry. Labor relations and labor markets were everywhere ordered by institutions of collective bargaining, with a strong functional split between white-collar workers (responsible for industrial planning) and blue-collar workers (responsible for routine manual labor). Keynesian welfare-statist legislation helped to smooth out aggregate demand and to secure high levels of social stability. In regional terms, this form of industrialization was seen by theorists of the day as a system of core–periphery relationships, within which core regions tended to grow at the expense of underdeveloped peripheral regions which in turn provided raw materials or agricultural products for the core at unfavorable terms of trade.

By the late 1970s, the processes of industrialization and regional development based on this "Fordist" mass production system in the United States and Europe were in crisis, most importantly because of cheap foreign imports and the increasing difficulty of achieving high levels of productivity gain and the limits this imposed on the ability of the system to keep wages – hence consumption – moving upwards (see chapters by Altvater and Itoh). The mass production system began to undergo major restructuring as producers searched for new models of industrial technology, organization, labor relations, and location (Coriat, Standing). In the core regions of mass

[1] In this introductory chapter, references to subsequent chapters are designated by the names of the author or authors in parentheses.

production, the effects of the concomitant plant closures and layoffs were aggravated by the intensified efforts of producers to respond to the crisis by decentralizing branch plants to low-cost peripheral regions. At the same time, after the 1960s, a series of new industrial regions based on flexible production arose outside of the heartlands of mass production.

Much of the theoretical reasoning of postwar social science portrayed the mass production system as the destiny of industrial society (Jessop). Its technologies, markets, institutions, politics, and geographical forms were seen as expressions of irreversible and univeral tendencies of modernization. Very few social scientists in the postwar period foresaw the processes of deindustrialization and decline which were dramatically evident by the 1970s in virtually all the major manufacturing regions of the United States and Western Europe, along with accompanying changes in politics, technologies, and markets. Even Marxian social theory, with its stress on the crisis tendencies of capitalist economies, proved largely inadequate to the task of understanding the historical changes that did take place (Jessop, Hirst and Zeitlin).

Social scientists in a variety of disciplines have now spent nearly two decades coming to grips with the effects of the breakdown of the mass production system, including the new processes of competition and global trade, profound redefinitions of occupational structures, the reshaping of labor markets and labor politics, and the rise of a series of new core regions. This book is evidence of the coalescence of an interdisciplinary project in the social sciences which is now generating powerful accounts of the decline of the Fordist mass production system, and the rise of a new flexible production order. More importantly, this book is in large degree focused on a new set of theoretical tools involving three interdependent dimensions of industrialization and regional development, i.e.:

a) their necessary institutional contexts,
b) their evolutionary dynamics, and
c) their geographical foundations and territorial specificity.

In particular, the book brings together work which draws on regulationist theory, institutionalist and evolutionary economics, and the new economic geography.

Regulationist theory provides us with a view of economic history as a chain of distinctive periods, together with a concept of the multilayered and political character of capitalist accumulation. These periods are defined by specific economic practices consisting of dominant sets of production relations (embodied in ensembles of leading industries), complemented by different political and quasi-political arrangements which coordinate the economy. Such practices are institutionalized when an effective coalition, or historical bloc is constructed which is capable of putting together and maintaining a specific set of social compromises. Regulationist theory identifies the specificity of the mass production system stretching from about the 1920s to the late 1960s as the marriage of a technological paradigm of production with Keynesian and Beveridgean institutions. Likewise, regulationist theory permits us now to pose the question of whether a new period of capitalist development based on the ascendance of the paradigm of flexible production and new forms of innovation-based competition is coming into existence; it also permits us to ask how production systems, institutions and markets are now different from those of the mass production period (Jessop, Altvater, Coriat, Leborgne and Lipietz, Dosi and Salvatore).

Institutionalist economics yields insights about the organizational logic of production. It has reopened the question of the nature of the firm and shed light on the many different possible articulations of markets and hierarchies that can efficiently coordinate production systems and labor markets in industrial capitalism. In so doing, institutionalist economics has replaced the view that there is always one single "best practice" for a given production activity, and shown that markets, technologies, and production organization are mutually interdependent systems embedded in relatively durable institutional contexts (Dosi and Salvatore). As a consequence, this body of theory indicates that sociopolitical organizations, far from constituting barriers to the operation of the market, are essential underpinnings of any efficient capitalist production system. The question is now what kinds of institutions will be built in the 1990s and how they will influence the organization of production, labor markets, and product markets in the early decades of the coming century (Hirst and Zeitlin, Saglio, Salais). In any case, empirical analyses of the advent of flexible production are proceeding with a sensitivity to the roles that institutions play in industrialization and regional development, and to the wide variety of institutional forms that may exist (Ganne, Saglio, Saxenian, Standing, Leborgne and Lipietz, Courault and Romani).

Evolutionary economics conceives of the development of technologies, markets, and institutions as pathways whose historical trajectory is governed by the complex interplay between prevailing rules of social order, and the probing and experimental character of much economic behavior in the context of prior states of the system (see particularly Dosi and Salvatore). The constitution of economic order is not a static equilibrium, but a constantly unfolding structure with many possible branching points; different pathways may be selected by the interactions among small events or by the regulatory effects of institutional order (Piore).

Economic geography deals with the territorial structures that flow from and sustain processes of industrialization and their regulation. It is through economic geography that the concrete synthesis of these other processes is realized as the question of regional development. Thus, each period of capitalist history tends to be marked by its own peculiar economic geography; just as the Fordist mass production era was rooted in a distinctive set of core regions, so with the rise of modern flexible production in the 1960s and 1970s, some sectors broke away from the old core regions and established themselves in new locational domains. In these domains, processes of institution-building – and thus of the eventual construction of a regulatory mode for the flexible production economy – are now occurring. In the same places, new kinds of technologies and labor relations which are critical to the new strategic competition and the evolutionary trajectory of production are being generated. It is thus here that the pathway of development of the contemporary period is, at least in part, being defined (Ganne, Saglio, Courault and Romani, Lorenz).

PERIODS OF CAPITALIST DEVELOPMENT AND THEIR GEOGRAPHICAL ANATOMY

Capitalism can be seen, in its most general form, as an arrangement in which commodity producers combine the physical means of production and labor in order to bring forth sellable outputs which generate profits. This general form, however, varies in its specific social configuration from time to time and from place to place; it emerges in practice as a series of historically and geographically specific *technological-institutional systems*. The latter involve:

a) an evolving technological and organizational structure of production;
b) labor markets and industrial relations, including industrial politics and mechanisms of the socialization of labor;
c) managerial cultures and norms;
d) market structures and forms of competition;
e) regulatory institutions at sectoral, regional, national, and international levels.

It is possible to identify relatively distinct periods of capitalist development in terms of dominant technological-institutional systems as they are embodied in ensembles of leading industries and complemented by different political and quasi-political arrangements which steer and coordinate the economy. In this sense, technological-institutional systems are much like what regulationist theory refers to as regimes of accumulation and modes of social regulation (Jessop, Hirst and Zeitlin, Itoh). Such episodes as the putting-out period of early capitalism, the classical mill-and-workshop economy of the mid-nineteenth century, the Fordist mass production era, and the currently emerging system of flexible production, all express distinctive historical and geographical tendencies of particular technological-institutional systems.

Even so, the detailed internal configurations of any given system – in matters of technology, industrial organization, labor markets, products, regulatory institutions, and so on – will often vary over space and time. Thus, the Fordist mass production system showed considerable variation between the different leading capitalist economies, as did corresponding forms of Keynesian welfare-statist regulation. Above all, the employment relation and its legal underpinnings were marked by the peculiarities of each national case, and this has had important repercussions for the ways in which the crisis of Fordism has been played out across North America and Western Europe.

If technological-institutional systems retain a sort of structural integrity over more or less long periods of time, they are nonetheless always susceptible to crisis and dissolution. Such a turn of events may occur for a variety of reasons, whether internal (saturation of markets, declining profitability, institutional failures, and so on), or external (intensified foreign competition, resource depletion, and so on), or a combination of both. At such times, the preexisting system of production relations and regulatory institutions can no longer absorb the tensions creating the crisis without significant restructuring. In the case of the recent crisis of Fordism, this was manifest in a prolonged drop of profitability in core mass production sectors, rising unemployment, the weakening of organized labor, and neoconservative onslaughts against the institutions of the Keynesian welfare state.

The transition from one system to another – which may be abrupt or prolonged over time – will often be associated with and brought about by a reconstitution of the geography of production, especially in cases where the ascending leading sectors of the new system are free of any dependence on the immobile resources basic to the old. Every shift of technological-institutional system hence opens up a period of restructuring and the possibility of a radical switch in the spatial bases of production from one set of core regions to another. In the new regions thus established, patterns of production and accumulation will proceed in markedly different ways from those that typified the old system.

The typical organizational and locational model of mass production industry was the growth center focused on vertically integrated lead plants with many backward linkages. On this basis, a small number of core mass production regions emerged in the mid-twentieth century, such as the upper Midwest in the United States (and nearby areas of Ontario in Canada), and the area stretching from the English Midlands through northern France, Belgium and Holland to the Rhine-Ruhr area of Europe, with outliers in northwest Italy and southern Sweden. The system of mass production was also associated with large-scale urbanization, due to the massing of the work force of the lead plants and their dependent satellites.

By contrast with mass production, flexible production systems are characterized by progressive vertical disintegration of production with numerous producers (of different sizes) caught up in tightly knit network structures (Piore). In these networks, groups of industrial establishments with especially dense interrelations tend to locate close to one another to facilitate exchanges of goods and information, and to take advantage of external economies in labor markets and infrastructure. The emerging economy of flexible production has brought into existence a series of new core production regions, and these are typically quite different from those of the mass production system. Three groups of contemporary regions of flexible production can be recognized.

First, craft-based, design-intensive industries such as clothing, textiles, furniture, jewelry, ceramics, sporting goods, as well as foci of precision metalworking and machine-building, may be found in two main types of location. One coincides with inner city areas in large metropolitan regions such as New York, Paris, Los Angeles, and London, with their large immigrant populations. The other coincides with old centers of craft production, as in Emilia-Romagna and Tuscany in Italy, parts of the Rhone-Alps and Mediterranean regions of France, and certain parts of southern Germany and Scandinavia (Ganne, Saglio, Courault and Romani, Bianchi).

Second, high technology industry has tended to locate above all at selected suburban locations close to major cities and in formerly non-industrialized areas such as Cambridge in the UK, or the French Midi, or especially the US Sunbelt with its major new high-technology growth centers in such places as Orange County, Silicon Valley, Chatsworth-Canoga Park, Dallas-Forth Worth, and so on (Saxenian).

Third, advanced producer and financial service agglomerations are found in or close to the central cores of large cities, such as Tokyo, New York, London, or La Défense in Paris.

In all of these new flexible production areas it has been possible to reconstitute production processes so that they are insulated from labor forces marked by the historical experience of Fordism, with its traditions of unionization, collective bargaining, and finely-delineated task structures.

FROM CORE AND PERIPHERY TO A GLOBAL MOSAIC OF REGIONS

From the 1930s onwards, the rapid growth of the core zones of the mass production system provoked considerable political concern about the question of regional development. In both Western Europe and the United States, various interventions were made to assist those regions which seemed unable to recover from the Great Depression; and then in the postwar years, an expanding series of programs was implemented in an effort to redistribute growth from leading to lagging regions. After the 1950s, many Third World countries also devised ambitious regional development programs aimed at promoting and diffusing growth.

These efforts at regional policymaking were all informed by a common intellectual perspective, namely, a conception of capitalist economic growth as a process which tends to create highly developed core regions on the one hand, and underdeveloped dependent peripheral regions on the other. To be sure, in regional development theory there are many different accounts of core–periphery relations. The traditional approach to this issue is built up around the investigation of comparative advantage (based on pregiven endowments), market exchange, and concomitant spatial flows of capital and labor. In this approach, the full liberation of factor markets is seen as the key to a developmental process in which regional income inequalities will become nothing more than temporary and self-correcting aberrations. For other theorists, the core regions and nations of the capitalist mass production system grew at the expense of underdeveloped peripheries, by locking the latter into a system of unequal exchange. Yet a third set of theorists identified polarization and trickle-down effects of the production system as the main factors underlying core–periphery inequalities, intensified by the low-wage labor surplus trap, and accordingly they called for policies to accelerate transfers from core areas to peripheries.

A more recent line of analysis revolves around the new spatial and international division of labor – mediated by the multiestablishment, multinational enterprise – in which the various phases of the production process are differentially allocated across space in relation to their varying technological and skill characteristics. In this view, advanced technical and managerial tasks are typically allocated to core regions, and routinized, low-skill, labor-intensive activities are allocated to the periphery. Trade then occurs between different regions but within the large firm.

Each of these theoretical advocacies has some validity, and indeed each (with suitable further qualification) captures something of current realities. That said, none of these viewpoints anticipated the processes of deindustrialization and decline which by the 1970s were dramatically evident in virtually all the core regions of the United States and Western Europe; none anticipated the perplexing rise of a series of new regions of flexible production; and by the same token, none had anything to say about the possibility that periodic shifts in the overall geographic configuration of the capitalist world could take place, including a redefinition of the structure of relationships between core and peripheral zones. The resurgence of flexible production organization in the modern world thus requires a reconsideration of the question of interregional trade, local economic specialization, and the broad configuration of the geography of production. Three points must now be made.

First, in contradistinction to the traditional theory of trade, comparative advantage is only rarely a matter of pregiven (still less natural) endowments. Comparative advantage is more frequently *humanly created* in the very process of trade, and one of the important ways in which this occurs is through a trajectory of regional development in which industrial agglomerations with their stocks of external economies are steadily brought into being. Striking examples of this process are cars in Detroit and aircraft in Los Angeles, where an early start and a series of technological breakthroughs hastened the domination of these regions in their particular sectoral specializations. This phenomenon of the "first mover" has recently been theorized in terms of evolutionary economics and the new strategic trade theory.

Second, whereas many kinds of commodity chains are spread out across the globe, as described by the theory of the new spatial/international division of labor, the most rapidly growing industries today are often strongly associated with particular industrial agglomerations. In any one of these agglomerations, semimanufactured outputs, subassemblies, and other kinds of inputs are made within the local industrial network, and are then passed on to plants in other locations and other agglomerations. In most industries today, a combination of agglomeration of key activities coexists with dispersal of other elements of the commodity chain; in brief, the "roundaboutness" of the division of labor is both functional *and* spatial. Accordingly, there can be no absolute opposition between the theory of agglomeration and the theory of the new spatial/international division of labor. Rather, each of these theoretical perspectives captures a different and equally valid facet of a single economic reality.

Third, in addition to established criticisms of the new spatial/international division of labor *qua* simple bipartite core–periphery structure, a further qualification must be put forward. Over the last couple of decades, there has been an increasing tendency for

a) large numbers of Third World workers to move into the burgeoning low-paying, unskilled jobs generated by the sweatshops, subcontract establishments, and low-grade service activities in the flexible production agglomerations of the core countries, and
b) high levels of technical competence and managerial control to develop in selected urban regions of the periphery, as in Brazil, Hong Kong, Singapore, South Korea, and Taiwan.

Indeed, two cities like Los Angeles and Hong Kong, despite their markedly divergent national backgrounds, probably have more in common with one another as centers of flexible production than do, say, Los Angeles and Detroit, which share a common national identity, but have their roots in different industrialization processes (cf. Leborgne and Lipietz, Standing).

These remarks also point in the direction of a needed rethinking of some aspects of Third World development theory. Until quite recently, much theory suggested that the most effective pathway to development was by means of large growth pole industries, complemented by upstream suppliers attracted in to the local area by policies of import substitution. Over the 1970s and 1980s, development strategies based on this theory ran into severe internal constraints (related to low aggregate demand) and external constraints (related to increased indebtedness). Meanwhile, forms of industrialization based on more labor-intensive and flexible production activities were proceeding apace in many of the Asian and some Latin American NICs and near-NICs. Production activities of these sorts characterize sectors such as textiles, clothing, shoes, plastics, metalworking and electronics. They are often sustained by super-exploitation of labor and state policies directed to export-oriented industrialization. In a few cases, especially in Asia, selected regions have been able to shift into relatively high value-added activities by means of product differentiation and technological upgrading. The experience of these regions, along with that of places such as the Third Italy, Jutland, or northern Greece in Europe, suggests that a flexible production base, combined with effective national and international marketing organizations, can generate significant rounds of economic growth, especially where producers are able to start climbing the price/performance frontier (Piore).

We propose a doubly faceted view of regional economic development in the new global context, and in contradistinction to the old core–periphery notion. On the one hand, the global economy is made up of a set of specialized regional production systems, each with its own dense system of intraregional transactional arrangements and local labor market activities. On the other hand, these individual regions are entwined in a worldwide web of interindustrial linkages, investment flows, and population migrations. At the global level, a number of critical institutional arrangements – the multinational enterprise, an emerging system of international subcontracting, interfirm strategic alliances, international agreements, and so on – play important mediating roles (Altvater). The nation-state remains a significant element in this structure of global production activities; however, the nation-state is also certainly less and less economically autarchic precisely because of the internationalization of structures of production (which puts increasingly severe constraints on national macroeconomic management) and the growth of international organizations (in which nations give up elements of their sovereignty in favor of coordination at higher territorial levels). It is in the light of these developments that we can see the economic geography of the contemporary world not so much as a core–periphery system, or even as an aggregation of nation-states, but as a *global mosaic of regional economies*.

THE IMPORTANCE OF INSTITUTIONS AND CULTURE

Capitalist relations of production and exchange are always embedded in wider sets of social relations and cultural norms. Indeed, in the absence of these, production and exchange could not occur as self-reproducing phenomena. We advance three reasons for this assertion:

1 At the level of individual decisionmaking and rationality, the economic system is rife with tendencies to market failure in the form of externalities, imperfect information, free-rider problems, and so on.
2 Capitalism is beset with inherent predicaments at the level of collective action and the macroeconomy, e.g. labor–management collisions, the economic cycle, and foreign trade imbalances, all of which threaten orderly production and exchange.
3 The domains of the social and the cultural (family life, education, forms of consciousness and ideology, and so on) underpin the economy but are not themselves subject to purely economic rules of order.

To these three points we may add a fourth observation, i.e. that markets are grounded in an infrastructure of institutionalized rules and norms, which vary over space and time.

The idea that successful reproduction of capitalist economic systems cannot proceed in the absence of institutionalized agencies and collective action holds not only at the level of the national economy but also at the level of the regional economy, where because of the specialization, agglomeration, and place-specific character of production, peculiar forms of institutional order often present themselves. We may illustrate this proposition by taking three examples of historical types of industrial agglomeration.

First, nineteenth-century mill-and-workshop complexes flourished on the basis of large numbers of low-wage workers within a patriarchal factory system. The residences of these workers formed dense urban neighbourhoods in which both social order and physical health were difficult to maintain. Thus, despite the prevailing *laisser faire* ideology, these problems became the targets of early urban planning interventions and urban reform movements. Second, in twentieth-century mass production regions, management and organized labor, together with municipal government, constituted an institutional framework within which workable compromises, agreements, and patterns of resource allocation were worked out at the regional level. Active urban planning, especially in matters of housing and transportation, was also part of this framework. Third, in flexible production agglomerations today, especially high-technology industrial agglomerations, market failures in technological innovation and transfer, and in the training of labor, are frequently in evidence. In some of these agglomerations, institutionalized efforts to address these problems have been made by the provision of publicly funded research activities and educational/training establishments.

We can analyze these institutional bases of regional economies by looking more systematically at their intersection with a series of major dimensions of place-bound economic and social life: i.e. interfirm transactions, technological innovation, local labor markets, and the organization of community.

Inter-firm transactions

Agglomeration is a strategy whereby producers ease the tasks of transactional interaction because proximity translates into lower costs and wider opportunities for matching needs and capabilities. But agglomeration alone does not necessarily lead to the formation of efficient transactional interrelations; indeed, there are powerful forces which, if not counterbalanced, may actively work against the formation of efficient transactions within agglomerations. We want to call attention in particular to three major issues. First, breakdowns of information exchange occur where one party holds privileged information that can be traded on opportunistically. If, for example, a subcontractor making subassemblies chooses to include a certain proportion of defective parts which can only be recognized when the final product has been in use for some time, then the subcontractor's information is asymmetric with that of the principal firm. Policing costs are particularly high in such cases and punitive monitoring is rarely fully effective. Second, failures of trust underpin and intensify this tendency; that is, in the absence of formal or informal means of ensuring that other parties to a transaction are likely to abide by a given set of standards, it becomes rational to be hesitant about doing so oneself (Lorenz). Third, where these problems prevail, fine-tuning of input–output flows is difficult to achieve and, in the absence of durable agreements at all levels of the production hierarchy, firms are encouraged to develop buffer stockpiles of critical inputs. All of these problems can be kept at bay by institutional infrastructures or social practices which increase information exchange and trust, and limit the probability that opportunistic behavior will benefit those who practice it. The Japanese *kanban* system and its associated practices of mutual aid represent one such type of institutional infrastructure; many other varieties can be found, in different industries and regions with their diverse historical trajectories and social structures (Lorenz, Bianchi, Saxenian).

As we shall now see, the same sort of infrastructure facilitates exchange of technical information which aids firms in the continuous and gradual refinement of products and processes.

Technological innovation

In the area of technological innovation market failures abound; since knowledge is extremely leaky and difficult for innovators to appropriate exclusively, profit-seeking firms tend to underinvest in many forms of product and process development. In many critical sectors, the basic research inputs either come from the nonprofit sector or are induced by public subsidies, as illustrated by the cases of the early semiconductor industry, aerospace, and now, biotechnology. Strategic planning and coordination, as with MITI in Japan, is an important way, too, of keeping industries focused on long-run developmental goals.

Technological innovations are frequently place-bound. That is, in industries characterized by rapidly changing products or processes, the stocks of knowledge and human capital upon which technological changes are based tend to be concentrated in the specialized labor forces which themselves are highly localized in a small number of places. In addition, in many innovative flexible production sectors, localized interfirm relations within the social division of labor are a critical means of developing and transmitting the information on which innovations are based, as when one firm identifies a need for specialized kinds of inputs or equipment which must be met by innovative activity on the part of a supplier firm within the network (Piore). In both of these localized dimensions of the innovation process, too, market failures can limit the industry's ability to carry out innovative activity, as when, for example, firms fail to invest in long-term research projects for fear that the workers involved in any project will use the knowledge thus developed in an opportunistic fashion, or when firms fail through lack of trust to communicate with each other their need for innovations. There is accumulating case study evidence that such failures are present in American high-technology agglomerations. Likewise, it appears that agglomeration-specific institutions are sometimes necessary to overcome some of these types of market failure (Saxenian).

Local labor markets

An additional area in which active institution-building is an essential element of successful regional development is that of local labor markets. Local labor markets depend for their effective operation on the development and circulation of information. This is particularly true in local labor markets associated with flexible production systems, for here the rapidly changing employment opportunities and resulting intensity of business transactions make firms and workers dependent upon access to large but ever-changing bodies of information. While spatial propinquity facilitates the functioning of these networks and allows the development of a fund of knowledge that helps participants screen and evaluate the information they receive, it is nevertheless frequently the case that the transmission of information about job vacancies and demands is severely hindered in the absence of collectivized channels of information transmission. Accordingly, we typically observe in centers of employment, public agencies and other organizations (employees' associations, labor unions, occupational guilds) devoted to the tasks of increasing the circulation of information.

In the same way, institutional support is a *sine qua non* of socially rational levels of skill provision and retraining activity. This predicament exists in a world marked by a complex division of labor and considerable future uncertainty. In these circumstances, employers will limit their investments in the development of skills for which an uncertain future demand exists, or in workers who may not be retainable once such investments have been made (whether due to voluntary quits or involuntary layoffs). Workers, for their part, will limit their own skill-building efforts in the absence of a reasonable degree of certainty that such efforts will be rewarded. Thus, in labor markets where training is not publicly subsidized it is probable that an undersupply of appropriate skills will occur. Moreover, in cases where an oversupply of skills occurs, e.g. because of a downturn in a particular sector, the social costs of retraining will often be less than the costs of long-term unemployment.

Community development and planning

Housing and transport are deeply implicated in the supply of labor. This is because local labor markets consist not just of an aspatial set of supply, demand, and wage-setting mechanisms, but also of an interlocking locational-*cum*-transport system of origins, destinations, and channels of access. Such systems are marked by manifold market failures, in part related to the size and durability of the fixed investments needed to sustain them and in part related to bottlenecks in their locational adjustment that result from private landownership, preexisting land use developments and, especially in the USA, municipal fragmentation. The consequence is that we often observe in such systems serious imbalances in the spatial distribution of jobs and housing and malfunctions of transport networks. At the same time, the household and the community are important foci of the social reproduction of labour and here again serious breakdowns often occur in the absence of instruments of collective order.

A politics of place

On the basis of the above discussion, we would argue that the making of a regional economy involves not just the development of a productive apparatus on the basis of the atomized decisions of firms and workers, but also a "politics of place," or in other words, the social construction of those institutional-regulatory structures that must be present in order to secure economic order and continuity. Some of these structures are within the domain of the national state (Leborgne and Lipietz). Others belong more properly to the level of the region as such. The regionalization of coordination occurs not only because of the prior existence of local government units, but also because the tendency of many economic activities to agglomerate in specialized districts fosters the emergence of place-specific problems of system-guidance (Bianchi). At the same time, localized cultures make their historical appearance and in many different ways (via business norms, workers' habits, even specialized languages), they shape the regional economy (Ganne).

REGIONAL DEVELOPMENT TODAY: ECONOMIC ORDER AND SOCIAL JUSTICE

In the new global–local economic reality we are describing here, the ability of the nation-state to regulate its own economic affairs is diminished; on the one hand, intensified international competition and interdependence have reduced the workability of national macroeconomic regulation (Altvater); on the other hand, the proliferation of specialized regional production districts as flexible production organization advances means that much policymaking and institution-building today are likely to be most effective when aimed specifically at the level of the industrial district with its particular production logic (Bianchi).

In the new flexible production regions that have emerged since the 1960s, we observe a wide variety of forms of institutions and market order, ranging from the intensely competitive economic relations that prevail in, say, the Los Angeles garment industry to the high levels of formalized coordination that exist in some regions in Germany (e.g. machine tools in Baden-Württemburg) and Japan (e.g. Toyota City) (Leborgne and Lipietz, Standing). It appears that regions in which coordination is weakly developed, and in which unregulated competition prevails, face many problems and predicaments that compromise long-run viability (Bianchi). These range from impediments to information exchange and interfirm collaboration to inadequate supplies of appropriately trained and socialized workers (Saxenian, Lorenz). Such regions are all the more vulnerable because in a world of contested markets they find themselves faced with competitors based in regions that do provide effective regulatory and coordinating services (Courault and Romani). Given such competition, regions with low levels of collective order are likely to find themselves squeezed in such a way that either they begin to lose market shares, or – as exemplified by the sweatshop segments of southern Californian high- and low-technology industry – they enter into a spiral of declining wages and working conditions and lowered rates of profitability combined with rising instability for secondary suppliers and subcontractors (Leborgne and Lipietz). Thus, it would seem that the viability of contemporary flexible production agglomerations depends on effective institution-building and policymaking at the regional level (Bianchi). That said, even economically successful and institution-rich agglomerations may veer into politically regressive configurations and there remains, in many of the growing production regions of the world today, an open question about how to marry economic efficiency and social justice.

There are, moreover, resurgent problems of uneven development on a world scale which cannot be resolved even by successful regulation within the growing nodes of the global mosaic of flexible production regions. The geographical distribution of high-wage, high value-added activity remains extremely uneven, and it is evident that the heightened interregional competition which has been stimulated by the increasingly open system of global production and exchange relations has had strongly negative effects in many places. At the same time, successful forms of local regulation and the social compromises upon which they are based are often subject to great stress exerted by the underregulated global system of markets and finance capital (Altvater).

We have only begun to confront the analytical and political challenges posed by the emerging new system of industrialization and regional development under flexible production. Probably the last time that a coherent overall social analysis and political program existed relevant to the then-current conjuncture was during the period of Fordist mass production and Keynesian welfare-statism in the late 1960s. The present book attempts to provide some of the conceptual underpinnings for addressing the new economic and regional realities in a systematic and analytically disciplined manner. This is essential if social science is to have a positive and progressive influence on the policies that will shape the pathways to industrialization and regional development in coming years. The task involves an insistent focus on the interactions between economic logic and political institutions at both the local and the global levels.

Part II

A NEW PERIOD IN CAPITALIST DEVELOPMENT?

2

FORDIST AND POST-FORDIST INTERNATIONAL DIVISION OF LABOR AND MONETARY REGIMES

Elmar Altvater

INTRODUCTION

Although there are *general* laws of capitalist development to be identified theoretically, an analysis of contemporary capitalism has to focus on the structure and the specific dynamics of the given *historical* reality of capitalism. "Fordism" is a descriptive category which summarizes such historical tendencies of capitalist accumulation in the last decades, in particular after World War II. Social and economic development take place within a given institutional, normative, and physical framework, which may be called a "regime" of accumulation: they are linked to the specificities of historical time and determined by the characteristics of geographical and social space. Thus, in order to understand capitalist dynamics it is not sufficient to identify the "general laws of motion" of the mode of production; rather, it is necessary to uncover historical and geographical specificities and particularities of capitalist development. This perspective motivated Aglietta's (1979) analysis of the "US experience" as well as the efforts of Antonio Gramsci (1967) to understand the impact of "Americanism" and "Fordism" on the growth (and expansion) of capitalist hegemony in the US during the 1920s and its repercussions on the political system in Europe, in particular in Italy.

The historical and spatial characteristics of capitalist development are not simply the expression of existing differences between national capitalisms – the US-American, the Western European, the Japanese, or the Latin American one. Each of these "capitalisms" has its proper trajectory, but also shares certain characteristics, e.g. business cycles, such as phases of prosperity and of depression. Capitalist societies also tend to generate crises which cannot be interpreted as a cyclical downturn in the course of a "normal" business cycle, but as a rupture of the institutional and structural form of economic accumulation and social reproduction. If there is such a "form crisis" (Altvater 1983a) or a "great crisis" (Boyer/Mistral 1978), we have to conclude that distinctive historical forms of capitalist reproduction, stages of development or distinctive modes of socialization do exist within the capitalist system in general. Marx himself made the distinction between the "manufacturing system" and "great industry," both forms of relative surplus production, but distinct in the mode of organizing the labor process and of moulding (national as well as global) capitalist reproduction structures.

Since Marx, it has become quite common (not only in Marxist writings) to identify different stages of capitalist development (for an overview: Altvater 1983b): Capitalism develops from early to late, from immature to mature, from competitive to monopoly and to state monopoly capitalism, from colonialism to imperialism – a sequence of stages or phases of development, each of them emerging from and resulting in a "great crisis" of transformation. The methodological problem involved, of course, is that of the relationship between historical stages and general laws, and that of the theoretical status of the category "crisis." Is each crisis a unique historical event, or are there general functions carried out by the dynamics of (historical) crises? It is not possible to answer that question here; it suffices to note that the analysis of a historical stage requires prior methodological deliberations about the relationship between "the general" and the "historical." The notion of Fordism calls attention to the historical forms of capitalist dynamics, structures, functions, and resulting tendencies or cyclical movements. It attempts a deeper theoretical understanding of historical forms which the capitalist system realizes in the course of its accumulation dynamics and of the efforts of the *bourgeoisie* to preserve social and political hegemony.

On the most abstract level, a *form crisis* is a double-sided process. It both destroys traditional social forms of social and economic regulation and it generates new forms of regulation. Whereas the process of destruction is more or less obvious and identifiable, it is much more difficult to deduce the generation of new social forms of regulation. Social stability can only be achieved by means of a complex set of complementary, compatible, and cohesive institutions. Since there is no single directing agent which defines and creates these institutions, complementarity, that is, a new mode of regulation or a new accumulation regime[1] can emerge only as the outcome of conflicting tendencies of progressive acts and regressive setbacks. In sum, a cohesive accumulation regime or mode of regulation is none other than a "discovery" (Lipietz 1987), a social form resulting spontaneously and not consciously from conflicting social actions.

Fordism was the outcome of just such a blind historical process, following a logic unknown to the participating actors, including governments as well as managers of big corporations or trade unions. In its very beginnings Fordism was merely the extension of Taylorist rationalization from the single labor process to the whole factory and via the mechanisms of competition to national industry as a whole. It proved to be a powerful machine of productivity increases and, therefore, it was historically successful insofar as it created a set of social forms complementary to technological and organizational rationalization on the factory as well as on the societal level. The increase of production per worker requires an increase in effective demand. Otherwise, a "realization crisis" becomes inevitable due to the "scissors" of overproduction and underconsumption.[2] Thus, one of the most important innovations in policy, politics and polity was the creation of the "Keynesian" interventionist state and of corporatist forms of bargaining and mediation of conflicting interests.

The condition of complementarity in the case of general productivity increases requires a rather simple correspondence: that of a growth rate of the national income which equals productivity growth. On the supply side, full employment of all factors of production only can exist, and on the demand side the gap of effective demand is filled, so long as this general correspondence is maintained. However, this outcome does not result from a clearing process of interests of individual entrepreneurs dominated by microeconomic rationality. Instead, entrepreneurs have to exert pressure on real wage increases in order to minimize labor costs and to increase their profits. It is a general contradiction in capitalist societies that each individual capitalist tries to reduce the nominal wage bill of his workers, whereas he needs general wage increases of all the other workers employed by other capitalists as the source of effective demand for his production. This general contradiction between micro- and macroeconomic rationality has played a tremendous role since the very beginnings of Fordism. Henry Ford himself responded to the double character of wages as costs and as the source of demand exactly as would a rational individual capitalist: during the crisis of the 1930s he reduced wages, fired workers, and tried everything to fight unions and workers' unrest (see Dobb 1966, Foster 1988, or de Jouvenel 1933). Only after a long period of "incubation" did the institutional complement to productivity increases come into existence by means of the politics of the New Deal in the United States and by means of the introduction of modern state interventionism of a Keynesian-Beveridgean type after World War II in nearly all developed Western capitalist countries. Therefore, only the period after World War II can truly be called the "Fordist" stage of capitalist development.

THE BASIC STRUCTURE OF FORDISM AFTER WORLD WAR II

Without repeating the details of a Fordist accumulation regime and mode of regulation (Lipietz 1986; Hübner 1989) I stress the importance of the relationship between productivity (Y/L) and labor costs (W/L). The one determines the growth of output per worker while the other defines the expansion of effective demand (by the workers). When both develop in accordance with each other the accumulation process may be balanced. There is no demand gap or situation of "underconsumption" and there is no oversupply or overproduction, *ceteris paribus*. Since W/L and Y/L express the portion of wages in national income W/Y, this formula implies also – negatively – the share of profits in national income ($P/Y = 1 - W/Y$).

In order to achieve this balance in the reproduction process, a set of institutions peculiar to and characteristic of Fordist social structures has to be established. Two important ingredients of that structure shall be discussed here: first, the labor and reproduction process and second, the components of the profit rate.

The labor and reproduction process

For our argument – which aims at identifying the relationship between the postwar international division of labor and the monetary regime – three facets of the labor process are important:

1 the Taylorist *rationalization* and its impact on productivity increases;
2 the importance of *monetization* (and commodification) of the reproduction of labor as well as of monetary compensation of (environmental) damages for human beings; and
3 the *regulation* of labor by the state by means of "Keynesian" interventions and the modern (social democratic) welfare state.

First, the rationalization of the labor process has become an increasingly global project. Its realization is due to the expansion of methods of scientific management from the United States – where they first were invented and implemented – to all other parts of the world where modern capitalism (or methods of rationalization like those used in the Soviet Union of the 1920s) has taken root. Yet, the diffusion of "scientific management" is an unequal and uneven process and it appears in very different forms in different parts of the world system. Therefore, "Fordism" does not describe an identical condition in all regions of the world. Whereas rationalization in developed countries is normally linked to the realization of a complete social process and with creation of complementary institutions in which productivity and wage increases are embedded, in less developed countries rationalization is only partial, affecting only segments of the labor process. This partiality of rationalization (and of modernization) is inherent to the international division of labor. Under the global aim of maximum productivity increases those sections of the production process are dispersed which promise optimal contributions to an overall rationalization process steered by transnational corporations. This is the crucial reason why in Fordist regimes the international division of labor is not only the result of market tendencies but of (micro)rational decisions of "scientific" managers with a global perspective.

Second, the form of monetization of labor-power and the monetary compensations for the destruction of the natural environment are very important to the development of social forms of conflicts. It is not by accident that the trade unions as well as the so-called traditional social movements normally are accustomed to fighting for higher wages and monetary compensation for all damages produced by the capitalist labor and reproduction process. Monetary claims define the form of social conflicts, e.g. the possible scenarios of workers' unrest, but also the institutions and the means of absorbing resistance, of the bargaining process and the settlement of its results, etc.

Third, the role of the Keynesian state fits into this institutional picture. The state supports rationalization via the increase of effective demand which helps to resolve the always threatening realization problem. The economic interventions of the state under the pressure of international competition are effective impulses for the dynamization and modernization of the system, for an increase of growth rates. And the welfare state is a very efficient vehicle for the mobilization of monetary funds for the compensation of unemployment, health problems, environmental damages, etc. But even within this comprehensive structure the Keynesian state has limits. It has already been mentioned that the rationalizing dynamics of the capitalist labor process go far beyond national frontiers and have a global reach. The Keynesian state by its very definition is a national state and is therefore limited to the national space in the exercise of its regulatory functions.

Components of the profit rate

It is not sufficient to interpret the development of capitalist economies after World War II as a systemically intelligent process of establishing an institutional network for balancing the "Fordist" equation of wage and productivity increases, that is for the long-run stability of unit labor costs: ${}^{W/L}/_{Y/L}$ (and of functional income distribution). Unit labor costs are not the "leading control variable" of capitalist dynamics, it is the *profit rate* on which the accumulation rate and therefore economic growth are dependent. Of course, labor productivity and wages per worker are components of the profit rate. But as Marx (contrary to Ricardo) pointed out, the profit rate π is also dependent on the development of the productivity of capital (i.e. the organic composition of capital):

$$\pi = \frac{Y/L}{K/L}(1-w)$$

This interdependent structure of variables (Y/L = labor productivity; K/L = capital intensity; w = unit labor costs W/Y; Y/K = capital productivity) taken into account, the profit rate and therefore the trajectory of economic growth depend on the development of capital productivity, labor productivity, capital intensity and wages per worker; under the assumption that profits are entirely invested, they determine the pace of growth. Many studies have shown that for the development of the profit and accumulation rate, unit labor costs play a very important role, but equally important are the dynamics of capital productivity (Armstrong, Glyn, Harrison 1984; OECD 1985; Altvater, Hoffmann, Semmler 1979; Stanger 1988). Moreover, capital productivity is quite analogous to the Marxian organic composition of capital, since Y/K and $v+m/c+v$ describe similar relationships in macroeconomic and in value terms. However, statistical tests by the OECD have shown that falling "rates of returns" may be caused more by a declining capital productivity than by an increase of unit labor costs:

> Overall, it appears that downward movements in capital productivity are the most important factor underlying falling rates of return; though declining profit shares also contribute to this outcome. This conclusion appears robust given the evidence of statistically significant time trends in almost all countries and sectors for which data are available.
>
> (Chan-Lee/Sutch 1985: 148)

Therefore the "Fordist equation" between productivity growth and the growth of wages per worker is not sufficient to explain the dynamics of capitalist prosperity after World War II, especially because the growth of capital productivity in the postwar period in all developed countries since the beginning of the 1970s exhibits a negative trend.

Table 2.1 Capital productivity growth (1950–84)

Country	*1950–73*	*1973–84*
France	1.50	−1.82
West Germany	0.57	−1.71
UK	−0.26	−1.45
USA	0.34	−0.47

Source: Maddison 1987: 656.

The dynamics of accumulation after World War II had three principal causes: the first was the complementarity of productivity growth and demand growth, which impeded the emergence of a realization crisis. Second, high capital productivity enhanced average profit rates and the dynamics of accumulation. But the third, and main, factor of capitalist dynamics came from the composition and interlinkages of the different components of the profit rate in the capitalist core countries. The uneveness of development, on the one hand a factor of backwardness, proved to be a driving force of the growth dynamics after World War II. Contrary to the assumption of Harrod-neutral technical progress, capital productivity in all countries of the Western world decreased (or the capital coefficient increased, respectively). But this occurred at a different pace of growth, as Table 2.1 clearly shows. On the other hand, the negative effect of an increase of the capital coefficient (that is the decrease of capital productivity) on the profit rate may be compensated by comparably low unit wage costs, that is either by a high growth rate of labor productivity or low wage-per-worker increases. There are different possible constellations of labor productivity, labor costs, capital intensity, and capital productivity; this is why the national average rate of profit may remain at a high level, even though one or more components may have a negative impact on it.

Positive effects on the average profit rate result from either high capital productivity due to high labor productivity or due to low capital intensity, or because labor costs are low. There are many possible constellations imaginable.

These abstract possibilities also describe the real situation of different countries of the Western world from World War II to the 1960s. In the United States real wages were comparatively high, but labor productivity was also relatively high, with capital intensity not growing enough to increase significantly the capital coefficient. In Western Europe, labor productivity was comparatively low, but capital intensity, and labor costs were also low. The same was the case in Japan. Therefore, for different reasons, profit rates of capital in all parts of the developed capitalist world were high in the first two decades after World War II. This constellation was the basis of high growth rates of GDP, of the spread of prosperity, and of the generalization of the institutional forms of this successful "growth machine."

Moreover, the most productive economy (the USA) was able to realize *superprofits* on expanding world markets: The growth of foreign trade during the postwar period was higher than ever before in modern times of the capitalist world system. In the OECD area it was nearly five times higher 1950–86 than 1900–50 (Maddision 1989: 26) and it exceeded the average growth rate of GDP in that period in six highly developed countries by more than 70 percent (Maddison 1987: 670). The rapid expansion of world markets facilitated the realization of permanently increased production of commodities and therefore of economies of scale. Compared to production costs in competing countries, the USA produced its export commodities at rather low costs. Sold at world market prices the difference between US costs and the international average cost level appeared as "windfall profits." However, this only could happen since the realization of extra profits by the most productive nation requires the expansion of the realization sphere, that is of the world market for commodities. The backward countries (Western Europe and Japan) on the other side experienced "the opportunities of backwardness" (Maddison 1987: 669). There are two reasons for this. First, they increased their competitiveness through a double process of holding unit labor costs at a comparatively low level and of undervaluing their currencies. Under the regime of fixed exchange rates this tendency occurred almost automatically due to uneven productivity increases. This especially played a role in West Germany and later in Japan. Second, these countries were able to fill the productivity gap surprisingly quickly and thereby succeeded in increasing the national wage level.

As a consequence of this constellation the United States had an inherent economic interest in promoting the world market and in establishing rules and institutions for the development of free trade after World War II. The accumulation model presupposed the institutionalization of world market structures which would facilitate international trade as well as international capital transfers. On the one hand, this was the framework in which the tendential convergence of productivity levels between the main trading partners took place. On the other hand, this was the precondition of the dominance of the leading country in the capitalist world market.

Moreover, the leading country did its best to promote productivity convergence via the internationalization of *productive capital* through private direct investments. At the end of the 1940s nearly the same portion of US direct investments abroad was directed to Latin America and Western Europe: 24.0 percent and 23.8 percent respectively. In the 1959–61 period Latin America only received 10 percent whereas Western Europe (EEC, EFTA, and Spain) got 46.2 percent of an amount of direct investments which was nearly nine times higher than ten years before ($4.54 billion compared to $621 million). Direct foreign investment in manufacturing (different from those in extractive industries, such as mining, which prevailed in Latin America) is a means of spreading technologies from one part of the world to another. In Western Europe, for instance, in 1965–6, discussion of the productivity gap vis-à-vis the United States was still predominant (Servan-Schreiber 1968; Mandel 1968). But only a few years later, this productivity gap was greatly reduced in many industries (for recent data, see Cuomo Report 1988; US Department of Commerce 1990). The rapidity of the convergence was surprising. As Maddison (1987: 678) clearly shows, between 1950 and 1973 the "rate of productivity convergence" with the lead country in the dynamic competing countries was in West Germany 3.42 percent, in France 2.59 percent, and in Japan 5.12 percent per year. Only in Great Britain was the convergence rate near zero, very likely due to the productivity decline in UK industries in this period (Glyn/Sutcliffe 1972). The catch-up effect explains more than 10 percent of the annual growth rate in West Germany, Japan and France (computed on the basis of data supplied by Maddison 1987, Tables 19 and 20). No wonder, since the growth of overall GDP and GDP per hour in the United States was far below the pace in France, Germany and Japan. The growth performance of the United

Table 2.2 Phases of growth in industrialized countries

	USA		*France*		*W. Germany*		*UK*		*Japan*	
	1950–73	*1973–84*	*1950–73*	*1973–84*	*1950–73*	*1973–84*	*1950–73*	*1973–84*	*1950–73*	*1973–84*
GDP	3.7	2.3	5.1	2.2	5.9	1.7	3.0	1.1	9.4	3.8
GDP/ hour	2.5	1.0	5.1	3.4	6.0	3.0	3.2	2.4	7.7	3.2
Y/C	0.34	−0.47	1.50	−1.82	0.57	−1.71	0.26	−1.45	1.39	−3.41

Source: Maddison 1987: 685.

Kingdom is an exception in this respect. The result of the converging development of productivity, of course, is the tendential equalization of productivity levels in the industrialized world, as Tables 2.2 and 2.3 clearly demonstrate.

But this development has two consequences for the accumulation process on a worldwide scale.

First, the catch-up process eliminated the extra (or windfall) profits of those US corporations that were highly integrated into the world market. This tendency has an impact on the US economy, since the overall profitability of US capital is improved by the sources of superprofits. In the case of their disappearance the tendency of the profit and accumulation rate to fall is enforced.

Second, the equalization of productivity levels is an economic and statistical expression of the propagation of "Fordist" working conditions and wage levels within the world of developed capitalist countries. By the same token the extension of Fordist institutional structures is also responsible for the equalization of economic conditions in all capitalist countries – although differences remain, of course, – which result in declining profit rates, as shown by the OECD (Chan-Lee/Sutch 1985: 127 passim). Therefore, we may conclude that the process of the internationalization of Fordist reproduction structures is a crisis-ridden process.

Further, the declining profit rates influence negatively growth rates of GDP as well as the performance of the labor market. Insofar as labor productivity increases more than GDP, unemployment is an inevitable consequence unless there is some outlet created for surplus labor, e.g. the reduction of working time. However, it is not possible to conclude that declining profit rates directly and immediately are causing an overall economic recession. Their effects are, rather, quite contrary: declining profit rates in the core countries may be an incentive for direct and portfolio investments in peripheral countries and by this mechanism stimulate the global accumulation process. Moreover, declining profit rates may also be an efficient stimulus for further investments by corporations in order to overcome a situation of increasing competition among capitalist corporations and countries. However, increased investment by the state and its financing by rising government debts *vis-à-vis* financial markets results in a "crowding out" of private investment, though the state as an investor only reacts to conditions created by the performance of private capital accumulation.

But there is one crucial and effective boundary to further accumulation when profitability tends to decline overall. That limit is defined by the interest rate on financial assets. Capitalists as any asset holder compare different rates of return to assets. If the interest rate exceeds the profit rate, financial assets are more attractive than productive investment. At this point it becomes clear that the monetary conditions of the accumulation process are of great importance. Therefore, we must examine the specific form of resolving monetary problems in the Fordist regime after World War II, before coming back to the consequences of the declining profit rate.

THE INTERNATIONAL MONETARY REGIME AFTER WORLD WAR II

One of the complementary conditions of postwar Fordism was the monetization/commodification of the labor–wage relationship as well as a mix of monetary and fiscal policy carried out by the state in order to fill the demand gap and to create full employment. The Keynesian solution to the problem of unemployment may be interpreted as a sophisticated social mechanism of monetization and (corporatist) class compromise. We have also stressed that the driving force of this system pushed it beyond national frontiers and created the postwar global capitalist system, which differs significantly from the preceding capitalist world system. Of course, capitalism since its very beginnings has been a world system (Wallerstein 1979; Amin 1975; Beaud 1987). But only after World War II did the capitalist tendency to transform the system into a world system become historical reality, as all functional forms of capital, i.e. commodity capital, productive capital, and interest bearing capital, were globalized. However, a basic and determining contradiction arose: although the economic reproduction process was globalized, the political forms of regulation remained national, due to the fact that the capitalist state is by its very origin a national state. The great achievement of the "Keynesian revolution" was the establishment of the "lender of last resort" and of "big government" on the level of the national economic system in order to avoid future crises of the 1929-type (Kindleberger 1986; Minsky 1986). But there was no equivalent institutional network capable of resolving this problem on the level of international economic relations.

The answer to this question, which arose immediately after World War II, seems today very evident: the regulation of the international monetary system in a way that is complementary to the regulated national systems. The absence of an international regulatory institution of last resort was overcome by the role which the "superpower" USA assumed in international political as well as economic affairs. Only the USA had the economic, political, and military power to assume the economic leadership in that system of international regulation. Therefore, the international monetary regime was at the same time an outcome of US dominance and a guarantor of its continued hegemony. But in order to transform dominance into hegemony, it was necessary to establish an international institutional framework for all actors (states, corporations, organizations, etc.) which fulfills the conditions of hegemony. According to Gramsci, hegemony is the ambiguous *ensemble* of power and consent. Therefore, as regime theory later defined it, the participants of the international system have to acknowledge or even to accept common explicit or implicit rules and habits, norms, and mutually established institutions (Keohane 1984; Cooper 1987). It is in their common interest to avoid free riders as well as to overcome the prisoners' dilemma and the frustration of "sour grapes." It is obvious that the consent required for an international regime must be *normative* and therefore prescriptive for the actions of participating actors. Any attempt at regime building is designed to achieve a coherent form of the mode of (international and global) economic reproduction and of (national) political organization on a world market level.

The "cooperative activity" of national states, that is the historical structure of a regime, is not fixed and stable for all times. Competition in all spheres, in particular in the economic system, is a very powerful driving force of institutional change and of shifts in the power structure of the hegemonic system. Therefore, hegemony – hegemony as the characteristics of a system as well as the hegemony of *one* state within this system – assumes a cyclical form and follows "long waves" of emergence, delegitimation, erosion, challenge of hegemony, and possibly "hegemonic war" (Modelsky 1981; Thompson 1983; Wallerstein 1979; Kennedy 1987). This statement becomes quite obvious when looking at the functions and means of hegemony, such as power, ideology, and in particular, money. Strange (1988) has devoted her analysis of "states and markets" to the different aspects of power (structural and relational power) and their means of exercise (military, economic, financial power, and disposal on knowledge), so that I will not deal with the emergence, decay, or – in general – the functions of power in the modern world system.

Instead, let us take a closer look at the monetary regime. What does money have to do with hegemony? From the viewpoint of neoclassical theories, a regime of flexible exchange rates, regulated by market forces, should resolve the basic problem of open economies (which is another expression for that basic contradiction between the globalization of the economic reproduction process and the national character of the state). This problem takes the form of a long-term equilibrium of the balance of payment, the most important indicator of the performance of a national state in a global economic environment. In the long run, exchange rates follow purchasing-power parities; in the short run, short-term capital flows, following interest rate differentials or expectations about exchange rates, may balance the payments of trading partners. Short-term capital flows serve as signal to the participating economies so that in the long run the balance of payments will be balanced: there are no losses and no gains in international reserves beyond their "natural" annual increases.

The balance of payments is composed of the balance of trade plus the balance of services minus the balance of long-term capital. Deficits and surpluses result in changes of reserves and in compensating flows of short-term capital.

$$Ex - Im - (K_{lim} - K_{lex}) = K_{sim} - K_{sex} + D$$

where Ex indicates commodity exports, Im commodity imports, K_{lim} and K_{lex} long-term Capital imports and exports, respectively; and K_{sim} and K_{sex} short-term capital imports and exports. D indicates the change in international reserves, such as gold and hard tradable currencies.

Neoclassical positions basically refer to the balance of trade, whereas Keynesians take into consideration all sub-balances of the overall balance of payments. Therefore, in the Keynesian discourse the exchange rate is not only dependent on purchasing power parities, but also on the elasticities of export and import prices, autonomous capital flows, the distribution of currency reserves, etc. In that case, flexible exchange rates in a multicurrency standard necessarily create economic instabilities via arbitrage speculation, attempts to promote the national currency by means of de- and revaluations, stabilization programs at the cost of employment etc., and due to the existence of free riders, who try to benefit from the system but do not want to pay for its costs. In order to prevent the success of free traders' attitudes, the monetary regime after World War II could not be based on flexible exchange rates and a multicurrency standard, but on the *institutionalization* of fixed exchange rates and one hegemonic currency as a means of regulation.

Since, first, international money as well as national money will be accepted by economic agents only if it is secure and if it therefore guarantees the value of assets; since, second, international fiduciary currency needs some kind of a collateral, as the experience of the 1920s showed; since, third, gold as an "eternal value" is not viable in modern capitalist economies which expand at high speed; and since, finally, a truly international money cannot be created – because there is no international state – the *international* monetary regime can only be based on the *national* money of the hegemonic state. Therefore, it was not by accident that the plans of institutionalizing an international fiduciary money (the "Bancor") failed, and that all attempts to reestablish the British pound as a world money after World War II were in vain.

The collateral of international money therefore is nothing but the economic power of the hegemonic state: economic strength measured in price stability and the level of productivity, a strong creditor position vis-à-vis the other partners of the system and political and military superiority. After World War II there was only one currency left which was – given these general conditions – able to exert the functions of a world money: the US dollar. But since the strength, i.e. the value, of world money depends on the performance of the hegemonic power, an erosion of the economic power of the leading nation immediately has repercussions on the functioning of the international monetary system and vice versa: crisis tendencies in the international monetary system affect the national money and thereby the performance of the national economy of the hegemonic power. The functioning of the international system and the economic performance of the hegemonic nation are intimately linked together, via a beneficial link thanks to the seigniorial advantages of the hegemonic power (Cooper 1987: 8), and via an unfavorable link due to the possible transfers of crisis tendencies from a national economy into the international system and vice versa.

Though the Bretton Woods System cannot be explained here in detail, suffice it to say that the system worked well in terms of growth performance and proved capable of integrating nearly the whole Western world into that system, even the recently decolonized countries of the (since the 1950s) so-called Third World.

Table 2.3 Growth rates of GDP, GDP per capita, and volume of exports

	1913–50	*1950–73*	*1973–87*
		GDP	
16 OECD countries	2.0	4.9	2.4
15 Developing countries	2.1	5.3	4.7
32 Countries	2.1	5.1	3.4
		GDP per capita	
16 OECD countries	1.2	3.8	1.9
15 Developing countries	0.7	2.7	2.5
32 Countries	1.0	3.2	2.2
		Volume of exports	
16 OECD countries	1.0	8.6	4.2
15 Developing countries	1.1	5.9	6.9
32 Countries	1.1	7.4	5.5

Source: Maddison 1989: 32.

The growth rates of GDP, productivity, and world trade, not to mention other indicators in the first two decades after World War II, were far above the average performance in this century. The growth of exports was also much stronger than overall production and production per capita (from 1950

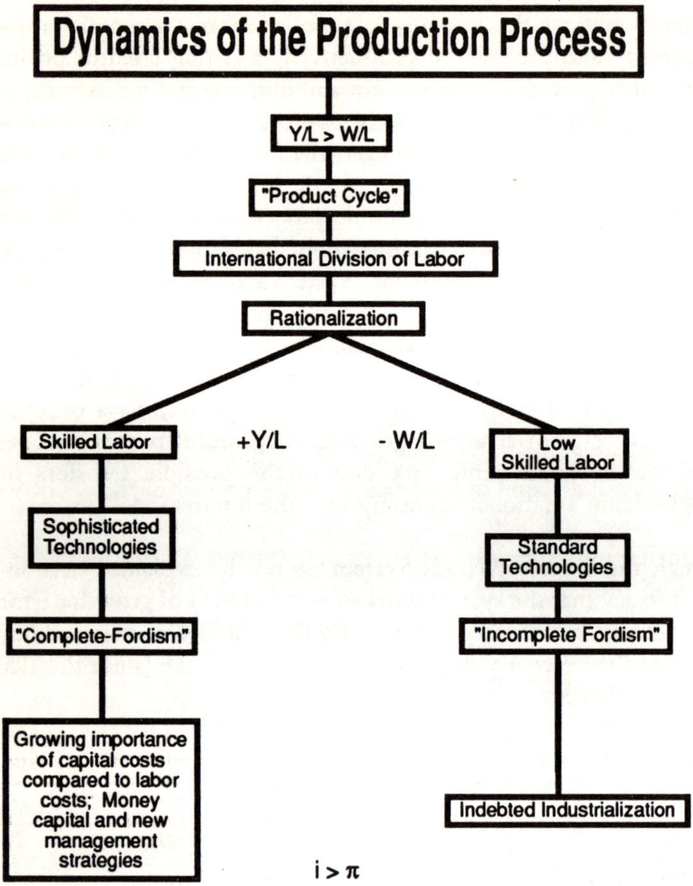

Figure 2.1 International configuration of national Fordisms in the "First" and "Third" Worlds

until 1984 it was about 1.7 times faster). This set of conditions has already been pointed out as a favorable precondition for growth in the industrialized world. But it had also consequences for the less developed countries, for in the environment of a fast-growing world market, the strategies of import substitution which were implemented after the breakdown of world trade following 1929 no longer made economic sense. The developing world, including both the formally decolonized countries and the long-independent countries, switched in the 1950s to an export-oriented and outward-looking strategy of opening internal markets to foreign capital. Though such strategies appeared to be the only way to industrialization and modernization, they also made the developing world dependent on the performance of the world market in general and on its leading economic powers (in the OECD-bloc) in particular. There was neither modernization nor just dependency but some kind of dependent modernization or modernized dependency. In any case, the countries of the Third World became subordinate partners within the international division of labor. An international configuration of national Fordisms in the "First" and "Third" Worlds arose, which may be characterized by Figure 2.1.

Since the dynamics of the production process in the industrialized world resulted in a slowdown of productivity increases (for the USA see Cuomo Commission Report 1988; Scherrer 1990 for an overview), on the one hand, and a palpable real wage increase per worker on the other hand, the "old" Fordist balance between productivity growth and demand growth eroded. The obvious expressions of this erosion were on the one hand growing inflation,[3] and declining profit rates and accumulation rates on the other hand. But thanks to the "product cycle," it now became technologically feasible, organizationally possible, and economically profitable to establish an international division of labor which in the industrialized countries saved expensive labor and halted the productivity slowdown by introducing new technologies into the production process, while sections of the Fordist production process with low skilled labor were transferred to some of the developing countries. Of course, in the 1980s there were counter-tendencies and further evolution of the international division of labor as a response to the debt crisis of Third World countries. But, in the 1970s, it did appear as if a global dispersion of the Fordist model was taking place as a strategy of transnational corporations in search of ways to reduce labor costs and to expand markets beyond the traditional countries.

Nonetheless, in both the core and the Third World, the "old" Fordist system has become incomplete. The necessity for the production and reproduction system to be complementary exists only on the level of the world market and no longer at the national level. The defects of Fordism are manifested in an uneven manner, and therefore present very different consequences in the industrialized and less developed worlds. Whereas in highly developed countries the first signs (technologies and social forms) of a new "post-Fordist" structure have already appeared, in the less developed countries, incomplete Fordism takes on a "bloody" (Lipietz 1987) and "bastardized" form. Most important for our argument, however, are the links between both sides of the coin: the "complete-Fordist"/"bloody-Fordist" division of labor has been more and more transformed into a creditor–debtor relation, hence into the buildup of an international debt-economy. The "debt regime" (Wood 1986) becomes regulated by the global process of interest rate formation on international financial markets. Therefore, it is necessary once again to turn our attention to the functioning of the monetary regime.

Its main characteristics – with regard to the question to be discussed here – may be summarized as a triangle (Figure 2.2):

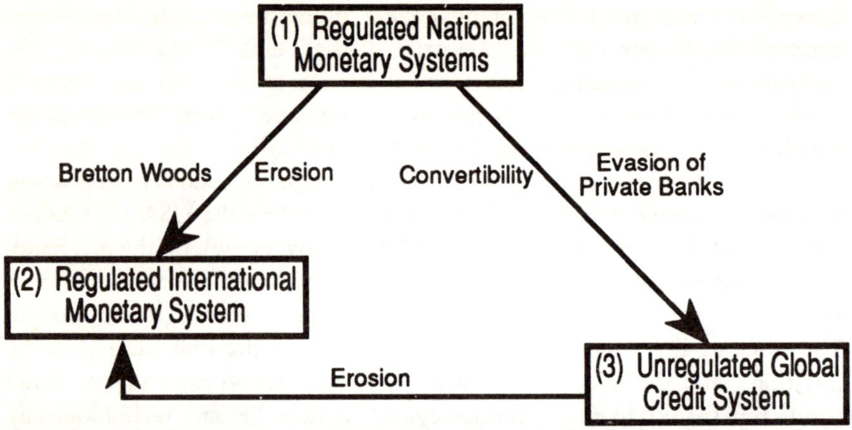

Figure 2.2 Monetary systems dynamics

The first two angles and their links already have been described: the compatibility condition for a regulated national monetary system in the case of an international reproduction process is fulfilled only in the case of the establishment of a regulated international monetary system. However, money exists in different forms and performs different functions. It serves as a *means of circulation* in order to facilitate *commodity* exchange, that is, world trade. The rules of the international monetary system of Bretton Woods, above all, proved adequate for regulating money as a means of circulation. It was a regime organized for the avoidance of a transformation of buyer–seller relations into debtor–creditor relations by means of short-term instruments and means to force debtors to pay off their debts. But money has other functions beyond facilitating commodity circulation. Money functions as a *means of payment* and therefore it assumes the form of *credit*. In this function it is indispensable for capitalist accumulation since loans finance investment of different types. Put another way, without money as a means of payment, the investment process would immediately come to a halt. Insofar as the capital accumulation process is becoming globalized, money as a means of payment (credit) also expands beyond national frontiers. The globalization of production is thus accomplished through the globalization of finance. The emergence and growth of transnational corporations after World War II necessarily generated the transnational banking system, albeit with a certain time lag. But the rules, norms, and institutions of the international monetary regime are only capable of controlling money as a means of circulation and not as the basis of international credit. The global credit system, therefore, emerged as an unregulated one, as a reality outside the regulatory competence of the international monetary regime. This deficiency of international regulation is not due to the absence of international political power, but to the inherent functions of money. The international monetary system becomes a "non-system" (Williamson 1988: 213) or – as Robert Triffin put it – an "international monetary scandal."

The emergence of the international credit system is thus a product of the successful performance of the capitalist world itself over the last few decades. But this success has now become a main problem for the regime, for (see third angle of the triangle in Figure 2.2) unregulated global credit was a factor of erosion of the (political institutional) regulation of the whole Fordist system. The reason for this is quite simple. First, the quantity of private liquidity since the beginning of the 1960s increasingly overtook that of official liquidity. One aspect of that process was the shift from the so-called "dollar shortage" of the 1950s to the "dollar glut" of the 1960s. Another expression was the growing difficulty of central banks in mobilizing official liquidity for the defense of fixed exchange rates against huge waves of private capital, which shifted without regulation from one currency to the other. Second, and more important than this quantitative aspect, is the "qualitative" logic behind these private capital movements; whereas official liquidity has been used to promote international trade and to support national (Keynesian/Beveridgean) policies of full employment, the flows of private liquidity followed the national differences of yields on capital. The consequence: a new "privatization of the world economy" (Czempiel 1989) in the form of transnational corporations and internationally operating private banks took place, undermining the regulatory capacities of nation-states and international institutions. The growing deficiencies of politically controlling international capital movements made room for the emergence of international capital markets. The formation of interest rates took place beyond the reach and influence of national monetary policies. Approaches like the Mundell-Fleming model (Mundell 1968, 1971; Fleming 1962) tried to formulate policy responses on this new international environment for national economic policy, but without real success (Herr 1989).

Transnational corporations have become the *dramatis personae* of the global economic system rather than governments and political institutions. On the level of the international credit system, the control of the interest rate as one of the most important target-variables of national economic policy has increasingly fallen into the realm of global financial markets. The underside of that development has been called the "loss of interest rate sovereignty" (Scharpf 1987). This tendency was one of the most important reasons for the breakdown of Keynesian economic policy and of the corresponding institutional structure of the class compromise on a national level. Moreover, as already mentioned the productivity lead of the United States has evaporated. In the course of this transformation process the USA has lost the advantage of unchallenged competitivity, and the dollar as the national money of the United States and as the world money of the global system has in part lost its collateral. The first consequence has been a remarkable devaluation of the dollar in the form of rising inflation rates; a deterioration of the exchange rate vis-à-vis other leading currencies (DM, SFR, yen, etc.) followed as well as a shift towards a multicurrency standard. In the second instance the devaluation of the dollar, due to its eroding collateral, took the shape of rising interest rates in order to compensate for the growing risks. Consequently, the hegemony of the United States in the global system has been challenged by competing countries.

THE CONTRADICTIONS BETWEEN REAL AND MONETARY ACCUMULATION

It has already been shown that the "Fordist equation" of balanced increases of productivity and wages per worker only partially reflects the dynamics of the capitalist accumulation process. In order to understand its contradictions, the development of the productivity of capital (and therefore of the profit rate) has to be taken into consideration. First, there is overwhelming empirical evidence that in the long run the profit rate has a tendency to fall, even in the recent period of "Fordist" reproduction structures; this has already been described above. Second, this tendency is accompanied by the deregulated interest rate formation on international financial markets. A third tendency already mentioned is the erosion of US competitiveness, that is, of the country supplying international financial markets with the world money. It is to be expected that money as a means of payment (money in the form of credit) becomes more expensive (in terms of the interest rates) the less collateral is available for the backup of credit money. Thus, the decay of US hegemony in the global system is a very powerful driving force behind the (real) interest rate increases.

But there are additional causes for interest rate increases: the loss of competitiveness of the US economy in general compared to other highly developed capitalist economies resulted in a trade balance deficit which had to be financed by capital imports. A cursory look at the development of the US balance of payment since 1950 shows clearly that, beginning in the 1970s, the United States was transformed from a creditor position into a debtor position in the world economy (Table 2.4).

The USA was transformed from a (trade) surplus country to a deficit country and from a capital-exporting country into a capital-importing country. This shift would not be problematic were the USA not playing the role of the hegemon in the world economy. But in the long run the exertion of hegemony and permanent capital imports are not compatible. These factors therefore destabilize the monetary regime. The tendency of eroding hegemony also is reflected in interest rate differentials between the leading countries (Table 2.5).

Beginning with the second half of the 1970s, interest rates for long-term

Table 2.4 The structure of the US balance of payment after World War II ($ billions; aggregated balances)

Period	*1950–9*	*1960–9*	*1970–9*	*1980–9*
Trade balance	29.3	40.8	−103.8	−919.3
Current account	6.0	33.3	−4.3	−808.3
Capital balance (long-term)	–	−27.8	−35.4	363.6
Exchange and gold reserves (as % of world reserves)	(1952) 50.0	(1962) 27.4	(1982) 8.3	(1987) 6.7

Sources: Economic Report of The President 1988; OECD 1988; Data for 1988 and 1989 are preliminary: Data for the capital balance comprise 1980 to 3rd quarter 1987.

Table 2.5 Long-term interest rates

Year	*Long-term interest rates*					*Interest rate differentials*		
	FRG	*GB*	*Switz.*	*Japan*	*USA*	*USA/Japan*	*USA/FRG*	*Japan/FRG*
1970	9.41	9.22	5.72	7.19	6.58	−0.61	−2.83	−2.22
1971	7.99	8.91	5.27	7.28	5.74	−1.54	−2.25	−0.71
1972	7.88	8.91	4.96	6.7	5.64	−1.06	−2.24	−1.18
1973	9.33	10.72	5.59	7.26	6.31	−0.95	−3.02	−2.07
1974	10.38	14.77	7.13	9.26	6.98	−2.28	−3.4	−1.12
1975	8.48	14.39	6.44	9.2	7	−2.2	−1.48	0.72
1976	7.8	14.43	4.98	8.72	6.78	−1.94	−1.02	0.92
1977	6.16	12.73	4.05	7.33	7.06	−0.27	0.9	1.17
1978	5.73	12.47	3.33	6.09	7.89	1.8	2.16	0.36
1979	7.43	12.99	3.45	7.69	8.74	1.05	1.31	0.26
1980	8.5	13.79	4.76	9.22	10.81	1.59	2.31	0.72
1981	10.38	14.74	5.57	8.66	12.87	4.21	2.49	−1.72
1982	8.95	12.88	4.85	8.06	12.23	4.17	3.28	−0.89
1983	7.89	10.81	4.51	7.42	10.84	3.42	2.95	−0.47
1984	7.78	10.69	4.58	6.81	11.99	5.18	4.21	−0.97
1985	6.87	10.62	4.7	6.34	10.75	4.41	3.88	−0.53
1986	5.92	9.85	4.24	4.94	8.14	3.2	2.22	−0.98
1987	5.84	9.5	4.03	4.21	8.63	4.42	2.79	−1.63
1988	6.11	9.36	4.02	4.27	8.98	4.71	2.87	−1.84
1989	6.77	9.25	4.75	4.55	9.19	4.64	2.42	−2.22

Source: Sachverständigenrat 1990; own estimations.

capital in the United States have been considerably greater than interest rates for long-term capital in Japan or the Federal Republic of Germany. This tendency, of course, reflects the loss of trust of asset holders in the strength of the US dollar. The increase in interest rates as well as in the interest rate differentials have important repercussions on the global process of *real* accumulation. Should interest rates for long term (and short term) capital be tangibly above or near the expected rate of return on real capital (the profit rate on productive capital), the microeconomic units may prefer to hold financial assets instead of investing liquid funds into the widening or deepening of real capital. The contradictory tendencies of profit rates (falling) and of interest rates (increasing) already have been described by Hilferding (1898) in his "Financial Capital," at that time he drew far-reaching conclusions concerning the further structure and development of capitalism from competitive to monopoly and "financial capitalism."

Even if we do not hold to Hilferding's conclusions, we have to take into consideration that rising real interest rates (Atkinson/Chouraqui 1985: 7 passim) in all countries and net interest payments which (over a considerable time period) grow faster than net operating surpluses (Chan-Lee/Sutch 1985: 152 passim) reflect structural changes in the accumulation process. When the market value of firms grows less than the net capital stock (Tobin's *q*), the incentive to invest in real capital becomes less effective (Tobin 1969). In the case of Tobin's *q* declining to less than unity, financial investment is more lucrative than real investment. Historical data show that Tobin's *q* has been falling in all developed capitalist countries, although the data basis is rather limited (Chan-Lee 1986: 205–32).

This tendency toward the declining profitability of capital expressed by means of different indicators (profit rate; interest rate; Tobin's *q*), is affecting the mode of regulation of the international monetary regime. Internationally formed interest rates have to be explicitly integrated into micro- and macroeconomic processes of decision making, much more so than before. The "Fordist equation" between wages, productivity, and profits now must be complemented by a new factor: interest rates. In Keynesian economics the interest rate played always an important role, but Keynesianism employed an assumption which is no longer valid; i.e. that the national state is capable of controlling the movement of interest rates "to that point relatively to the schedule of the marginal efficiency of capital at which there is full employment" (Keynes 1936: 375), at least by means of the "euthanasia of the rentier, and, consequently, the euthanasia of the cumulative oppressive power of the capitalist to exploit the scarcity-value of capital" (Keynes 1936: 376). Today, a reduction of interest rates in order to correspond to lower schedules of the marginal efficiency of capital (or of profit rates) seems to be impossible: either it has to take place on an international and not on a national level or the level of interest rates has to be taken as an exogeneously given variable for national economic policy and therefore also for decision making on the corporate level. The outcome of this change is that capital costs (reflected in the interest rate) become much more important for business decisions than labor costs. This is the reason for new management strategies, which are often viewed as exploring territories "beyond Fordism."

Since each financial investment (loan owned by the creditor) is at the same time a process of indebtedness (debt due to the debtor) the delinking of monetary accumulation from real accumulation results in the build up of an international debt economy. This has taken place via financial innovations and the spread of financial institutions all over the world. A financial network has been superimposed upon the already existing network of trade relations and direct investments of TNCs. Whereas the task of official institutions such as the IMF in the "old" monetary system of Bretton Woods was the financing of balance of payment deficits, in the reprivatized global economy this task has been fulfilled by private financial institutions in a world of tendentially growing current account imbalances (as a percentage of GDP; Atkinson/Chouraqui 1985: 32).

Credits and debts even on an international level do not generate any economic or political problem so long as debtors are capable of paying interests (and amortization on the principal) so that they convert after a certain time into creditors – and vice versa: creditors into debtors. But some conditions have to be met, before the "debt cycle" (World Bank 1985, Chenery/Strout 1966) can function. In the first place we must remind ourselves of a distinction which Kindleberger made in order to explain the causes of the contemporary debt crisis of Third World countries. He distinguishes between development loans, consumption loans, recycling loans (Kindleberger 1981) and – we may add – capital flight loans. Only (foreign) loans used for the financing of development projects can be serviced out of the returns of the development project. In the case of all the other types of (foreign) loans the debt service burdens (and overburdens) the national revenue of the indebted country. If economic difficulties of servicing the debt arise, the interest rate immediately will rise for the compensation of increasing loan risks (e.g. country risks). Hyman Minsky has derived his thesis of "financial instabilities" in capitalist economies from the inherent difficulties of servicing debts due to rising interest rates which overburden the economic capacities of debtors, be they hedge finance units, speculators or Ponzi units (Minsky 1982).

In response to this inherent mechanism of financial systems, the international banking community has invented "financial innovations" in order to respond to growing loan risks, especially in the international financial markets, via rollover techniques, the transformation of short-term deposits into long-term loans, the securitization of loans, and the transformation of interests into spreads and fees. But these financial innovations have extremely asymmetrical effects. They only serve as instruments for the international banking "community" to secure at least parts of their outstanding loans and – much more important – of interest payments by debtors. But they have not relieved debtors of their debt burden. This tendency has been further aggravated by the failure of many development loans due to miscalculations by the project management or declining terms of trade on the world markets for agricultural and mineral primary goods. This event is also an outcome of the delinking of real and monetary accumulation. The reason is quite simple. Since development loans are only effectively used if development projects are profitable enough to cover interest costs and amortization payments, any decline of their profitability has an immediate impact on the capacity of servicing foreign debts. The performance of development projects in countries which pursue a strategy of indebted industrialization depends on international markets while markets themselves are dependent on the economic stance and economic policy of OECD countries (Balassa 1984). Therefore, the success of development loans became questionable because of the accumulation crises in the industrialized world, which transferred the negative growth effects via market mechanisms (on product markets as well as on financial markets) into the developing countries. But these countries were, compared to the industrialized countries much worse off, not because of their comparably low levels of private and public welfare, but because of their indebtedness. The mechanism of the "debt cycle" did not work at all, because debtors did not succeed in decreasing the debt overhang and – the mirror-image – creditor countries did their best in order to remain creditors vis-à-vis the Third World. The IMF, originally an institution of regulating money as a means of commodity-circulation (fixed exchange rates), has increasingly been transformed into an institution in support of the international credit system: by enforcing debtors to undertake adjustment measures and by impeding the efforts of single financial institutions to behave as free riders via a bailing out.

A bitter conclusion has to be drawn. The same cause – declining profit rates and rising interest rates – transformed the world economy into a "casino capitalism" (Strange 1986), a worldwide debt economy. The decoupling of monetary from real accumulation created for developing countries a kind of a cul-de-sac and for the industrialized countries an outlet for liquid funds into risky financial assets (Altvater/Hübner 1989). The original international monetary regime – ruled by fixed exchange rates, the dollar as the world money and the USA as the hegemonic power, and last but not least by equilibrating mechanisms in order to avoid fundamental imbalance – is over.

THE FORDIST STALEMATE

We now come back to the argument depicted in Figure 2.1. In the industrialized "Fordist" countries, the increasing importance of monetary flows as compared to categories of real accumulation, provoked a change in management strategies. I will not discuss the spread of speculative centralization battles, hostile takeovers, etc. (Report on Leveraged Buyouts 1987 and 1989; Altvater/Hübner 1989), but rather stress the growing role of capital costs (interest rates) for decision making on the national as well as on the corporate level. For the nation-state, increasing real interest rates have resulted in the already mentioned "loss of interest-sovereignty," and consequently in the impossibility of continuing traditional Keynesian/

Beveridgean policies of demand management for full employment thus provoked the growing influence and attractiveness of monet recommendations and of supply-side economics for the formulatio… economic policy. At the corporate level, the increased importance of interest rates compared to the profit rate is reflected not only in a declining Tobin's *q* and, therefore, a growing attractiveness of financial investment; insofar as real, productive investments are undertaken, their character and strategic targets are changing: in the direction of relocating industries, of implementing just-in-time strategies, of flexibilizing the whole production process and the administration of corporations. All these methods basically serve one objective, that is to decrease capital costs and to reduce all other costs in order to increase profit rates on a level comparable to interest rates. The driving force of capitalist modernization, therefore, is the dynamics of monetary markets more than that of real accumulation and the distributional conflict between capital and labor. The traditional "Fordist equation" between growth rates of productivity and wages has been substituted by a more complex relation between interest rates and profit rates on the one hand, and the reduction of labor costs in order to increase profits on the other hand. Where does the effective demand come from? With the declining importance of industry in national income, mass consumption of workers as the decisive component of effective demand in "Fordist times" is increasingly going to be replaced by the consumption of employees of the service sector and by those who spend their interest income: not only financial capitalists but also the growing number of employees of the financial industry. In Germany, for instance, the share of the wage bill paid in manufacture declined from 43.3 percent in 1960 to 38.0 percent in 1988, whereas the share of wages paid in the service sector increased from 7.1 percent to 12.9 percent and the share of wages in the public sector grew from 14.3 percent to 19.3 percent (Sachverständigenrat 1989: 252). In the US the trend was quite similar: the contribution of the manufacturing sector to the national product declined from 28.0 percent in 1960 to 19.7 percent in 1985, whereas the contribution of the financial and insurance sector increased from 14.1 percent to 15.9 percent, and that of other services from 10.0 percent to 16.1 percent (Economic Report of the President 1990: 306). Of course, these indicators only show a tendency on the demand side from the Fordist form of generating mass consumption to a new regime compatible with the new conditions of profit production under the pressure of international financial markets.

The result is a deep change of the international economic system: the decline of the rate of real investment and the shift from widening to deepening strategies has caused the well-known problem of structural unemployment in the developed countries. The impact of the growing weight of capital costs and the reduced importance of labor costs for modern corporate planning on the international division of labor is quite obvious:

the management of large corporations is more concerned with keeping capital costs low than looking for favorable production sites where labor is comparably cheap. Therefore, for many industries it makes no more sense to follow the strategy of the "new international division of labor" (Gordon 1988), i.e. a strategy which was based on use of a combination of skilled and unskilled labor in different parts of the world in order to minimize labor costs. For, compared with capital costs labor costs play a minor role in corporate calculations.

Even more important is the fact that due to the privatization of monetary and financial markets the steering capacity of nation-states and international institutions has been significantly reduced. The strategic options are therefore determined by market processes more than by political forces as embodied in institutions and rules. The consequence on the plant and enterprise level is enforced flexibilization not only of the production process, but of all stages of the whole circulation process of capital: new production concepts (Kern/Schumann 1984) in order to rationalize the production process as a technological and social system, new management strategies in order to minimize capital outlays for stocks and inventories ("just-in-time"), corporate financial strategies in order to economize monetary capital and to use financial markets as "parking lots" for liquid funds. Therefore, modern industrial corporations behave increasingly like financial institutions ("Toyota-Bank"; "Siemens is a bank which holds a small electronic workshop"), where cash management is more important than labor management. Of course, tendencies of this kind have always been present in capitalist economies, but under the regime of high interest rates which cannot be controlled by the nation-state, the driving forces to adapt real processes to the tendencies of monetary accumulation become much more important than ever before.

On the national level, adjustment and adaptation to world market structures becomes a powerful political imperative. Consequently, neither the old Fordist equation nor Keynesian policy models fit into this economic environment at the end of the century. The trajectory of Fordism has come to a halt. The rupture with the "old" model of accumulation and mode of regulation is quite obvious, but although there are signs of "post-Fordist" structures, so far no new stable, complete, compatible, and cohesive system has emerged. First, interest rates that are permanently higher than profit rates (and Tobin's q less than unity) are not compatible with real accumulation, economic growth, and full employment. Even in a world characterized by high real interest rates there may be some growing points, but these regions (today Japan and some countries of Western Europe, especially Germany) are accompanied by the expansion of lagging areas of poverty and even misery in other parts of the world. Second, no new institutional framework for the regulation of reduced labor and expanded leisure time has evolved. Third, the regulation of the metabolism of human beings with nature is deficient; the ecological question – which has not been discussed in this chapter – remains unsolved. And, last not least, in the aftermath of national Keynesianism, neither monetarism nor supply-side economics resolved the urgent problems of unemployment on a national level. Post-Keynesianism, however, makes sense only on a global scale. In order to achieve this, the reconstitution of a new hegemonic order is necessary. But none of these deficiencies, at present, can simply be removed. The first function of the "great crisis" already has been fulfilled: the "de-formation" of the old Fordist structure. Thus far, it is possible to speak about "post-Fordism", but only in a negative sense. For, the second phase, i.e. the "trans-formation" to a new form of accumulation and regulation, has yet to occur.

3

FORDISM AND POST-FORDISM

A critical reformulation

Bob Jessop

As the language of Fordism and post-Fordism has entered everyday discussion it has also been vulgarized.[1] This reduces its utility for theoretical understanding and empirical analysis and generates many confusions and controversies. This chapter critically reworks the conventional terminology of Fordism and post-Fordism by distinguishing four levels on which they can be analyzed. It also notes a fundamental analytical asymmetry between the two terms and calls for more cautious and critical use of the notion of post-Fordism.

THE FOUR LEVELS OF FORDISM

The concept of Fordism was popularized in the USA by Henry Ford himself and was already part of social scientific and popular consciousness in North America and Europe in the 1920s. Even then it had different nuances and these have since become disturbingly diverse. We can introduce some order into this confusion by distinguishing four levels of Fordism: the labor process, the regime of accumulation, and its modes of regulation and societalization.

The labor process

As a distinct type of capitalist labor process, Fordism refers to a particular configuration of the technical and social division of labor involved in making long runs of standardized goods. Fordist "mass production" is typically based on a technical division of labor which is organized along Taylorist lines, subject in its immediate production phase to mechanical pacing by moving assembly line techniques, and organized overall on the supply-driven principle that production must be unbroken and in long runs to secure economies of scale. The assembly line itself mainly exploits the semiskilled labor of the "mass worker" but other types of worker (craft or unskilled manual workers, foremen, engineers, designers, etc.) are employed elsewhere. In addition Fordism ideally involves systematic control by the same firm of all stages of accumulation from producing raw materials through to marketing (cf. Siegel 1988: 5). This complex technical division of labor is sometimes related to a complex regional division within or across national economic spaces: in late Fordism, for example, one might find "engineering and conception in region I, skilled production in region II, unskilled production in region III" (Lipietz 1982: 37). Thus the dominance of mass production in a given enterprise or sector does not exclude other labor processes or types of worker. Instead it subjects them to its own logic. For the dominance of mass production means that, by virtue of its impact on productivity and productivity growth, it is the main source of dynamism in a firm or sector; and that other processes and activities will be organized to support, enhance, or complement it.[2] This level of analysis is basically microeconomic.

The regime of accumulation

As an accumulation regime, i.e., a macroeconomic regime sustaining expanded reproduction, Fordism involves a virtuous circle of growth based on mass production and mass consumption. Many studies assume that the Fordist regime and its reproduction are autocentric, i.e., that the circuit of capital is primarily confined in national boundaries. On these assumptions Fordism's virtuous circle involves: rising productivity based on economies of scale in mass production, rising incomes linked to productivity, increased mass demand due to rising wages, increased profits based on full utilization of capacity, increased investment in improved mass production equipment and techniques, and a further rise in productivity. The apparent harmony among the steps in this virtuous circle does not guarantee its realization. Thus some analysts note the role of various margins of flexibility[3] and/or "built-in stabilizers" that help to keep the circle virtuous despite the inevitable tendencies towards instabilities and disproportions (e.g., Boyer and Coriat 1986). Others note that these virtues themselves require at least two key proportionalities to be met. These are that:

1 increased productivity in Department I (the capital goods sector) offsets the rising technical composition of capital[4] if the capital/output ratio is not to increase and thereby depress profits; and
2 the growth rates in wage earner consumption and in productivity in Department II (the consumer goods sector) move in a similar range – countering tendencies toward a crisis of underconsumption due to insufficient demand and toward a wage-induced profits squeeze (e.g. Lipietz 1982, 1985).

Not every firm or branch of production must be dominated by Fordist techniques for this mode of growth to occur as long as the leading sectors are Fordist. Indeed, if mass production is to find a mass market, there must be matching growth in output of other types of goods (such as steel, oil, roads, family housing, and electricity) and services (such as retailing, consumer credit, and the servicing of consumer durables). Thus some authors note the dominance of an "auto-industrial complex" under Fordism to highlight how other production sectors complement mass production in the vehicle sector (cf. Perez 1983: 369). Even with such refinements, however, work on the Fordist accumulation regime is firmly rooted in economic analysis. To consider its institutional and organizational supports we move to a third analytic level.

Its mode of regulation

Fordism can also be examined as a mode of regulation, i.e., as an ensemble of norms, institutions, organizational forms, social networks, and patterns of conduct which sustain and "guide" the Fordist accumulation regime and promote compatibility among the decentralized decisions of economic agents despite the conflictual character of capitalist social relations (cf. Lipietz 1985: 121). Fordism can be specified through the forms assumed by different moments in the circuit of capital, the ways in which these forms get reproduced, and their articulation with each other. Thus one could explore the distinctive features of the Fordist wage relation (the skill profile of the collective laborer, the organization of labor markets and wage-effort bargaining, the nature of the wage form, and the balance between the private and collective reproduction of labor power); the Fordist enterprise (its internal organization, the source of profits of enterprise, forms of competition, other ties among enterprises, links to banking capital); the nature of money (its dominant form and its emission, the banking and credit system, the allocation of money capital to production); the nature of commercial capital (especially in mass consumption and distribution); and the links between the circuit of capital and the state (the forms of state intervention in different moments of the circuit of capital). This third level of analysis entails a mesoeconomic, institutionally focused analysis of the circuit of capital. It gives a middle-range link between the labor process and the basic features of macroeconomic reproduction and must be carefully defined for each moment in Fordism:

- The wage relation is organized around the key role of semiskilled labor in large plants or concerns;[5] management recognizes unions for collective bargaining, unions in turn concede management's right to control the labor process and corporate strategy; wages are indexed to productivity growth and prices and are downwardly sticky despite fluctuations in demand for labor; collective bargaining and/or minimum wage legislation spread rising wages to employees in non-Fordist sectors and thereby maintain relativities and demand; and indexed welfare benefits financed from progressive taxation generalize mass consumption norms to the economically inactive. Even in a largely autocentric Fordist regime this pattern is compatible with dual labor markets and/or non-unionized firms or sectors – as long as there is adequate overall social demand for mass-produced goods.

- The ideal-typical Fordist enterprise is a large concern in which ownership and control are separated. It has a distinctive multidivisional, decentralized, market-oriented organization overseen by a central board which engages in long-range planning. This pattern was pioneered at General Motors and became the norm for other large US industrial firms in the 1920s and 1930s. Elsewhere mass production was often coupled with trusts and cartels. The main source of profits of enterprise in Fordism is relative surplus-value based on continual improvements in productivity and economies of scale. Those firms at the forefront of Fordist process and product innovation can also earn "technological rents." In both cases the main form of capitalist competition is monopolistic rather than liberal. Rather than participating in a flexi-price system in which prices vary with demand, firms engage in cost-plus pricing, price leadership behavior, and competition through advertising.
- Money is fiduciary and national in character rather than an international commodity money; private credit is supplied by a hierarchically organized banking system overseen by a central bank; corporate expansion depends on access to private credit as well as reinvestment of profits of enterprise; consumer credit is a major factor in households' ability to purchase housing and major consumer durables and thus sustain demand; and state credit policies are targeted on aggregate demand and full employment.
- Commercial capital has a key role in establishing the links between mass producers and mass demand via mass advertising, mass retailing (multi-branch retailing, supermarkets, mail order), mass credit (installment-plan transactions, mortgage finance), consumer research, etc. The mass media have acquired a crucial role in diffusing mass consumption norms (Beniger 1986). Marketing needs also feed back into design (even at the expense of strict pursuit of the logic of mechanized mass production) to encourage marginal product differentiation, annual style changes, etc. Thus design becomes a key factor in linking mass production and mass consumption (Sparke 1986: xxi).
- The Fordist state is a Keynesian welfare state which has two key functions in promoting the virtuous circle of Fordism. It manages aggregate demand so that the relatively rigid, capital-intensive investments of Fordist firms are worked close to capacity and firms have enough confidence to undertake the extended and expensive R&D as well as the subsequent heavy capital investment involved in complex mass production;[6] and it generalizes mass consumption norms so that most citizens can share in

the prosperity generated by rising economies of scale. Where the latter function involves only limited state provision for collective consumption, the state must ensure adequate levels of demand through the transfer of incomes. More intense intervention is likely both as Fordism rises to dominance and during its declining years.

Analysis of the mode of regulation is concerned with the economy in its integral sense, i.e., the social context in which expanded economic reproduction occurs.[7] It specifies the institutional and organizational conditions which secure Fordism as a national accumulation regime and is especially helpful in defining the peculiarities of different Fordist regimes. But it should not be divorced from work on the more general dynamics of capitalism. For the latter defines the basic tendencies and counter-tendencies, structural contradictions, strategic dilemmas, and overall constraints which inevitably shape modes of regulation, which find a provisional, partial, and unstable resolution in the latter, and whose continued presence and even development eventually undermine any given institutional and organizational solutions. In short, while this level is distinct from Fordism as an accumulation regime, it cannot be properly understood without considering how modes of regulation modify and yet remain subject to the generic laws of capital accumulation.

Its mode of societalization

Other analysts approach Fordism in terms of its overall social impact. They go beyond its microeconomic base, its meso-level institutional preconditions, and its macroeconomic effects to explore its general repercussions on other institutional orders (such as the political system or cultural life) and/or other axes of societal organization (such as its spatial patterning). Such concerns involve not only the direct impact of Fordism but also its indirect effects. Among the latter one could explore measures adopted to deal, for example, with its adverse effects on social cohesion and institutional integration. The recursive pursuit of these effects and their further repercussions has no inherent limits. Thus one could study the impact of growing social welfare expenditures on the labor process and employment, the structure of demand, the dynamic of central–local government relations, or the rise of social movements oriented to welfare issues. With these direct and indirect effects we reach the fourth level of analysis.

Considered as a generic mode of "societalization," i.e. pattern of institutional integration and social cohesion, Fordism moves social relations further towards a mass *société salariale* in which the vast majority of the population depend on an individual and/or social wage to satisfy their needs from cradle to grave (Aglietta and Brender 1984). This contrasts with the pre-Fordist period, when workers were involved in capitalism primarily as producers and met their consumption needs mainly from petty commodity and/or subsistence channels. Fordism itself promotes two complementary trends in consumption: first, growing private consumption of standardized, mass-produced commodities in nuclear family households and, second, provision of standardized, collective goods and services by a bureaucratic state. The first trend is tied to the strategic marketing of "ideological commodities" (such as cars, televisions, washing machines, refrigerators, or mass tourism) whose individualized consumption becomes a mechanism of permanent self-normalization as consumers adopt the "American way of life" (Haug 1986; Luescher 1986; Wolf 1987). The second trend reflects the growing socialization of the social reproduction of labor power as well as efforts to manage the individual and social costs of the Fordist model (cf. Jessop 1986).

Thus, in addition to the state's role in such activities as general education and vocational training, unemployment and retirement insurance, health care, or housing provision, it often deals with such side effects of Fordism (or Fordist forms of modernization) as drastic falls in the rural population, destruction of traditional working class milieux, the privatization of family life, the depopulation of inner cities, or the environmental and social impact of the automobile. One major effect of Fordism here was to expand the local state as the vehicle for socializing consumption, coping with the side effects of the Fordist mode of growth, and managing its local crises – regardless of whether local accumulation itself was mainly Fordist or not. Closely related to this was the growth of the new middle classes (including state employees) who serve Fordist expansion and/or deal with some of its social repercussions. In turn this growth prompted crisis to the extent that such services were less amenable to productivity boosting Fordist techniques.

Another dimension of societalization is its spatial patterning. Fordism involved the growth of core industrial regions comprising large metropolitan areas surrounded by networks of smaller industrial cities. These regions were dominated by the leading Fordist firms and their suppliers; drew in raw materials and, on a growing scale, migrant or foreign labor, from the rest of the world; and churned out mass-produced goods for global markets (Harvey 1989: 132; Storper and Scott 1988: 10). As the Fordist regime developed, firms allocated activities, sourced supplies, and sought markets on an increasingly global scale. Urban life also assumed Fordist characteristics. This can be seen in suburbanization (especially in the USA) and/or high-density urban renewal based on industrial construction techniques, especially in Europe (Florida and Feldman 1987; Harvey 1989). There was also a Fordist "politics of place" centered on "a consumerist representation of urban life as manifest in ideals about the nuclear family, suburban residence, and private car ownership" (Storper and Scott 1988: 30).

Somewhat further removed from its economic logic are the specific forms of political life associated with Fordism. State intervention was reorganized to facilitate the Fordist mode of regulation through economic and social programming and increased administrative discretion as well as an expanding role for the local state in collective consumption and social welfare. Forms of representation and the social bases of the state were also modified. Thus trade unions and business associations had a key role in economic management and political bargaining over social welfare; and, as the parties of government accepted the Fordist compromise between capital and labor, they tended to become "catch-all," people's parties. In northwest Europe this pattern was closely linked to the dominance of social democratic or labor parties but similar policies were pursued elsewhere by regimes with more conservative governing parties (cf. Hirsch and Roth 1986; Roobeek 1987).

So far we have not questioned the common assumption that the Fordist accumulation regime, whether or not it is autochthonous, is at least essentially autocentric in character. But small, open economies (such as Denmark, Sweden, Austria, or Canada) actually moved towards a mass consumption society after 1945 by occupying growing non-Fordist niches in an emerging supranational productive system. In short, under global Fordism not all economies had to be Fordist in all respects. Rather, in a global division of labor whose dynamic was mainly determined by the leading Fordist sectors in the leading economies, economic success could occur in at least two ways. First, national economies could themselves assume a mainly Fordist dynamic, with growth being largely based on an expanding home market; or, second, they could occupy one or more niches which allowed them to enjoy rising standards of mass consumption based on growing export demand and profits in non-Fordist sectors (small batch capital goods, luxury consumer goods, agricultural goods, shipping or financial services, and/or raw materials). Where an economy is not itself primarily Fordist, however, its mode of growth must complement the dominant Fordist logic. In this way it can still be involved in the Fordist growth dynamic rather than being (increasingly) excluded from it.

STRUCTURE AND STRATEGY IN THE ANALYSIS OF FORDISM

Besides these different levels of Fordism we can distinguish their structural and strategic moments. The former refer to the actual organization of the level in question, the latter to the strategic perspectives and discourses which are currently dominant on that level. Structure is the legacy of a complex historical process and often embodies major structural contradictions, strategy could well be short-term or ephemeral and involve irrational means and objectives. Indeed, structures rarely have a simple, unequivocal relation to a single strategy and they often prove recalcitrant in the short term to attempts to recompose them. Conversely, even when strategies are long-term and organic in character, they may not yet have been institutionally embedded in structures nor have constituted new "worlds of work" in practical consciousness.[8] New conceptions of work usually emerge ahead of their time and become hegemonic and embedded in common-sense practices only through complex historical practices. The same holds for accumulation regimes and modes of regulation. In addition, of course, many strategies are "irrational, arbitrary, and willed" and so will sooner or later fail if they remain unmodified (for good illustrative studies of these points, see Elam and Borjeson 1989; Jenson 1990; Kristensen 1990; and Williams, Haslam, Wardlow, and Williams 1986). This suggests that we must establish whether the Fordist era really involved the predominance of Fordist structures. Or was it just marked by the hegemony or dominance of Fordist strategies which lent coherence and direction to what would otherwise have remained rather inchoate and disconnected economic, political, and social changes? It is essential but hard to disentangle Fordist myth from Fordist reality.

It is indeed problematic that Fordist concepts were often hegemonic even when national economies were less often clearly Fordist in their organization and dynamic. The history of many national economies both before and after World War II could well be written in terms of successive attempts to impose Fordist principles on non-Fordist economies. This led to some interesting hybrid economies which have placed national economies in very different positions to exploit post-Fordist opportunities. The vitality of the "Third Italy" is often cited here – although both its reality and relevance are often contested too. There are also many more general questions about the heuristic value of the whole debate about Fordism and post-Fordism. Indeed, as the ensuing remarks might suggest, there is a risk of getting bogged down in mainly taxonomic discussion.[9] Whereas this might help in clarifying some of the issues at stake in disputes over Fordism, it does not as such advance our understanding of the dynamics of postwar growth, the origins or nature of the crises which emerged in the mid- to late 1960s, or the way out of the crisis. For this we must move beyond taxonomy to consider the causal mechanisms and connections implied in different approaches to Fordism.

PROBLEMS WITH THE CONCEPT OF FORDISM[10]

The labor process

Fordism à la Henry Ford was actually quite limited in diffusion and was never fully realized even in Ford's own plants in North America – let alone those in Europe. Indeed two common criticisms are that only a small part of manufacturing output is produced in Fordist conditions and only a small proportion of the labor force is employed in Fordist manufacturing. Both charges hold even for the supposed golden years of Fordism. In part this reflects basic technical limits of assembly line techniques for automated and continuous-flow processes, jobbing or small batch production, or even potentially high volume goods whose composition and level of demand are too varied to justify using product-specific techniques. This means the potential for Fordism depends on the weight of specific sectors and the nature of their products. Thus, before judging the utility of "Fordism" for understanding the labor process in a given economy, we should examine both its industrial profile and its competitive performance. Its explanatory force would be cast in doubt where an economy managed to compete successfully in supposedly Fordist sectors without itself being Fordist. Its utility would be confirmed where a failure to adopt Fordist techniques is linked to manufacturing decline in open competition. Britain's combination of "flawed Fordism" and continuing economic decline would seem to fit the Fordist model. So does Germany's ability to translate a strong capital goods sector and high value-added consumer durable sectors into continuing export success. But Japan poses problems. Indeed Sayer has claimed that "while Japan has forms of organization not unlike those described as 'flexible specialization', it also has other characteristics which cannot usefully be subsumed under western versions of fordism or postfordism" (Sayer 1989: 667).

The limited spread of Fordism also reflects certain social limits making it impractical or unacceptable even where it is applicable. For existing labor market institutions and social traditions could well prevent or block penetration of Fordism even for that "classic" Fordist sector – cars. Several recent studies demonstrate how Fordist mass production has been blocked here by such factors as skill structure, managerial capacities, market size, labor market conditions, and union organization (e.g., Tolliday and Zeitlin 1987).

We might also ask whether studies of the Fordist labor process have been too concerned with manufacturing or, indeed, in this sector, with mass consumer durables generally or even cars and trucks alone.[11] For this diverts attention from the growth generated since 1945 by other sectors (such as aerospace, petrochemicals, or synthetic materials) and biases studies of mass production itself towards some, possibly atypical, manufacturing sectors. A less restrictive definition of Fordist mass production helps us find it elsewhere. Defined as the production of standardized goods or services through a Taylorized technical division of labor, the use of dedicated machinery, and a focus on economies of scale,[12] for example, Fordism occurs in sectors ranging from battery farming to tax collection, mass retailing to mass incineration, multiple choice testing for the mind to mass screening for the body, fast food to mass transit (on US agriculture, see Kenney *et al.* 1988; on agriculture in general, Kamppeter 1986; on office work, Beniger 1986: 432f).

Regarded as a labor process, then, Fordism can be variously defined.

Descriptively this does no harm as long as each definition is relatively precise and different definitions are clearly distinguished.[13] But problems occur in causal explanation because more inclusive definitions entail different chains of argument and/or causation from less inclusive definitions. Here too clarity is essential. For each definition, given certain assumptions about the capital–labor relation and forms of capitalist competition, one could generate arguments about the dynamics of class conflict in the enterprise and/or the course of capital accumulation. We also need a more complex set of distinctions which does not residualize all but a small subset of labor process relations. Even the dichotomy between "flexible specialization" and "mass production" with its two positive terms is not much better. A multi-valued typology would generate rich descriptions and explanations.

Regime of accumulation

To put the historical cart before the conceptual horse, we can ask whether there was ever a Fordist regime of accumulation and how one might identify it? Emphasizing the virtuous circle of mass production and mass consumption in an autocentric national economy means that few national economies could be described as Fordist. Only the American economy had the continental scope and range of resources to be virtually self-sufficient and develop something like a true Fordist accumulation regime. In these terms it might be better to argue that Fordism occurs mainly on a local or regional scale in the form of Fordist industrial districts or else panregionally or supranationally in the form of Fordist circuits of capital.[14] Fordist regional economies, pockets of flexible specialization, and other local regimes can all be found inside national boundaries. This poses problems about their role in the national mode of growth and their insertion into the dominant mode of regulation. A further complication arises because advanced capitalist economies were significantly involved in foreign trade from the formation of Fordism through its heyday into its declining years. This holds above all for small, open economies. Thus, whatever the virtues of an ideal-typical, autocentric Fordist regime, ideal and reality were often far apart.

The idea of a Fordist accumulation regime could be rescued if one agreed that its virtuous circle need not actually occur through a close coupling of mass production and mass consumption *within* national economies but could be secured instead where a national economy has the following features. Its dynamism would be based on intensive accumulation in one or more leading sectors, rising productivity due to economies of scale and/or other sources of relative surplus value, rising wages indexed to rising productivity and profitability, a corresponding growth in mass consumption, rapid domestic expansion in the production of mass consumer goods and/or the various complementary goods and services needed to enjoy them, and, to close the circuit, sufficient export earnings to finance the import of mass consumer goods and other inputs needed to keep the virtuous circle in operation. Thus a given national economy need not itself produce complex mass consumer durables as long as it generates enough export earnings to finance their importation and has a mode of regulation which generalizes both mass consumption norms and effective demand. Even so the overall economic dynamic would differ from that in other regime types: for example, from those based on extensive accumulation and competitive regulation or intensive accumulation without mass consumption. For mass consumption would still require a supporting infrastructure and complementary service sectors and this would modify occupational and sectoral patterns.

Treating Fordist accumulation regimes in this way involves some complex theoretical maneuvers with somewhat ambivalent implications for the utility of Fordism as a concept. Positively it eschews analyzing national or regional economies in isolation and brings into focus the complementarities among different national accumulation regimes. It bans assuming that Fordism emerged through the simple diffusion of the US model and requires closer attention to its different modes of regulation. It also highlights the need to examine how international regimes (or modes of international regulation) served to block further exploitation of dependent peripheries and encourage trade among the metropolitan economies. Negatively we are led to ask whether work on Fordism involves anything more than putting old growth models into a new terminological bottle. Sometimes it seems that a Kaldor-Verdoorn type of explanation (stressing the virtuous effects of a shift from low-productivity agriculture to high-productivity manufacturing and/or the potentially self-reinforcing effects of rapid growth in productivity and output) has been repackaged and sold (in some cases misleadingly) as Fordism.[15] If this type of explanation holds, the key to growth in mass consumption would be rising mass incomes from rapid economic expansion rather than mass production as such. In turn this would direct attention to the general conditions of postwar economic growth. Perhaps Fordism could then be brought back as an accumulation strategy or organizing myth giving shape and coherence to economic growth. But this would be to emphasize strategic issues rather than merely describe the actual growth dynamic. The origin and spread of Fordist ideas, projects, and strategies would still be important fields of enquiry but should not be mistaken for actual modes of growth in different postwar economies. Equally important would be studies of the real driving forces behind growth – contrasting the roles of military spending,[16] Keynesian welfare spending, and so forth.

A longer historical perspective is valuable here, as it shows the experimental, accidental development of the dominant Fordist paradigm. For, if Fordism's history is limited to the domestic consolidation of Americanism and its diffusion, one ignores attempts to develop alternative models of Fordism in the interwar period. Taylorist and Fordist paradigms were significant in the Soviet Union, for example; and, voiced in various productivist, technocratic, and futurist projects, they also enjoyed wide cultural and political appeal in Western Europe – especially "where representative government was deemed to be working badly" (Maier 1970: 29). Nazism and Fascism gave Taylorism and Fordism a distinctive political gloss. They proposed state-sponsored "rationalization" of all economic and social life as a prelude to winning mass markets elsewhere through bilateral trading agreements and/or force of arms.[17] As a resource-rich, continent-wide, and still expanding economy, the US could plausibly adopt an isolationist stance and aim for autocentric growth: German and Italian policymakers felt their economies had to look abroad to secure the potential of Fordist rationalization (cf. Siegel 1988). European fascination with Americanism and Fordism was undermined by the Great Depression, however, reviving only after Allied victory in its turn undermined faith in block policies such as the Nazi *Grossraumwirtschaft* or Japan's "Asian Co-Prosperity Sphere." This helped the US to hegemonize postwar reconstruction and assert the virtues of Fordism and the American way of life. In promoting its own model of Fordism the US was advantaged, of course, by its economic and military dominance. This enabled it to influence institutional reform in West Germany and Japan and establish international regimes which encouraged Fordist accumulation – most notably through installing an international oil regime which secured increasingly cheap and plentiful supplies of an energy resource essential to Fordist expansion (Bromley 1990). Thus, even accepting that early postwar growth regime was mainly autocentric, various favorable international conditions were needed for this domestic takeoff to occur.

Mode of regulation

The key problem here is the wide variation in the modes of regulation compatible with Fordism considered as an accumulation regime. Metropolitan Fordism has occurred with the most varied labor market institutions. The mass Fordist worker and the mass union were not always the pivotal forces in industrial relations and collective bargaining. The M-form (or Sloanist divisional company) mostly took root in Europe in the late 1960s rather than in the take off to Fordism (Franko 1974); and Japanese company forms are far from Sloanist (Sayer 1989). Many European economies as well as Japan have retained a sizable and politically significant petty bourgeoisie in the commercial sector long after a Fordist model would have seen a shift to mass retailing. Conversely mass consumption emerged in the late nineteenth century in both Britain and the United States (Beniger 1986; Tedlow 1990). In addition metropolitan Fordism is associated with liberal, corporatist, and dirigiste forms of political regime. Keynesian demand management developed rather late in most countries and was applied rather ineffectively in almost all economies. Likewise the timing and pattern of welfare state developments assumed very different forms across all the advanced capitalist economies.

Thus, rather than insist on one particular institutional configuration as comprising *the* Fordist mode of regulation, we should search for family resemblances. Some common patterns can be discerned: a linkage between wages and productivity; the spread of collective bargaining; monopolistic competition; the growing role of state credit in investment and consumption; state involvement in generalizing mass consumption norms to significant groups – peasants in Japan, farmers in Europe and the USA, the growing army of state employees, and welfare recipients. There was also a growing strategic concern with economies of scale, productivity, planning, "growthmanship," etc., which affected social and political life as well as forms of economic activity.

Mode of societalization

The Fordist mode of regulation, if such there is or was, has proved so varied that its consequences for societalization also vary massively. Moreover, since the starting points for Fordist expansion differed and sociopolitical and institutional structures are hard to transfer across societies, we can find quite different "national Fordisms." But it is often unclear how far these differences are due to a generic Fordist logic working in different circumstances or to preexisting structural differences fundamentally modifying the generic logic to produce different modes of growth. Thus we turn to another issue, namely, periodization.

Periodization

Since each of the elements in Fordism has its own prehistory and many elements will survive beyond the Fordist epoch, crucial questions arise about the exact timing of this epoch. If there is no discontinuity or rupture as the labor process, accumulation regimes, modes of regulation, or social structures of accumulation evolve, are we justified in talking about a Fordist epoch? Even if we concede that mass production makes a difference, it might not be the key innovation in explaining postwar growth. Research on long waves reveals that more periods of economic expansion have occurred than is implied in the simple contrast between pre-Fordist and Fordist regimes. It also suggests that postwar growth was based as much on abundant cheap energy, petrochemicals, synthetic materials, other process industries, and aircraft as on mass production of consumer durables. Likewise, we could ask whether the crucial divide in the transition to post-Fordism is the demise of the Fordist assembly line (or even mass production as a whole) or the penetration of microelectronics into an ever-growing range of economic activities (including mass production). Without satisfactory answers to such questions, discussions about Fordism and post-Fordism have little point.

Nor can satisfactory answers be found in the genesis of individual elements of Fordism (however this is defined) but only in discontinuities in their overall articulation. Thus we should try to identify new configurations which have acquired a distinctive and emergent dynamic as a result of this articulation. At least two problems arise here. Firstly a great deal of effort must go into tracing the long gestation of a rather short-lived phenomenon – since Fordism's heyday did not even cover the full thirty-year postwar boom.[18] Secondly, the takeoff into Fordist growth seems to depend on a long series of chance events with the end of World War II providing only a convenient but stereotypical reference point for later periodizations.

My own conclusion is that any serious discussion of Fordism as a distinct phase in capitalist expansion must proceed from a focus on the mode of regulation. Analyses of the labor process are too limited – especially since there are real problems in assessing the significance of the Fordist labor process in its strict meaning. Emphasizing the Fordist regime of accumulation is undermined by the problematic character of the claims about autocentrism and the virtuous circle of balanced growth generated by mass production and mass consumption. But stressing societalization, i.e., the ramifications and repercussions of Fordism on the pattern of institutional integration and social cohesion, assumes the unassumable – that we already agree how to define Fordism. By a process of elimination we must either define it in terms of a specific mode of regulation or else reject Fordism as a useless concept. Many critics would welcome the second alternative but a reasonable case can be made for the first. Thus I recommend that Fordism be defined in terms of a core mode of regulation whose minimum features comprise:

a) a wage relation in which wages are indexed to productivity growth and inflation,
b) the state has a key role in managing demand, and
c) state policies help to generalize mass consumption norms.

Working with this basic definition we can examine the economic preconditions of the Fordist mode of regulation in specific modes of growth (or national accumulation regimes and their insertion into the world economy) and forms of organization of the labor process; we can examine the crisis of Fordism in terms of changes in the circuit of capital and/or the modalities of class struggle which help undermine the effectiveness of the Fordist mode of regulation as well as in terms of the distinctive, sui generis dynamic of this mode of regulation itself; we can examine the attempts to maintain this mode of regulation in the face of its crises and their eventual failure; and we can examine its broader preconditions and effects in the organization of state, economy, and civil society. In each case we must be sensitive to national variations among different modes of growth and regulation as well as to the changing pattern of complementarities and tensions among them.

THE TRANSITION TO POST-FORDISM

What conclusions follow from this concerning post-Fordism? An adequate account must treat it like Fordism – distinguishing its different levels and adopting the same critical spirit. During the current economic turbulence, however, even its generic features are uncertain. The asymmetry between Fordism and post-Fordism in this regard is even suggested by the simple chronological prefix in the latter term. So let us be wary. If such prefixes really conveyed much information, terms like "postliberalism" or even "pre-post-Fordism" would tell us much about Fordism.[19] Yet even a detailed specification of the latter's different dimensions leaves a lot unanswered. Thus a minimum condition for mentioning post-Fordism would be to show why it emerged after the Fordist era. Otherwise we should talk of a variant of non-Fordism.[20]

If the prefix is to convey real meaning rather than just provide a chronological marker akin to the initial French usage of "*après-fordisme*" (literally, "*after* fordism"), further arguments must be adduced. A case might be made along one or both of the following lines. Either one could show how post-Fordism has emerged from tendencies originating within Fordism but marks a break with it; and/or one could indicate how the ensemble of old and new elements in post-Fordism resolves or displaces one or more of the contradictions and crises which decisively weakened Fordism.[21] If post-Fordism is said to emerge from Fordism, major discontinuities must also be demonstrated: otherwise it would be better to talk of high Fordism, late Fordism, or neo-Fordism.[22] Alternatively one could justify the concept by showing how post-Fordism resolves or displaces the distinctive contradictions and crisis-tendencies of Fordism. This does not imply that post-Fordism would lack its own contradictions and crisis tendencies. These can be taken for granted. But it would still make sense to "post" this new regime if it somehow overcomes (or is widely held to do so) the problems typical of the Fordist era.[23]

Serious analysis of post-Fordism must go beyond noting that it occurs after Fordism and show how it relates to specific developmental tendencies and crises of Fordism. Doing so would not as such prove that "post-Fordism" is the best rubric for exploring and explaining current changes in advanced capitalist economies and/or the capitalist world economy as a whole. This is all too often assumed in post-Fordist perspectives, however, which simply ignore other bases of periodization with no attempt at justification (cf. Bromley 1988).

Apart from these basic conceptual problems, there are sound empirical reasons for casting doubt on post-Fordism. There are clearly many different starting points and trajectories en route to post-Fordism and most evidence is contradictory, incomplete, or provisional – especially in relation to post-Fordist accumulation regimes, mode of regulation, or societalization. Even assessing whether the American, Japanese, or German model will be the new hegemonic paradigm risks marginalizing other possibilities. These include an emergent *pax trilateralis* based on a complementary, tripolar global division of labor or a variety of paradigms based on elements drawn from each model and adapted to local circumstances.[24] Moreover, since it is doubtful whether Germany or postwar Japan are best seen as Fordist, it might prove misleading to treat the eventual dominance of a "Fujitsuist" paradigm and/or a recharged Modell Deutschland as post-Fordist.

Faced with these difficulties, prudence might well be the better part of valor. But, as Marx reminds us, "*hic Rhodus, hic salta!*" Having written elsewhere on post-Fordism and having just spent some time here looking about me, I might just as well leap. Taking as my springboard the basic assumption that a post-Fordist accumulation regime will be based on the dominance of flexible production in combination with differentiated, non-standardized consumption, I will specify the various levels of post-Fordism.

As a *labor process*, post-Fordism can be defined as a flexible production process based on flexible machines or systems and an appropriately flexible work force. Its crucial hardware is microelectronics-based information and communications technologies. These are relevant to "manual and non-manual work, to small, medium and large businesses, at corporate, divisional and workplace level, to managements and unions, and so on" (Clark 1989: 6). They can also be used "to steer, control and provide immediate feedback on a wide range of human and machine operations. When linked into electronics-based telecommunications systems, these 'real time' technologies can also effect enhanced information links and flows across space, integrating activities across departments and sites, and between individuals and organizations in different countries" (Clark 1989: 6). This would allow new or enhanced flexible specialization by small firms or producer networks even in small-batch production and, indeed, outside manufacturing, could promote flexibility in the production of many types of services in the private, public, and so-called "third" sectors. Hence the scope for the post-Fordist labor process to shape the dynamic of the emerging economic system is far greater than was true of Fordism. In some areas post-Fordism will see a further extension of Taylorism (e.g., lower grade clerical work and some aspects of design) or its further intensification (e.g., sweated manufacturing or assembly work). Elsewhere there may be a bilateral convergence towards diversified quality production from both mass production and handcraft production (Streeck 1987).

This pattern could properly be labeled post-Fordism insofar as it emerges from the Fordist labor process itself and/or is seen as responding to the crisis of Fordism. While not denying the partial validity of the former link, it is easier to defend the latter. Stressing how post-Fordism emerged from the Fordist labor process makes it hard to distinguish it from neo-Fordism. And, since post-Fordism can be applied to branches which were not previously organized along Fordist lines, further doubts are raised over the validity of the prefix. By looking at how post-Fordism evolves in response to crises of Fordism, however, we can include process and product innovations which emerged outside any immediate Fordist context. Flexible specialization complexes which have long coexisted with Fordist mass production and now seem to have won a new lease of life in both material and ideological terms could be included here; so could the key role of new technologies (such as microelectronics, biotechnology, and new materials) in overcoming some of the problems of Fordist control. In seizing on these new or recharged sources of flexibility, capitalists hope to overcome the alienation and resistance of the mass worker, the declining quality of products, the relative stagnation of Taylorism and mass production, the competitive threat from low-cost "peripheral Fordist" or "bloody Taylorist" producers in the Third World, and the relative saturation of the markets for standardized mass-produced goods; and/or to meet the growing demand for more differentiated products, for measures to break the rising costs of non-Fordist service sectors (especially in the public sector), and to boost productivity in other manufacturing sectors. The new technologies may also help resolve more general problems in Fordism such as its excessive use of fuel, energy, and raw materials and its damage to the natural and built environment (cf. Roobeek 1987).

As a stable *mode of macroeconomic growth*, any virtuous post-Fordist circle would reflect the newly dominant form of labor process as well as changes in international economic relations. An ideal-typical national post-Fordist accumulation regime would have the following dynamic. It would be based on flexible production, growing productivity based on economies of scope, rising incomes for polyvalent skilled workers and the service class, an increased demand for differentiated goods and services favored by the growing discretionary element in these incomes, increased profits based on technological rents and the full utilization of flexible capacity, reinvestment in more flexible production equipment and techniques and/or new sets of products, and a further boost to economies of scope. Compared with the ideal typical Fordist accumulation regime, post-Fordist growth need not involve generalizing core workers' rising incomes to other workers and/or the economically inactive. Moreover, as Fordist expansion is said to have been largely based on a growing home market and post-Fordist modes will be more oriented to worldwide demand, global competition could further limit the scope for general prosperity and encou. age a market-led polarization of incomes.

One possible solution to this latter problem would be international Keynesianism to manage global demand and generalize high consumption norms together with more local supply-side policies to enhance the structural competitiveness of the productive systems engaged in the race for post-Fordist modernization. Another solution would be more "extensive" in kind: reintegrating the former Soviet bloc and Communist China into a single world market to generate demand through the decomposition of the crisis-ridden relations of state socialist production and new forms of East–West trade.

Besides its emergence from and organization around genuinely post-Fordist labor processes, this new accumulation regime could be treated as post-Fordist insofar as it resolves (or is held to do so) crisis-tendencies in its Fordist predecessor. These were the relative exhaustion of the growth potential which came from extending mass production, the relative saturation of markets for mass consumer durables, and the disruption of the virtuous circle of Fordist accumulation due to internationalization and the problems this created for national regulation. In these respects post-Fordism transforms mass production and goes beyond it, segments old markets and opens new ones, and is less constrained by national demand conditions.

As a *mode of regulation*, post-Fordism would involve commitment to supply-side innovation and flexibility in each of the main areas of regulation. Given current uncertainties about this aspect, I offer only a brief and tentative account of some possible features.

- The post-Fordist wage relation could involve a basic recomposition of the collective laborer (with a tendency towards polarization between polyvalent skilled workers and the unskilled as compared to the Fordist tendency towards homogenization around semiskilled "mass" labor); the organization of internal and external labor markets around different forms of flexibility (functions and skills, duration and form of labor contract, wage package, etc.); a shift towards enterprise- or plant-level collective bargaining; and new forms of social wage. Industrial relations strategies could focus on integrating core workers into the enterprise and on mobilizing the production intelligence of workers by dissolving the Taylorist distinction between conception and execution. There may also be more intensification, marginalization, and insecurity for peripheral workers. They will often be poorly paid, unorganized, and recruited from politically marginalized social groups such as ethnic minorities, rural–urban migrants, and illegal immigrants.
- The post-Fordist enterprise system could see a shift from the primacy of the hierarchical, well-staffed, bureaucratic "Sloanist" form of corporate structure towards flatter, leaner, more flexible forms of organization. New forms of organization between hierarchy and market will become more important in managing strategic interdependencies both within and among firms and in responding quickly to changing demands. More use will be made of outside consultants, specialists, and subcontractors as well

as revolving teamwork and greater internal competition; firms will also turn to joint ventures, licensing or contracting of technology, strategic alliances, collaborative R&D, design partnerships, and so on. Profits of enterprise will depend on: the capacity to engineer flexible production systems and to accelerate process and product innovation; the search for technological rents based on continuous innovation in products and processes; and economies of scope. Competition will turn on non-price factors such as improved quality and performance for individual products, responsiveness to customers and customization, and rapid response to changing market conditions. Some commentators also expect a polarization between giant transnational firms offering a full range of goods and/or services in generic and multifaceted fields of technological competence and a multiplicity of smaller (but often transnational) firms aiming at specific niches within global or other markets. Others predict a new hierarchy of industrial–financial relations with its peak occupied by global players in high value-added and expanding product markets and international banks oriented to the needs of transnational firms (cf. Grou 1984; Amin and Robins 1990).

- On current trends the money form will be dominated by private, rootless bank credit which circulates internationally; more flexible forms of credit will be developed linked to a growing range of financial instruments; state credit will be subject to limits set by the logic of international money and currency markets. Whether these trends are sustainable is hotly debated.
- With the increased emphasis on differentiated forms of consumption, commercial capital will be reorganized to create and serve increasingly segmented markets. The hypermarket, the shopping mall, and the boutique are often cited as archetypal post-Fordist forms of consumption and contrasted with the supermarket and department store.
- State intervention will shift from a Fordist concern with managing national demand through Keynesian and welfare state measures. For the irreversibly international character of post-Fordism has the paradoxical consequence of reinforcing the state's role in promoting competition – not just of individual firms or national champions but of the overall productive system and its sociopolitical supports. If this marginalizes the state's role in managing national demand, it increases its role in the constant and continuous restructuring of the supply side (Kundig 1984: 60). Welfare policy will also be closely integrated into this restructuring process. Moreover, as individual nation-states may well lack the means to organize competition, this implies an enhanced role for continental or panregional states (such as the European Community). During the transition to post-Fordism this will not only involve rolling back the frontiers of the Fordist type of state but also rolling forward those of a new type of state. The post-Fordist world will be structured through the interaction of national or regional rivalries in the race for societal modernization and the dynamic of a global production system.

Taken together these forms comprise a distinctive ensemble of regulatory practices. They also seem to emerge out of tendencies inherent within Fordism and to resolve at least some of its crisis-tendencies. Some of these new structural forms and regulatory practices developed from attempts to manage the crisis of Fordism, others from attempts to escape it; some are primarily defensive, others offensive. Among the problems they help solve are: the collapse of Fordist incomes policies and the crisis of Fordist labor market institutions, the contradiction between Fordist wage forms and the post-Fordist need to promote responsible autonomy, rising R&D costs, rapidly changing and shortening product life cycles, greater risks of market failure, the availability of technologies permitting greater task integration and easier communication between divisions, and so on. Politically, the new forms of state intervention respond to Keynesian stagflation, the fiscal crisis of the state, slower productivity growth in the welfare state compared to the private sector, the rigidities and dysfunctions of bureaucratic administration and planning, the growing resistance shown by class forces and new social movements toward the forms and effects of the Fordist state, and so forth.

The post-Fordist "*mode of societalization*" is especially uncertain because, in contrast to the postwar dominance of the American model, there is now strong competition between Japanese, German, and American models. At best we can describe the societalization effects of the uneven transition toward post-Fordism. There are already clear signs of reorganization in the spatial division of labor in and across national systems. For flexible production seems to be avoiding the old Fordist production centers and is typically located in the suburban extensions of Fordist metropolitan areas, in relatively nonindustrialized hinterland areas, and, at least in services, in central business districts (Storper and Scott 1988). These new sites of production are rearticulated into the global circuit of capital and only its central nodes (the primary milieux of innovation) can function as locally integrated, agglomerated, self-generating growth poles; other sites are becoming more fragmented and are being inserted at various lower points in the global hierarchy (Amin and Robins 1990).

PROBLEMS WITH POST-FORDISM

There is a risk that building a model or paradigm of post-Fordism might encourage teleological and/or functionalist analysis. At worst this involves assuming that an inevitable, preordained transition is under way from Fordism to post-Fordism – impelled by the changing logic of the productive forces and/or competitive pressures imposed by the strongest capitalist forces. The theoretical fallacies here are well known since there is much scope for social forces to resist and/or shape the technical and social innovations. Even if we avoid teleology, functionalism may yet ensnare us.

The risk here is that, having constructed a paradigm of post-Fordism, we then assess everything in terms of its role in advancing (or else blocking) the transition to post-Fordism. But, if we cannot yet tell what the final form(s) of a post-Fordist labor process, accumulation regime, or mode of regulation will be, it is both foolhardy and fallacious to argue that specific structures or strategies must prove functional or dysfunctional in the transition. This helps explain the fundamental asymmetry between the concepts of Fordism and post-Fordism.

The labor process in post-Fordism

Let us ignore the many scientific, technical, and financial limits making the fully automated factory impractical and examine the more general shift towards flexibility in manufacturing.[25] This can take two main forms: static and dynamic (Coriat 1990) see also Coriat, in this volume, chapter 6. The former depends on a firm's ability to adjust its product mix "instantly" to demand fluctuations and thereby operate at or near to full capacity. This is only effective where the goods produced have a short life or will soon become obsolete so that economies of scale are limited: otherwise a larger firm equipped with flexible machinery would outperform small or medium-sized firms (Coriat 1990: 157–9, 163). By contrast dynamic flexibility operates on a longer time horizon and involves "production lines able to evolve rapidly, in response to changes in engineering of products or of processes" (Coriat 1990: 167). It is ideal for new products with growing demand and/or products with stable volumes of demand but periodic shifts in the features offered or demanded (Coriat 1990: 169; cf. Elam 1989). Lying even further beyond the reach of flexible specialization is the production of large, lumpy investment goods such as public telecommunications switching systems (Sayer 1989: 675). These may benefit from computer-integration of different stages in production but their actual manufacture will be beyond the scope of small or medium-sized firms even when organized flexibly in industrial districts. Nonetheless the changes occurring in the labor process, even if limited in scope and often realized only partially, do seem to involve significant departures from Fordist practice. Even if one is not justified in talking about a novel post-Fordist accumulation regime or mode of regulation, the evidence does indicate certain genuine post-Fordist trends in the labor process. The real problem is to assess their significance in relation to other trends and alternative accounts.

The accumulation regime

It is still too soon to define the macroeconomic regime of accumulation other than in abstract terms. This is especially true given the complex and still uncertain changes occurring in Eastern Europe and the former Soviet

Union as well as the continued development of the European Community and its associates. Coupled with the rapid changes occurring under the aegis of transnational companies and their strategic alliances, this makes it easy to forecast that there will be no return to the Fordist *status quo ante* and hard to anticipate the precise forms of any new regime.

Mode of regulation

Since objects and modes of regulation are mutually related, it would be foolish to forecast the main outlines of any future post-Fordist modes of regulation. Objects of regulation are not fully constituted prior to struggles over their regulation but are partially constituted in and through such struggles (Jessop 1990a). Chance discoveries and trial and error experimentation also play a major role in consolidating modes of regulation. Given the uncertainty about the elements of post-Fordist accumulation regimes, their articulation into durable moments of a stable post-Fordist mode of regulation is doubly uncertain. This explains the conflicts over key aspects of post-Fordism such as flexibilization of the wage relation; the meaning, scope, and importance of flexible specialization; the nature and significance of industrial districts; and the relevance of small, flexible firms in many still crucial areas of production. It is also much debated whether the international monetary system has become too flexible for its own good – let alone that of productive capital (see Altvater, in this volume, chapter 2). Niche marketing strategies often end up entombing their protagonists. And, while there seems to be a "hollowing out" of nation-states as functions are transferred upward to supra- or transnational bodies and/or downward to new forms of local and regional state, the forms of the state and state intervention in promoting and managing the transition to an uncertain post-Fordist future are too many and too varied to allow easy prediction. Nor is it clear how far current changes in state forms and functions are due to the transition to post-Fordism and how far they result from other trends, dilemmas, and contradictions.

Societalization

Neither a viable post-Fordist accumulation regime nor a coherent post-Fordist mode of regulation has yet arrived – if they ever will. This suggests that discussion of post-Fordist societalization must be restricted to three areas of enquiry:

a) how shifts in the labor process affect class, gender, and other social relations;
b) the social problems due to Fordism's crisis and the search for alternatives; and
c) the political processes involved in this search process.

At most we can describe some of the social forms involved in the transition to post-Fordism but equally in the transition to other structural forms. It is too soon to talk of a post-Fordist mode of societalization.

Periodization

Regarding the labor process we can raise similar doubts on periodization to those involved in accounts of Fordism. The question naturally arises whether the computer-integration of production from design to marketing really marks a crucial break in industrial organization or whether other shifts are more important. This is especially problematic since there is clearly a continuing role for mass production in many product areas. Conversely, many Japanese solutions now adopted in the West (such as JIT production) actually originated in the 1940s and 1950s – before the crisis of Fordism (cf. Sayer 1989: 670). This does not rule out inserting them into a distinctive post-Fordist system, but it does require a more nuanced and balanced analysis to show exactly what is new and/or "post" about this system.

CONCLUDING REMARKS

But few positive substantive results have emerged from this review. This is mainly due to the intellectual raw material on which we have been working: the distinction between Fordism and post-Fordism. Thus my conclusions largely consist in methodological reflections and involve few substantive comments. In particular I have suggested that there are three main sources of confusion involved in this distinction:

1 a failure to differentiate among possible sites where Fordism and post-Fordism may develop;
2 a failure to distinguish between the strategic or rhetorical affirmation of one or another aspect of Fordism (or post-Fordism) and its actual instantiation in specific structural features of a social formation; and
3 a failure to recognize the basic asymmetry between the concepts of Fordism and post-Fordism which is due to the closing of the Fordist era and the uncertain future of post-Fordism.

This said, if one or other concept is to be used, I prefer Fordism. In part this reflects the opportunity the passing of the Fordist era gives to study both the transition to Fordism and the regularities in and across different areas of consolidated Fordism. "Fordism" serves both as a heuristic concept when establishing the historical specificity for particular formations and, in its guise as an accumulation regime, it is useful in defining the basic tendencies, counter-tendencies, and crisis forms of postwar capitalism. Even here it has been necessary significantly to qualify the nature and meaning of Fordism, however, by giving rather loose definitions of its manifestation on different levels and/or by admitting a multitude of subtypes or hybrid forms. This is what prompts complaints about taxonomic fury overwhelming both detailed histories and dynamic causal analysis of tendencies and counter-tendencies. However justified these complaints may be in some cases, it still seems worthwhile to deploy the notion of Fordism as part of a broader conceptual system.

These problems are reinforced in dealing with post-Fordism. It is hard to define a coherent post-Fordist accumulation regime or mode of societalization as an ideal type or as a heuristic construct. There are also real doubts as to whether the path of capitalist development leads to a "post-Fordist" future. Quite specific theoretical and empirical conditions must be satisfied before we can reasonably talk of *post*-Fordism. Moreover, even if this particular concept were abandoned in favour of one lacking a chronological prefix (such as flexible accumulation, Fujitsuism or "Spaethkapitalismus"),[26] good grounds would still remain for doubting that it adequately describes the future of capitalism. For this will be determined by the class struggle and capitalist competition in all their manifold forms as well as by forces rooted in other institutional orders; it will also depend increasingly on global forces rather than those confined to particular nation-states or plurinational productive systems. This inevitably introduces problems both for those trying to find solutions to the problems of Fordism and those trying to study them. The debate has only just begun and I have tried to present some conceptual guidelines and some more substantive reflections to help it along.

4

FLEXIBLE SPECIALIZATION VERSUS POST-FORDISM

Theory, evidence, and policy implications

Paul Hirst and Jonathan Zeitlin

There is widespread agreement that something dramatic has been happening to the international economy over the past two decades: rapid and radical changes in production technology and industrial organization, a major restructuring of world markets, and consequent large-scale changes in the policies of economic management at the international, national, and regional levels. At the same time there is a great deal of confusion about how to characterize these changes, the mechanisms at work, and the policy implications for different groups of economic and political actors. One way of accomplishing these tasks is to postulate a change of basic manufacturing organization from a "Fordist" pattern that prevailed in the years of the long post-1945 boom to a "post-Fordist" successor in the later 1970s and 1980s. Many people habitually conflate three approaches to industrial change under this heading: flexible specialization, regulation theory, and a more diverse body of explicitly "post-Fordist" analyses. The resulting problem is that significant differences of approach are concealed by a superficial similarity between the proponents of flexible specialization and a set of apparently similar but underlyingly divergent ideas. The purpose of this chapter is to examine systematically the differences between flexible specialization, regulation theory, and other variants of "post-Fordism."

The chapter is organized as follows. In the first three sections we outline the basic arguments of the three approaches, focusing on their fundamental assumptions and theoretical architecture. In a fourth section we deal with the problem of evidence, first by examining the general methodological approach of each theory, and second by examining the substantive issues raised by their different attempts at evidentialization. In the fifth and final section, we consider the policy implications of the three theoretical approaches for the manufacturing sector.

FLEXIBLE SPECIALIZATION: TECHNOLOGICAL PARADIGMS AND POSSIBLE WORLDS

Despite their apparent similarities, flexible specialization and post-Fordism represent sharply different theoretical approaches to the analysis of industrial change. Where post-Fordism sees productive systems as integrated and coherent totalities, flexible specialization identifies complex and variable connections between technology, institutions, and politics; where post-Fordism sees industrial change as a mechanical outcome of impersonal processes, flexible specialization emphasizes contingency and the scope for strategic choice. The distinctiveness of flexible specialization as a style of analysis can best be appreciated by examining the way in which its theoretical architecture builds upwards from simple ideal types to a complex and multileveled system of concepts applicable to a diverse range of empirical cases.[1]

The central building block of this approach is its distinction between mass production and craft production or flexible specialization as technological paradigms, ideal-typical models or visions of industrial efficiency. Mass production for these purposes can be defined as the manufacture of standardized products in high volumes using special-purpose machinery and predominately unskilled labor. Craft production or flexible specialization, conversely, can be defined as the manufacture of a wide and changing array of customized products using flexible, general-purpose machinery and skilled, adaptable workers.[2]

Neither mass production nor flexible specialization on this view is inherently superior to the other. Each model is theoretically capable of generating a virtuous circle of productivity improvement and economic growth. Under mass production, subdivided labor and dedicated equipment can reduce unit costs through economies of scale, extending the market for standardized goods and facilitating new investments in special-purpose technologies, which further reduce costs, extend the market and so on. Under flexible specialization, conversely, versatile labor and universal equipment can reduce the cost of customization through economies of scope, extending the market for differentiated goods and facilitating new investments in flexible technologies, which narrow the price premium for customized products, extend the market and so on. But the practical realization of either possibility depends on a contingent and variable framework of institutional regulation at the micro-level of the firm or region and the macro-level of the national and international economy. Hence the technological dynamism of each model and its potential for sustained development cannot be evaluated outside of definite institutional and environmental contexts. Thus, just as there may be technologically innovative forms of both mass and craft production, so too are there stagnant variants of each model in which firms compete through squeezing wages, working conditions, and product quality – practices as common in large, declining enterprises as in small sweatshops.

The structural properties of each technological paradigm define a set of micro- and macroregulatory problems whose resolution is crucial for their long-term economic success.[3] In each case, however, similar problems may be solved in different ways, and a plurality of institutional frameworks can therefore be observed within both mass production and flexible specialization alike. For mass production, the crucial microregulatory problem is that of balancing supply with demand in individual markets: coordinating the flow of specialized inputs through the interdependent phases of production and distribution; and matching the output of productive resources that cannot easily be turned to other uses with the normal level of demand for each good. But as Piore and Sabel argue in *The Second Industrial Divide* (1984), these common goals may be pursued through a range of individual strategies, such as market segmentation, inventory variation, and superficial product differentiation, while the institutional framework provided by the large, hierarchical corporation likewise varies considerably both within and across national economies. Thus the organization of mass production firms in the United States, Germany and Japan, to choose some notable examples, differ significantly along key dimensions such as levels of administrative centralization and vertical integration, relationships with financial institutions, and systems of shop floor control.

For flexible specialization, by contrast, the crucial microregulatory problem is that of sustaining the innovative recombination of resources by balancing cooperation and competition among productive units. Two major types of institutional framework may be identified for the performance of these functions: "industrial districts" of small and medium-sized firms; and large, decentralized companies or groups. In the industrial districts, geographically localized networks of firms subcontract to one another and share a range of common services which are beyond the capacity of individual enterprises to provide for themselves, such as training, research, market forecasting, credit, and quality control. Successful districts are also typically characterized by collective systems of conflict resolution which encourage firms to compete through innovation in products and processes rather than through sweated wages and conditions. Within any particular district, however, there may be substantial differences in the roles played by specific institutions, from trade or employers' associations and cooperative banks or credit unions to trade unions, churches, and local government; and the political complexion may also vary sharply from "red" regions such as Tuscany or Emilia to "white" ones such as the Veneto or Baden-Württemberg.

In large, decentralized companies, on the other hand, the relatively autonomous productive units often resemble small, specialized firms or craft workshops, while obtaining services such as research, marketing, and finance from other divisions of the parent enterprise. As in the case of mass production corporations, however, large, flexible firms may also differ significantly from one another – for example, in their relationship to banks or trade unions – depending on their individual histories and the national institutional context. There are signs, too, that the extended period of volatility in international markets since the mid-1970s is giving rise to what Sabel calls a "double convergence" of large- and small-firm structures, as small firms in the industrial districts build wider forms of common services often inspired by large-firm models, while the large firms themselves increasingly seek to recreate among their subsidiaries and subcontractors the collaborative relationships characteristic of the industrial districts.[4]

As the Great Depression of the 1930s demonstrated, the market stabilization strategies of large corporations by themselves could not solve the central regulatory problem of mass production: how to ensure a continuing balance between consumption and production across the national economy in order to amortize lumpy investments in product-specific equipment. While the Keynesian welfare state emerged as the dominant form of macroregulation during the postwar period, here too the differences among national economies remained striking: differences, for example, in the methods of managing budgetary aggregates, in the commitment to countercyclical deficit finance and public welfare provision, and in the role of collective bargaining agreements and other "private" means for relating purchasing power to productivity growth. Like mass production itself, Keynesian macroregulation was as much a project as an accomplished fact; and nowhere was this more true than at the international level. Despite Keynes's own postwar proposals, no effective institutional mechanisms were created to ensure a steady expansion of global demand in line with productive capacity or recycle purchasing power from surplus to debtor countries in the world economy.

If the macroregulatory requirements of mass production are relatively well defined, those of flexible specialization remain the least developed aspect of the model. Thus Piore and Sabel argued that the superior capacity of flexibly specialized firms to accommodate changes in the level and composition of demand makes macroregulation less vital than in mass production, giving the price mechanism a greater role in equilibrating supply and demand as in the nineteenth-century competitive economy. But Piore and Sabel also emphasized the microregulatory need for such an economy to take wages out of competition and maintain welfare services in order to avoid debilitating breakdowns of solidarity among economic actors, distinguishing possible low- and high-consumption variants of a flexible specialization regime through the contrasting images of Bourbon Naples and a Proudhonian artisan republic.

More recently, Sabel has developed these ideas by treating macroeconomic regulation as a problem of reinsurance: whereas for mass production, the key problem is that of reinsuring firms against unpredictable fluctuations in the level of demand through macroeconomic management, the problem for flexible specialization is that of reinsuring regional economies against large-scale shifts in its composition by establishing interregional mechanisms to facilitate structural adjustment. On this basis, in turn, he distinguishes two possible futures for the welfare state under a regime of flexible specialization: an exclusive or dualist variant in which regional economies increasingly opt out of the national welfare state while remaining vulnerable to unpredictable external shocks as well as to disruption from those left outside the system; and an inclusive variant which would integrate firms and industrial districts into national systems of training and reinsurance, extend flexible specialization to less successful regions and social groups, and build on existing trends towards the decentralization of the welfare state itself.[5]

A final macroregulatory issue concerns the implications of flexible specialization for the international division of labor. In *The Second Industrial Divide*, Piore and Sabel suggested that one possible scenario might be the emergence of new forms of interdependence in the world economy, as mass production migrated to underdeveloped countries, while advanced economies increasingly shifted over to flexible specialization. Under these conditions, the First and Third Worlds might also come to share a common interest in a new institutional framework of multinational Keynesianism to regulate world demand and ensure macroeconomic stability (Piore and Sabel 1984: 279–80). But flexible specialization might also be conceived as an alternative development strategy for parts of the Third World itself. Such a strategy might build on existing forms of small-scale enterprise concentrated in the substantial "informal" sectors of many developing economies; and it might also build on the unavoidable flexibility of preexisting forms of mass production imposed by the constraints of narrow markets and shortages of appropriate skills and materials. Either way, flexible specialization might offer an attractive route to economic development for such countries in which "appropriate technologies" were not necessarily inferior and modern forms of industrial organization could more easily be adapted to local conditions.[6] Like any development strategy for the Third World, flexible specialization would clearly be advanced by the creation of effective mechanisms of international macroeconomic coordination, but unlike mass production it could also be successfully pursued under the more likely conditions of continued volatility in the world economy. Which of these possible worlds may in fact be realized, and to what extent, like the institutional frameworks of micro- and macroregulation, cannot be derived from flexible specialization as an abstract model, but depends instead on the outcome of strategic choices and political struggles.

From this account it should be clear that flexible specialization is at once a general theoretical approach to the analysis of industrial change, and a specific model of productive organization whose micro- and macro-regulatory requirements may be satisfied through a variety of institutional forms. But in no sense can this general approach be understood as an evolutionary teleology in which the triumph of flexible specialization as a specific model is a necessary consequence of some immanent logic of economic or technological development. Much of the debate over flexible specialization has in fact missed the mark by construing the latter as a similar type of theory to post-Fordism in its many variants.

Contrary to what many critics have supposed, for example, mass production and flexible specialization are ideal-typical models rather than empirical generalizations or descriptive hypotheses about individual firms, sectors, or national economies.[7] As the original formulations made clear, neither model could ever be wholly predominant in time or space. Thus mass production requires a continuing role for skilled workers and craft production both inside and outside the large firm, to design, set up and maintain special-purpose machinery on the one hand and to manufacture goods for which demand is too small or unstable to justify investments in dedicated equipment on the other. Conversely, some standardization of intermediate goods and components is a necessary condition for the flexible manufacture of diversified final products.[8] Hence the persistence of firms, sectors, and even whole national economies organized on alternative principles does not in itself undercut the notion of a dominant technological paradigm in any given period.

At a deeper level, moreover, the analytical distinction between mass production and flexible specialization is also compatible with the empirical finding that hybrid forms of productive organization are the rule rather than the exception. As historical research conducted within this framework shows, firms in most countries and periods deliberately mix elements of mass production and craft or flexible production because they are acutely aware of the dangers involved in choosing an unalloyed form of either model. Thus economic actors' understanding of the pure models paradoxically leads them to hedge against risks in ways that blur the lines between them. The resulting interpenetration of elements of flexible and mass production also means that firms often find it easier to shift strategies from one pole to another than an abstract consideration of the two models might lead one to expect.[9]

Contrary to another widespread misconception, therefore, flexible specialization is neither a technological nor a market determinism.[10] Just as trajectories of technological development in this approach are shaped by competing visions of production, so too are patterns of demand shaped by competing visions of consumption. Thus, for example, the realization of either virtuous circle between investment, productivity and the extension of the market depends not only on the creation of an appropriate institutional framework, but also on the relative success of flexible and mass producers in persuading consumers to accept or reject a price premium for differentiated goods over their standardized counterparts. This dynamic interaction between production and consumption means that the flexible specialization approach regards market structures not as fixed parameters which impose a uniquely appropriate form of conduct on economic actors, but rather as contingent historical constructs which reflect the competitive strategies of the actors themselves. Hence, for example, current trends towards the diffusion of flexible specialization as a productive strategy result not only from the pervasive volatility of demand but also from the conscious efforts of firms organized along these lines – most notably in Japan – to fragment the mass market still further through the constant introduction of new speciality products.[11]

While flexible specialization strategies may be pursued within a plurality of productive and institutional forms, the range of variation is neither infinite nor arbitrary. Thus, for example, the regulatory requirements of flexible specialization are incompatible with a neoliberal regime of unregulated markets and cutthroat competition. In each of its institutional forms, flexible specialization depends for its long-term success on an irreducible minimum of trust and cooperation among economic actors, both between managers and workers within the firm and between firms and their external subcontractors. And as we have already noted, such cooperation depends in turn on the establishment of rules limiting certain forms of competition such as sweated wages and conditions, as well as on collective institutions for the supply of nonmarket inputs such as technological information or trained labor. Hence flexible specialization should not be conflated with opposed conceptions of "flexibility" as labour-market deregulation which have become common currency not only among businessmen, policy-makers, and trade unionists, but also among post-Fordists and their critics.[12]

If flexible specialization depends on trust and cooperation, finally, this does not imply the absence of any conflict. On the contrary, flexible specialization, like any system of production, is prone to potentially debilitating conflicts among economic actors, not only between employers and workers, but also between firms and their subcontractors or subsidiaries. The reproduction of social consensus within these systems, though it may build on formative experiences in the past, can only be sustained in the longer term through the creation of institutional mechanisms for the resolution of disputes whose operation is broadly satisfactory to all the parties concerned. While the maintenance of consensus is always provisional and contingent, so too is the crystalization of particular conflicts into durable antagonisms between social groups: neither outcome is predetermined by flexible specialization as a technological paradigm without reference to a definite social and institutional context. For flexible specialization, unlike post-Fordism and its Marxist antecedents, social and political identities cannot be derived from the structure of production through the ascription of objective interests to abstract categories or classes of actor; and therein lies another fundamental difference between these two contrasting approaches.

"POST-FORDISM": TOTALITIES AND SOCIAL TYPES

The term "post-Fordism" has such a wide currency and is used so indiscriminately that some demarcations are necessary if different approaches grouped under the term are not to be confused. It has often been used, along with "Fordism", to characterize aspects of the work of the flexible specialization and regulation schools. Here it will be used to specify a body of work quite distinct from both of them, although in many cases drawing eclectically on their concepts.

One of the most popular and influential advocates of the concept of post-Fordism in Britain was the political magazine *Marxism Today*, which attracted a politically varied readership right up to its closure in 1990. The edited collection *New Times* (Jacques and Hall 1989) is largely drawn from the pages of that journal. Here we will take it as a typical exposition of the rise and fall of Fordism as an economic and social system and its replacement by post-Fordism. This popular account is similar in most respects to other, more academic versions such as Harvey (1990) and Lash and Urry (1987), which we will ignore here for reasons of economy.

Marxism Today depended heavily on a stereotyped version of the technical, economic and social relations that preceded "new times" – typified by the concept of "Fordism." As its leading members were mainly neo-Gramscian Marxists, this concept owed much to Gramsci's essay "Americanism and Fordism" in the *Prison Notebooks* (1971) and also to works like Harry Braverman's *Labor and Monopoly Capital* (1974), which emphasize Taylorism as a form of production and labor organization characteristic of the Fordist era. Ford's name was used to sum up a series of innovations in manufacturing introduced in the first two decades of this century in the USA and supposedly generalized as a model of industrial production worldwide thereafter. Fordism is mass production on the assembly line model, using special-purpose machinery and mainly unskilled labor in a division of labor based on the increasing fragmentation of tasks. The Fordist era is characterized by the dominance of mass markets and long runs of standardized goods. Fordism is based on the technological efficiency of planned production arising from the separation of conception and execution, and on the economic efficiency of large-scale plants. It comes to dominate by economic logic, the logic of competitive advantage and market performance. However, it was only when Fordist production was coupled with appropriate macroeconomic policies by national governments that this production system could find the stability to exploit its full potential; Fordist growth was possible when the mass purchasing power to sustain the mass markets necessary for mass production was assured. Fordism and

Keynesianism were responsible for the great postwar boom from 1945 to the first oil price crisis of 1973.

The Fordist system implied a definite type of society, industrial society based on a homogeneous, male, full-time working class, concentrated in large plants in large industrial cities. The assembly line, the concentration of labor in large plants and a full-employment economy promoted the central role of the unions and workers parties in politics. Traditional social democracy was underwritten by Fordism and with it the primary role of the welfare state.

What is wrong with the Fordist stereotype? First, that it ascribes the dominance of Fordism to economies of scale, to a narrowly economic explanation without reference to actual markets, plant sizes or specific forms of production organization. Taylorism is taken for granted, without reference to the actual complexities of work organization or the role of labor. The division of labor has never been a purely management prerogative: strongly institutionalized patterns of industrial relations mean that it is impossible to read off work organization from stereotyped visions of management objectives. Moreover, these objectives have never been so stereotyped or free of other complex constraints of economic calculation as the notion of Taylorism supposes. Distinct national patterns and phasings of adoption of standardized mass production are ignored.[13] Second, it takes Keynesianism for granted – ignoring the complex questions of what macroeconomic policies were pursued in different national economies. Thus, for example, neither West Germany nor Japan followed "Keynesian" policies in their major periods of postwar expansion, adopting instead orthodox fiscal and monetary policies.[14] Close attention to the different national experiences in the great postwar boom reveals quite different manufacturing strategies and institutional underpinnings to growth. *The Second Industrial Divide*, far from conforming to the Fordist stereotype, has the great merit of carefully considering these different national routes to economic organization.[15]

"Fordism" involves a simplified view of manufacturing and macroeconomic management, homogenizing the postwar world to overstress the social and political differences after 1973. "Fordism" was the coherent concept in *Marxism Today*'s analysis; it provided the rigid point of departure against which to project the changes from the early 1970s onwards. "Post-Fordism" is a much less coherent concept. It is a way of bundling together a series of economic and social changes. Just as Fordism is supposed to have prevailed throughout the industrial world – other countries following the American model – so "post-Fordism" is a worldwide phenomenon, with, it appears, Japan leading the way. *Marxism Today* argued that Britain exemplifies the new "post-Fordist" pattern of economic relations.

Not only are national and regional complexities obliterated here, but there is no real explanatory core to the post-Fordist case other than the decomposition of Fordist structures. The post-Fordist case involves, by and large, borrowing and radically simplifying the flexible specialization approach to manufacturing. However, this borrowing is accompanied by very little attention to the wide range of forms and hybrids of flexibly specialized production and their social and institutional conditions. The primary role in explaining change is assigned to the collapse of mass markets. Market differentiation and volatility are taken as purely external factors enforcing new patterns of manufacturing. The switch from Fordism to post-Fordism is seen in conventional economic terms as market-led. Classic capitalist entrepreneurs of the conventional type and individual firms respond with new intrafirm strategies to the new market conditions. Strategies are reactive and firm-initiated, decisions draw production techniques, product ranges and labor market changes after them. This violates the basic assumptions of the flexible specialization approach, that effective strategies are anticipatory more than reactive, and that not merely the firm but interfirm and collective regional and national patterns are crucial in the balancing of competition and cooperation necessary for their more progressive institutionalizations.

Marxism Today characterized Britain as a "post-Fordist" society without enquiring further into the extent to which British firms have, indeed, adopted new forms of production technique and manufacturing organization. The evidence for such adoption is scant. British firms appear to be reproducing many of their worst faults of the postwar decades: inattention to training, poor delivery times, misuse of technology, misconceptions of competitors' strategies, etc.[16] How then can Britain's economy be typified as "post-Fordism"? Only by our realizing that its use by *Marxism Today* was not as a rigorous concept organizing a detailed empirical account of British manufacturing but as a loose sociological metaphor.

Most of *Marxism Today*'s authors were Marxists or post-Marxists. The use of a concept drawn from production organization to characterize a whole era would thus seem to be following on the traditional Marxist methods of assigning a determining role to the forces of production. But technological determinism was one of the main things *Marxism Today* and its key authors like Stuart Hall strived to avoid for over ten years. The use of features of the production system to characterize the wider society is no longer a rigorous, if rigid, causal determination, rather it has slipped to the level of a casual metaphor.

The analysis of post-Fordism in *New Times* is little more than pop sociology combined with a tendency derived from classical Marxism to think of societies as coherent types. The "post-Fordist" concept is linked to that of postmodernism to produce a view of modern society as fluid and changing, dominated by a shift from collectivism to individualism, from production towards consumption and the service sector, from substance toward style, and to new political issues and new social movements.

Pointing out that a traditional collectivist socialism based on the assumption of the electoral primacy of a mass, homogeneous manual working class is dated can hardly count as either politically innovative or sociologically novel; it has been a commonplace since the 1950s. The characterization of the "Fordist" era in these terms is an absurdity, no one attentive to the politics of European labor movements could ever have believed in this unitary working class, this hegemony of collectivism or this assumption of previous political and electoral success.

When we turn from this fantastic history of the era before 1973 and the illusory post-modernist contrast with the world afterwards we can see how unlike "new times" are the most successful regional and national economies that have followed flexible specialization strategies.[17] Neither the Third Italy nor Japan are models of postmodernist social and cultural fluidity. In Germany and Italy flexible specialization has been based on legacies of labor skill and institutions of regional and social cooperation that persisted into the supposedly "Fordist" era. In each of these countries conservatism in social relations and attitudes has provided the basis for social cooperation and economic coordination necessary to develop innovative strategies in manufacturing. Social continuity has provided the stability and support to absorb and promote radical change in technology and economic organization. One can overstress the role of conservatism and traditional "bourgeois" attitudes in these cases, since all three societies were radically disrupted by the experience of fascism. Nevertheless, all three rebuilt their societies and economies after the war, combining innovation and conservatism, and drawing strength for cooperation from the experience of the antagonisms and divisions of the fascist era. Other regions have found different routes to the patterns of cooperation and institutional collaboration necessary to promote flexible specialization, but none of the major industrial districts looks remotely like "new times." The best thing one can say for *Marxism Today* is that it sought to discard obsolete socialist shibboleths, but the means it chose to do so were almost as damaging as the illusions from which it sought to escape. In particular it portrayed Britain as a rapidly changing and innovating country, with Mrs Thatcher's Conservative Party in the lead because it recognized these changes and sought to exploit the new individualism consequential upon them. Not only has British manufacturing not changed in the way *Marxism Today* supposed, not only is the British economy rerunning on a new basis the macroeconomic failures of the 1960s, but British social attitudes stubbornly refuse to conform to the stereotype of "new times."[18] *Marxism Today*'s analysis comes close to celebrating Thatcherism: seeing it as breakthrough when Britain is teetering on the edge of economic failure.

The writings of Christopher Freeman and Carlota Perez offer another distinct and far more academically sophisticated variant of the use of the terms "Fordism" and "post-Fordism."[19] Their work in contrast to the above is marked by two key differences, a sophisticated attention to economic theory (particularly Joseph Schumpeter's work on investment, technological innovation and growth cycles, and Kondratiev's work on long waves of technological innovation linked to periods of upswings and downswings in growth) and a more detailed and specific attention to technology.

Freeman and Perez identify five successive waves of innovation linked to Kondratiev cycles. In each of these waves the upswing is dominated by the generalization of a new "technological paradigm," which emerged from the crisis in the preceding downswing and which in turn leads to a new downswing as the impact of technical innovations and the investment cycle falters. Each successive wave involves new institutional forms which support the phase of economic and technical innovation; these institutions then fall into crisis during the period of downswing, and then involve a period of institutional transition or crisis as the conditions for the next wave are prepared.

The five waves are set out schematically in Freeman and Perez (1988):

1 the period from the 1770s–80s to 1830s–40s, divided into an upswing called the "Industrial Revolution" and a downswing called "Hard Times" and characterized by early mechanization;
2 the period from the 1830s–40s to the 1880s–90s, divided into an upswing of "Victorian Prosperity" and a downswing of the "Great Depression" based upon steam power and railway technologies;
3 the period from the 1880s–90s to the 1930s and 1940s, divided into an upswing called the "Belle Epoque" and a downswing of the second "Great Depression" in the 1930s and based upon electrical and heavy engineering technologies;
4 the period from the 1930s–40s to the 1980s and 1990s and divided into an upswing called the "Golden Age of Growth and Keynesian Full Employment" and a downswing called the "Crisis of Structural Adjustment" and based upon Fordist mass-production technologies; and
5 an emerging fifth period beginning in the 1980s and 1990s based upon new information and communications technologies, basically a "post-Fordist" phase.

The work of Freeman and Perez involves major and crippling theoretical problems despite its seriousness and its attempts to avoid some of the major pitfalls of explaining growth through cycles of innovation. They are careful to try to specify the technologies underpinning the major waves, and to look for evidence for their periodization and diffusion. Unlike other "post-Fordists" they are careful to remain at the level of techno-economic evolution and not to generalize from this into social types, which are ensembles of economic, social, and cultural relations. They are also careful to avoid drawing simplistic and overblown political conclusions from their

successive waves of technological innovation. However, they still subscribe to the "Fordist/post-Fordist" dichotomy and to roughly the same periodization as the other two bodies of work considered here.

The first major theoretical problem is that Freeman and Perez's analysis is closely tied to the cycles of long waves of technological innovations proposed by the Russian economic statistician Nikolai Kondratiev (Kondratiev 1979). Kondratiev's cycles do not merely illustrate their thesis, they are central to its periodization and to its explanation of successive phases of growth, crisis, and innovation. If Kondratiev's claims are false then so is the framework of Freeman and Perez. Kondratiev's work is statistically based and depends on inferences from economic data on growth – particularly on relations between prices and output. As such it offers no convincing causality for a 50–60 year cycle, or why the pattern of technical innovation should "bunch" in this way. It is essentially a work of statistical economic history and its periodizations are dependent on certain relationships being shown to be statistically significant and to correspond to the chosen cyclic periods. It has stimulated a great deal of debate and it has been comprehensively tested and found wanting by Solomos Solomou (1987). Solomou found that there was no significant correlation between production trends and Kondratiev's waves in Britain, France, Germany, or the USA, and no correlation of innovation clusters in the way Kondratiev supposed. In the absence of a 50–60 year cycle divided into roughly equal periods of upswing and downswing, shorter and more contingent cycles of growth or technological innovation will not sustain the Freeman–Perez framework. In the absence of a rigorous statistical proof of the Kondratiev cycle Freeman and Perez must fall back on a weaker version of Schumpeter's account of the investment cycle.

The second major theoretical problem is that although Freeman and Perez are careful to refuse a simple teleology of autonomous technical change, recognizing the role of economic and conjunctural factors in the reception and diffusion of technologies, they still tend to see change in productive systems as driven by technology rather than by business organization and strategy. As we have seen, the flexible specialization approach – far from foregrounding technology – emphasizes the adaptation of techniques to distinct forms of productive organization. Thus the adoption of microelectronics, the core technology of Freeman and Perez's fifth ("post-Fordist") wave, does not of itself lead to flexible specialization strategies and production methods. Perez's discussion of "flexible production," despite its relative sophistication, remains predominantly at the level of technical processes and potentialities and does not give due weight to production organization, market transformations and institutional changes at the suprafirm level (Perez 1985).

Third, the concept of a "techno-economic paradigm" is very different from the "technological paradigm" of flexible specialization, despite the apparent similarity of the terms. A "techno-economic paradigm" is a diffusible pattern of technology linked to changes in cost-structures. It remains the case that this paradigm is the core of an evolutionary phase of a technical-economic system that comes to dominate every branch of the economy, and it prevails because of the intrinsic gains of the technologies involved and the availability of generalizable low-cost inputs, corresponding to classical economic ideas of competitive efficiency. It is not an ideal-typical model of a new form of productive efficiency, effective because adopted by key economic actors; rather it is a new form of technological efficiency *per se*.

Fourth, technical change is evolutionary in Freeman and Perez's analysis and prevails because of its inherent adaptational advantages, after giving due allowance for delays and crises. It is, despite qualifications, much closer to a form of technological-economic determinism than flexible specialization theory. Evidence of this is the tendency to view successive waves of innovation as necessary and to view the corresponding institutional adaptations of the economy to the new technologies in a functionalist manner. Technical change is conceived much in the manner of Marx's inherent dynamic of the forces of production. The "socio-institutional framework" needed to accommodate such technologies is seen much in the manner of Marx's relations of production, as a social linkage that is functionally necessary to a certain phase of technology and which then moves into a crisis of functional mal-adaptation when the potential for innovation in the prevailing technology is exhausted. This can be illustrated by a passage from Carlota Perez which could be a translation from *Capital*:

> the prevailing pattern of social behaviour and the existing institutional structure were shaped around the requirements and possibilities created by the previous paradigm. This is why, as the potential of the old paradigm is exhausted, previously successful regulating or stimulating policies do not work. In turn, the relative inertia of the socio–economic framework becomes an insurmountable obstacle for the full deployment of the new paradigm.
>
> (Perez 1985: 455)

Freeman and Perez are more careful in building their social types than *Marxism Today*: the types are techno-economic systems not ensembles of all possible sociocultural relations, and they offer a more coherent and general causality than do the other "post-Fordists," but the result is an evolutionary theory with a strongly technical necessitarian cast. As such, although their theory is not classically Marxist, it has many of the same faults as the technological determinism explicit in the dynamic role of Marx's forces of production. The flexible specialization approach has been constructed to avoid just these faults, and so, despite some similarities of terminology and despite some very interesting points about technical change and the specific technologies involved, Freeman and Perez's views are fundamentally opposed to its main theoretical thrust.

REGULATION THEORY: MIDDLE WAY OR BLIND ALLEY?

These theoretical contrasts emerge most starkly from a comparison of flexible specialization with the more simplistic versions of post-Fordism put forward by *Marxism Today* or Freeman and Perez. But between these two poles stands a large but elusive "middle ground" which seeks to combine the openness and contingency of the flexible specialization approach with a continuing insistence on the systematic nature of capitalism as a mode of production and the centrality of class struggle in its development. Undoubtedly the most important and developed representative of this approach is the French "regulation school" in its "Parisian" variant.[20] Despite the many subtle differences among themselves, the Parisian regulationists share a conceptual framework and methodological approach which appears to mark them off from flexible specialization and post-Fordism alike. This group of heterodox economists, many of whom are associated with the French state planning apparatus, draw their inspiration from a distinctive combination of Althusserian structuralist Marxism, post-Keynesian macroeconomics, and *Annales*-school historiography of the *longue durée*. On this basis they are out to execute a "slalom between the orthodoxies" of neoclassical general equilibrium theory and classical Marxism to produce a rigorous but nondeterministic account of phases of capitalist development which leaves considerable scope for historical variation and national diversity.[21] As in the case of post-Fordism and flexible specialization, this section will examine the theoretical architecture of the regulation approach before going on to consider critically the relationship between this conceptual edifice and its putative methodological foundations.

Like post-Fordism, but unlike flexible specialization, the regulation school takes as its point of departure the concept of capitalism as a mode of production. Capitalism, in this view, is a contradictory and crisis-ridden economic system which requires some form of institutional regulation for its continued reproduction; but in contrast to orthodox Marxism, the operation of these crisis tendencies and their resolution is "underdetermined" by the abstract properties of the mode of production iself. Social and political struggles therefore play a crucial role in the creation of the regulatory institutions which sustain each new phase of capital accumulation; and like post-Fordism, but unlike flexible specialization, the central actors involved in these struggles are conceived essentially in class terms.

At a lower level of abstraction, the regulation school analyzes successive phases of capitalist development in terms of a series of modes of development based on a combination of regimes of accumulation and modes of regulation.[22] A regime of accumulation is a relatively stable and reproducible relationship between production and consumption defined at the level of the international economy as a whole. Each regime of accumulation encompasses a number of interrelated elements, from the pattern of productive organization and the time horizon for investment decisions through the pattern of income distribution and effective demand to the relationship between capitalist and noncapitalist modes of production. Within any regime of accumulation, however, each national economy may have its own distinctive mode of growth depending on its insertion into the international division of labor. Thus the regulation school distinguishes four major regimes of accumulation in the history of capitalism since the eighteenth century: extensive accumulation; intensive accumulation without mass consumption (Taylorist); intensive accumulation with mass consumption (Fordist); and an emergent post-Fordist accumulation regime whose contours have yet to be fully determined.

A mode of regulation, on the other hand, is a complex of institutions and norms which secure, at least for certain period, the adjustment of individual agents and social groups to the overarching principles of the accumulation regime. Like regimes of accumulation, modes of regulation are complex ensembles composed of a number of interrelated elements: the form of monetary and credit relationships; the wage–labor nexus; the type of competition; the mode of adhesion to the international regime; and the form of state intervention. On this basis, the regulation school likewise distinguishes four major modes of regulation over the past two centuries: old regime regulation (*regulation à l'ancienne*); competitive regulation, monopolistic regulation, and an emergent semiflexible mode of regulation whose contours again remain to be determined. Within any national economy, however, each mode of regulation may be realized in very different ways depending on the preexisting institutional context and the outcome of domestic social and political struggles.

When a regime of accumulation comes together with an appropriate mode of regulation, the resulting mode of development makes possible a sustained period of technological progress and economic growth. Under these circumstances, crises are essentially cyclical and perform an equilibrating function for the economic system as a whole. But each mode of development also contains its own internal limits – reflecting the deeper contradictions of capitalism as a mode of production – and eventually comes to decay over time. The consequence is the onset of a structural crisis in which regulatory institutions no longer perform a positive function for the regime of accumulation, and the economic system loses its self-equilibrating character. Only the creation of a new relationship between accumulation and regulation can break the impasse and revive the growth process, but such a breakthrough in turn depends on the strategic choices and political struggles of the major social actors.

Despite differences of emphasis between individual writers, for example, the regulation school attributes the crisis of Fordism during the 1970s and 1980s primarily to two structural tendencies within this regime of accumulation. First, the progressive exhaustion of productivity gains from Fordist forms of work organization undercut the virtuous circle between consumption and investment, and precipitated a clash between corporate profitability and the institutional mechanisms which sustained aggregate demand within each national economy. Second, the gradual erosion of American hegemony in the world system undercut the role of the dollar as an international currency, destabilizing the implicit mechanisms which had ensured a steady expansion of global demand during the postwar period despite the absence of more formal institutions of international macroeconomic coordination. The effects of these tendencies were then amplified by conjunctural shocks such as the oil price rises of 1973 and 1979 and by the strategic responses of the major actors: internationalization of production and marketing on the part of large firms; defense of real wages on the part of trade unions; and adoption of deflationary policies on the part of the state. If the roots of the crisis of Fordism for the regulation school are at once structural and conjunctural, there is nothing automatic about the emergence of a new mode of development, and a number of future configurations are conceivable, from neo-Taylorism plus neo-liberalism through flexible specialization in the narrow sense to flexible mass production plus international Keynesianism. Which of these possible outcomes will in fact be realized is determined, on this view, both by the compatibility between the various components of the emergent regime of accumulation and mode of regulation, and by the strategic choices of large firms, trade unions and the state.[23]

At first glance, therefore, the regulation school appears to avoid the theoretical pitfalls of the more simplistic versions of post-Fordism while maintaining a close link between empirical analyses of industrial change and the structural dynamics of capitalism as a mode of production. Like flexible specialization, the regulation school avowedly rejects both determinism and functionalism, while enthusiastically embracing historical contingency and national diversity. Neither the onset of structural crises nor their resolution are automatic processes dictated by the laws of motion of capitalism itself; regimes of accumulation do not secure their own regulatory requirements; and neither the capitalist class nor the state plays the role of a "system engineer" consciously ensuring the dynamic stabilization of the economy. The emergence of an appropriate mode of regulation is thus always "miraculous," provisional, and potentially unstable: at best, one can speak of an *a posteriori* functionalism in which a particular set of regulatory institutions, once constituted, prove compatible with the demands of a viable mode of development.[24] Finally, as we have seen, different national modes of growth may coexist within the same international regime of accumulation, while each mode of regulation may be realized through a variety of institutions, giving rise to a wide range of distinctive national experiences in any historical epoch.

But how consistent are these methodological principles with the theoretical architecture of the regulation approach? On closer examination of the key texts, a series of problems can be identified which highlight the tensions between such methodological declarations of intent and the practical application of regulation theory to the explanation of empirical phenomena.

The first set of problems concerns the nature of regulation itself. What precisely is to be regulated within this approach? Is it the general contradictions of capitalism as a mode of production – redefined in neo-Keynesian terms as the need to maintain a dynamic balance between production and consumption – or the specific dilemmas of an individual regime of accumulation? What is the relationship between the various elements of the mode of regulation and which level is most crucial to the achievement of sustained growth? Despite formal declarations that modes of regulation are complex structures each of whose elements must be compatible with the others, most regulationist analyses in practice tend to privilege a single component of the system – usually the wage–labor nexus or the form of competition – and use it to characterize an entire phase of development as Fordist or post-Fordist, competitive or monopolistic. How far do the objects of regulation such as the wage relationship or the form of state intervention precede the emergence of a particular mode of regulation, and how far, as Jessop suggests, do the objects and mode of regulation form a structural coupling whose contours can only be determined under a definite set of historical circumstances? Despite their methodological reservations about functional explanations, most regulation school analyses reason in practice as if the persistence of the objects of regulation could be accounted for by the development of a smoothly functioning mode of regulation.[25]

A second set of problems concerns the relationship between higher and lower levels of analysis or between theoretical abstractions and empirical cases. How far can the historical succession of regimes of accumulation and modes of regulation be separated from the structural properties of capitalism as a mode of production? To what extent is the existence of a plurality of national modes of growth genuinely compatible with the postulation of a dominant international regime of accumulation during any given period? How far can divergent national configurations of institutions such as collective bargaining or the welfare state be legitimately considered as limited variations of a single mode of regulation?

When the regulation school seeks to characterize broad historical periods in terms of dominant regimes of accumulation and modes of regulation, its exponents tend to elide the diversity of national experiences acknowledged in more detailed case studies.[26] Despite the many empirical caveats scattered through their work, therefore, regulation theorists, as other critics have charged, systematically overstate the dominance of Fordist modes of regulation during the postwar period, whether in terms of the pervasiveness of Taylorist work organization, institutionalized collective bargaining, Keynesian demand management, or the welfare state.[27] Conversely, the application of a regulation approach to national case studies, typically involves not only a severe "stylization of the facts" to fit its theoretical categories, but also ad hoc modifications of the categories themselves to accommodate observed variations. The result is the multiplication of hybrid formulations poised uneasily between theory and empirical description, such as "flex-Fordism" (Germany), "blocked Fordism" (UK), "state Fordism" (France), "delayed Fordism" (Spain, Italy), "peripheral Fordism" (Mexico, South Korea, Brazil), or "primitive Taylorization" (Malaysia, Bangladesh, the Phillipines).[28] When regulation theorists discuss the possible shape of an emergent post-Fordist mode of development, finally, they often fall back illegitimately on general tendencies of capitalism as a mode of production – such as the concentration and centralization of capital or the development of economies of scale – to privilege the likelihood of certain outcomes over others. Thus, for example, in arguing against the viability of flexible specialization as the basis for sustained growth, Boyer revealingly asks: "Is it plausible that the trends that have operated since the Industrial Revolution could be reversed? Does it not imply a complete shift in the balance of power and social and economic structures? Is it conceivable that all the indivisibilities that characterize contemporary society could thus be reduced . . . ?"[29]

A third set of problems concerns the role of classes as social actors in the emergence and decay of successive modes of development. At a methodological level, regulation theorists reject the idea of classes as collective subjects whose interests can be derived from the abstract structure of capitalist relations of production, recognize the existence of a plurality of social actors, and assign considerable autonomy to the state as something more than a bulwark of capitalism. When it comes to more specific regulationist analyses, however, social forces are typically analyzed in class terms with little attempt at empirical justification. Thus, for example, the declining rate of productivity growth under Fordism is largely attributed to class struggle at the point of production, assuming without argument that particular disputes between managers and work groups can be unproblematically aggregated into more general conflicts between capital and labor. Similarly, national modes of regulation are characteristically treated as class compromises, with little specific attention to the identity or organization of the actors concerned: to what extent, for example, can trade unions, business firms, or employers' associations in countries as different as the US, France, and Japan all be treated as direct representatives of wider social classes? Finally, despite occasional references to the importance of "new social movements," regulation school analyses of the political forces and social constituencies which may shape a possible post-Fordist future are overwhelmingly couched in class terms.[30]

Each of these problems highlights the inherent tensions between the totalizing ambitions of regulation theory – taken over from classical Marxism as well as from neo-classical economics – and its putative commitment to the recognition of national and historical variation. The regulation school set out to discover a middle way between general theory and empirical analysis, but their approach has run into a blind alley in which its conceptual holism must alternatively override or be undermined by the diversity of particular cases. Despite its apparent methodological sophistication, therefore, the theoretical architecture of the regulation approach is ultimately little different from the more simplistic versions of post-Fordism, while contrasting quite sharply with that of flexible specialization.

PROBLEMS OF EVIDENCE

Methodological questions

So far we have considered the theoretical structure and major substantive claims of the three main bodies of work which address the problem of changes in methods of manufacturing organization, and the approaches to technology and business strategy that accompany them. We have shown that each of these approaches stems from a distinct theoretical background and that each makes quite distinct claims about the type and level of changes that are taking place in contemporary manufacturing industry. These very real differences tend to be masked by widespread assumptions in political and academic debate that the theories are all saying something similar about the same basic, if changing, reality. This is a common-sense mistake that it is essential to rebut if we are to be able to assess the respective claims of the theories to validity and to consider the adequacy of their means of demonstrating the very distinct knowledge-claims that they advance.

It is an illusion to assume that each of them can be tested by reference to some common set of changes occurring "out there" in the real world. Each of these bodies of work sets up a very different world, each preconstructs it in theory, and each specifies in a particular way what sort of evidence and research strategy will count in showing what they claim to be happening is taking place. Theory, in the sense of a set of concepts, the construction of a set of objects on the basis of such theory, and the postulation of a domain of evidence relevant to such concepts and propositions are all particular to the style of work in question. We cannot construct a theory-neutral domain of evidence that will suffice to adjudicate between the claims of the three styles.

Rather we must look first at each of the styles of work and see what theory-evidence relation it seeks to establish. Then we must question the adequacy of that relation. Is the process of evidentialization consistent with the claims of the theory? Does the evidence proposed and gathered actually justify these claims? Is it practically possible to build up a body of evidence? If not, are there substitute forms of evidentialization that will provide weaker but indicative or persuasive support for the theoretical claims advanced?

All these questions relate to the relationships possible between theoretical concepts, the objects they construct and the type of evidence considered appropriate. These relationships are complex. We assume that there is no one "correct" set of epistemological doctrines or methodological protocols that will cover them all. We are far from wishing to legislate prescriptively about epistemological doctrines or styles of research. Clearly, we have little time for the naive empiricism or positivist phenomenalism that would refer each of these theories to some common body of given facts or constructed operationalizations of observable phenomena. Equally we are by no means satisfied that a realist epistemological premise actually enables one to capture "reality" more effectively than conventionalist premises about the theory-dependent nature of the objects of investigation. In practice, all demonstration and evidentialization is more complex for any definite body of work than general epistemological protocols or methodological textbooks suppose. We are skeptical about the merits of generalized doctrines in epistemology and the philosophy of science as guides to social-scientific research. But we are not indifferent to questions of validity and evidence. Our concern with these questions is governed by two suppositions.

The first is that justice must be done to the distinct claims of a theory about the evidence relevant to it. It is illegitimate to seek to "refute" a theory solely on the basis of evidence external to the theory and in disregard of the arguments it may make about why that evidence is inappropriate. However, serious contradictions between theory and evidentialization – exemplified by inconsistency, uncertainty of claim, and constructing ad hoc arguments to make evidence fit ambiguous propositions – all tend to diminish the possibility of taking the tests proposed by a theory seriously or to accept its evidence at its own value. We shall claim that there are such contradictions between theory and evidence implicit in regulation theory, and that they stem from its basic objectives and intellectual starting points. We shall also claim that the "post-Fordist" view as advanced by *Marxism Today* in particular, is too loose theoretically to be evidentialized in any rigorous way.

The second supposition is that ultimately the only way to determine the validity of these theories and their very distinct constructed objects is not to refer each of them to a common external "reality," but to relate each of the theories one to another and judge them, in terms of the plausibility of the arguments that can be advanced within one of the theories – flexible specialization – against the others. This claim is not theory-neutral and its outcome rests not on some general methodological protocol, but on the explanatory power of the specific argument advanced. Ultimately the test of a theory is its intellectual productivity, not its elegance, nor its conformity to established methodological canons, nor its relation to pregiven political expectations or desirable policy outcomes. That productivity rests solely on its capacity for arguing and showing what it claims is the case with a reasonable degree of probability.[31]

We shall now consider each of the main styles of work in turn in order to examine the general theory/evidence relation it constructs before going on to consider substantive questions about that evidence.

Flexible specialization

The flexible specialization approach takes its start from a criticism of social theories which assume that society is a "totality," a set of relationships governed by a single general principle and consistent in their character with such a principle. It also entails the criticism that such theories frequently presume a process of necessary social development or evolution based on certain fundamental "tendencies" operative in such a totality.[32] These criticisms apply to both Marxist and non-Marxist general sociological theories alike. Flexible specialization emphasizes the contingency and complexity of the connections between social relations; it insists on the distinctiveness of the national and regional routes to the establishment of such connections between social relations; it recognizes the crucial role of strategy and bodies of ideas in constructing such routes, and it is aware that things could have been otherwise. It is, therefore, alert to the specific conditions producing certain outcomes and to the possible coexistence of several distinct sorts of outcomes. The variety of possible outcomes that can be constructed from the basic ideal–typical concepts of flexible specialization is, therefore, considerable and each establishes a different relation between concepts, constructed social objects, and type of evidence that will demonstrate whether or not the social outcomes connected to such objects are operable.

To illustrate this complexity we shall outline three kinds of relation between theory and evidence in flexible specialization.

First, flexible specialization can be used in a mode we call the normative–empirical. Flexible specialization emphasizes that each social "world" contains a number of possibilities. A prevalent technological paradigm and the typical modes of social organization connected to it arises for a complex variety of factors, and has predominance over other possible outcomes for reasons that fall far short of social-structural or historical logics or necessities. This means that we must be alert to competing strategies and assess outcomes in terms that do not predetermine which of them will prevail. Hence the attention given to historical alternatives to mass production and the search for other reasons for the saliency of mass production strategies than an assumption of their inherent efficiency due to economies of scale.[33] Part of the role of evidence here consists in showing that other alternatives were possible, that they coexisted with the dominant paradigm, and that they offer distinct routes to innovation and change should the specific complex of conditions favouring the dominant paradigm cease to apply.

Thus flexible specialization is concerned to rewrite history in order to show that the complexity of the past helps us to recognize that there are a variety of options in the present. The relation of theory and evidence in such historical work is complex, in particular such claims cannot be refuted by pointing to the importance of mass production; rather they depend crucially on the conditions under which it came to prevail and national and regional variations in the forms of mass production strategies themselves. The same complexity of evidence about the coexistence and possibility of a number of worlds relates to present debates as well. Part of the role of theory is to identify certain instances or cases of progressive flexible specialization strategies, to show that such things are socially possible and to investigate whether they can be generalized given appropriate policy commitments and satisfactory conditions. If flexible specialization strategies are possible, if their conditions are not too difficult to satisfy, and if certain of their policy consequences and social outcomes are attractive from a certain normative standpoint then the role of evidence here is to serve as a support for advocacy and a means of generalizing the process of learning from certain national, regional, or enterprise experiences.[34] Simply showing that flexible specialization strategies have not been generalized, that they exist only in certain cases, and that they do not exist in a pure ideal-typical but in a hybrid form thus does not constitute a refutation of flexible specialization as advocacy. All the advocate of flexible specialization as a normative approach has to do is to show that such strategies are possible and that they can be expanded beyond given cases, even if in a hybrid form. Thus much of the "empirical" criticism of flexible specialization analysis is beside the point, since the use of such concepts is not confined to the hypothesis that flexible specialization is the prevailing or generalized mode of manufacturing organization.

Second, flexible specialization serves as a positive heuristic. Thus flexible specialization theory includes a battery of concepts drawing attention to a number of distinct ideal-types of production systems, progressive and stagnant variants of the same, possible forms of hybridization, and also ways in which these various forms can be combined in large and small companies, national and regional economies. The result is a very large range of possible situations and complex cases, a wide variety of types and hypotheses. Thus it will not do to select one of these types or hypotheses and seek to "refute" it without reference to the others. This is a common failing of the critical literature which tends to operate as if flexible specialization were a theory which gives necessary prominence to small firms over large ones, or which supposes that the industrial district based on small firms is the sole or major form of flexible production. The ideal-type is not to be taken as an empirical generalization, and, therefore, it should not be treated as if it consisted in a proposition that the majority of firms in a given national economy would conform to its features. Moreover, the simple ideal-type is just a part in a more complex and multilayered process of theorization. This process emphasizes the importance of social context, the complexity of coexisting strategies and structures of manufacturing, the contingent nature of their conditions of existence, and the variety of possible outcomes. Flexible specialization cannot be reduced to a few simple hypotheses. At the same time, this theoretical complexity is not the result of ad hoc argumentation and incoherence.

Third, flexible specialization serves as a negative heuristic. Flexible specialization is a theory about the nature of manufacturing as a form of social organization. Even if flexible specialization were not widespread, the concept would still be valuable. For the analysis of mass production and its conditions of existence could not remain the same, the specific social routes to its generalization as a paradigm and the variant forms of its institutionalization would gain in saliency as against traditional claims that mass production prevailed because of technological necessities and economies of scale. Flexible specialization is thus not merely an hypothesis about one type of production but is part of a much wider theory about production systems in general and their socio–political conditions of existence.

"Post-Fordism"

The advocates of "post-Fordism" are trying to respond to certain perceived social changes, however much this may clash with Marxist orthodoxy. Yet "new times" remains haunted by the Marxist concept of mode of production. "Post-Fordism" is created by generalization from a bundle of actual or supposed social facts, and the concept of the new social system is then used to explain all the phenomena of "new times." Thus the presence of certain changes in the composition of the labor force is taken as indicative of the development of productive flexibility in UK manufacturing, whereas no such claim is warranted by the evidence. The vices of Marxist schematization of social relations persist, even though metaphor and ad hoc generalization have replaced rigorous concepts.

"Post-Fordist" arguments like that of *Marxism Today* simultaneously under- and overtotalize. They overtotalize by aggregating together phenomena into a social type: deriving social relations from a metaphor based on production organization. "Post-Fordism" is a new type of society and despite claims about fluidity of social relations, the co-present elements are not permitted to vary except to the most limited degree. Thus the inability to see that those areas in which the most "advanced" production exists (the most "post-Fordist" manufacturing organization) are frequently socially and institutionally conservative (the least "postmodernist" social relationships).

At the same time these arguments undertotalize because they are incapable of postulating consistent causal relationships between the phenomena in this new social type, its elements are merely co-present. For the post-Fordists, economic determinism is passé, and so is the essentialism of the mode of production and its necessary tendencies. But no new explanatory schema has been developed consistent with the level of generality of the social typology. Like regulation theory, "post-Fordism" is a retreat from classical Marxism and classical sociological theory that has not fully settled its accounts with the latter.

The result is a form of argument that is ad hoc in its basic substance, rather than just forced into ad hoc arguments to reconcile specific contradictions and ambiguities as is the case with regulation theory. It is difficult to see how the "post-Fordists" could construct a domain of evidence since there is no coherent theory to be evidentialized. Merely pointing to certain phenomena, showing things have changed, or showing that the bits of a list do indeed exist will not do. "Post-Fordism" is simply not coherent enough as a theory to merit more specific concern about the evidence that might justify it. We shall not, therefore, go on to discuss it when we consider the specific issues about evidence below.

For Freeman and Perez, on the other hand, definite questions of evidence do arise. The most important of these is the question of statistical proof of the existence of Kondratiev waves. As we have covered this issue in our theoretical discussion of the "techno-economic paradigm" approach we will not take it further in the following section on substantive problems of evidence, and as this issue is the one on which their analysis stands or falls we will not consider lesser issues of evidence further here.

Regulation theory

The regulation theorists share the same basic concepts, though their specific methodological pronouncements differ. At the same time two methodological positions are common to most regulationists: first, a desire to move away from the more rigidly economic determinist, historicist, and conceptual realist features of classical Marxism; second, a willingness to use forms of evidence, such as econometric tests, quite alien to classical Marxism.

Alain Lipietz expresses this methodological opposition to conceptual realism particularly forcefully in *Mirages and Miracles* (1987). He argues that concepts are only valuable insofar as they enable us to grasp phenomena, and that one should not make a fetish of them. Conceptual realism, the hypostatization of concepts into entities, is a major fault of much Marxist social analysis and an epistemological obstacle to concrete analysis. Lipietz uses Umberto Eco's nominalist monk William of Baskerville to make his point.[35] The rejection of conceptual realism has been a commonplace, at least since Whitehead's day.[36] But conceptual realism has been a persistent feature of Marxism precisely because of its strong Hegelian legacy. Marxism views society as a totality driven toward certain states of affairs by tendencies inherent within it. This totality can be apprehended in general concepts because it exists as a determinate generality; concepts thus partake of the nature of this social reality. Conceptual realism in Marxism is not misplaced concreteness: it is rather the apprehension of the concrete generality in thought.[37]

Conceptual realism makes methodological sense only within a strongly structuralist or historicist conception of Marxism. Yet this is exactly the sort of determinism Lipietz, Boyer, and the other regulationists wish to avoid. They want to insist on the concrete variability of capitalism, on its crisis-ridden nature, and on the absence of historical inevitability. Yet they wish to retain the concept of capitalism as a mode of production and the existence of certain fundamental tendencies within it. How can this be done? The answer, it appears, is to postulate levels of analysis, and to use the higher levels of generality to explain the specific conditions which are the only real states of existence of the capitalist system; capitalism exists not as generality but as specificity.

In fact it is much more difficult than the regulationists think to be a nominalist and yet to retain some major elements of classical Marxism. Three main problems present themselves. First, conceptual nominalism prevails in regulation theory only at the formal level, and conceptual realism is overcome in practice by injecting elements of classical Marxist concepts into the phenomena. Nominalism is compatible with Marxism only by the "Marxianization" of the phenomena to be explained. The very problem of regulation is of this nature – it is a consequence of "Marxianizing" the economy. Regulation theory involves taking a version of the "problem" of reproduction outlined in Volume II of *Capital* as a necessary economic problem, and seeking to show how production and consumption can be interconnected by specific institutional forms. Concrete economic relations are treated as if they must actually resolve at least temporarily the contradictions in the capitalist mode of production. This is only a "problem" if you read *Capital* in a certain way, and believe that concrete economies actually have to answer in practice the problems posed by this reading.

Second, there is a theoretical failure to explain how the more general concepts and abstract levels of analysis derived from classical Marxism can actually be carried over into and made compatible with the more specific concepts and levels of analysis sought by regulation theory. This problem arises from rejecting as unacceptable the classical Marxist conception of the capitalist mode of production as a totality (a concrete generality) which subsumes particular capitalisms as mere instances of the workings of its laws. Levels of abstraction are needed because this direct presence of the totality in the concrete simply does not work: it leads to the postulation of certain necessary states of affairs, and yet these states of affairs stubbornly fail to materialize. Capitalism, far from collapsing, survives and proves capable of institutional innovation and national variation within a prevailing international regime of accumulation. Hence the withdrawal of the totality to a higher level of abstraction, and the postulation of the concrete as the specific domain in which the general "problems" of the capitalist mode of production are worked out in specific spatio-temporal forms. The problem is not merely one of levels of abstraction, how capitalism is present in and as its specific spatio–temporal conjunctures, but also involves that of concealed conceptual realism. The more general concepts are supposedly merely a means to get at a reality that is always and inevitably concrete, and yet somehow these concepts get "behind" reality and particular national economies become exemplars of variant forms of phases in the evolution of capitalism.

Third, there is the problem of the status of social actors, and social classes in particular, created by the complexity of the regulation school's relation to the classical concept of the capitalist mode of production. In classical Marxism classes are both categories of economic agents, defined by their relationship to the means of production, and social forces active in politics. The evolution of the capitalist mode of production, its contradictory tendencies, and the emerging crises of capitalism lead the social groups active in politics to crystalize around the opposed social poles of bourgeoisie and proletariat. Marxism has been in retreat from this simplistic social teleology of economic and political class relations ever since the *Communist Manifesto* was written. But regulation theory has a special problem since for it classes as categories of economic agents can only relate to capitalism at the highest level of abstraction. At more concrete levels, economic and political actors are seen to be specific and contingently constituted agencies. However, regulation theory has devised no means to relate classes at the different levels of abstraction any more than it can relate any other social phenomena. Thus it tends to continue to treat classes as stereotyped entities and to refer to them as active social agents, alongside or underlying more specific social groupings and organizations. The result is schematic analysis and the tendency to overgeneralize the political consequences of and responses to forms of economic relations both nationally and internationally – treating capital and labor as if they were a coherent social actors.[38]

Thus the main methodological problem of regulation theory is not some specific difficulty with evidence, but the much more basic issue of whether the basic entities and processes it seeks to evidentialize exist at all. As these entities are of a more general nature than specific bodies of evidence, such as statistical series, the latter cannot serve to demonstrate their existence but presuppose it if they are to count as evidence. Indeed, the use of evidence in regulation theory presupposes the general truth of its conceptions of capitalism.

Unlike flexible specialization, regulation theory is tied to certain general concepts that must be present in reality; it cannot operate in an ideal-typical mode. Flexible specialization does not suffer from this problem of different levels of generality; from a need for the device of concretizing abstractions through the construction of specific states of their realization in definite conditions. Flexible specialization neither needs to posit a totality nor does it need to limit the range of variation of regional and national economies to fit in with some prevailing phase of its actualization. We should be clear that the postulation of a social totality (in however complex and qualified a form) is not inherently more virtuous in explanatory terms than theories like flexible specialization which stress the variation, contingency and complexity of phenomena. We should also be clear that anti-totalizing theories are not more "empiricist" than theories that seek to make totalizing general concepts operative in reality. There are other routes to conceptual rigor than totalization or quasi-totalization: flexible specialization uses concepts as ideal-types to organize analysis and explain how in the concrete case contingent phenomena are causally connected one to another.

The problem with totalization is that it stakes too much on the validity of concrete generalities. It is forced to limit possible phenomena to fit in with the general principles of social organization that make social relations consistent wholes. The problem with regulation theory is not its commendable drive to break away from the worst features of totalizing Marxism; rather it consists in the belief that they can be purged by using the device of levels of abstraction while retaining elements of the concept of mode of production. Like most halfway houses, or third ways between alternatives that are perceived as unacceptable, it suffers in practice from the faults of that from which it seeks to retreat. The problems of evidentialization in regulation theory are thus effects of its basic theoretical strategy, and it is difficult to see how they can be overcome or that strategy reformed. Only a complete break with classical Marxism would resolve these problems of evidence, but that would explode the general problem of "regulation" in terms of which these questions of evidence arise.

From theories to evidence

Regulation theory

Of all the competing approaches, regulation theory has undoubtedly devoted the most explicit attention to the problem of empirical verification. Among its central methodological objectives is the reconciliation of Marxist analytical concepts with modern methods of formal economic modeling and econometric testing. Hence the regulation school characteristically seeks to verify its hypotheses by translating key concepts into appropriate micro- and macroeconomic models, which can then be assessed both for their internal consistency and for their ability to explain comparative and historical variations in statistical data sets. What were the mechanisms of wage formation in particular countries during different periods? What were the central causes of the crises of the 1930s and the 1970s for individual countries and the international economy as a whole? What are the prospective relationships between demand, investment, productivity, and growth under different possible configurations of regimes of accumulation and modes of regulation? These are the types of question which the regulationists' methods of empirical testing are designed to answer.

There can be little dispute about the ingenuity deployed by the regulation school in constructing such models, and the assumptions involved are often more realistic than those of more conventional economic analysis. But a number of fundamental problems can nonetheless be discerned. The first concerns the nature of the modeling exercise itself. However sophisticated the model, the theoretical conclusions that follow are no better than its initial assumptions. Thus, for example, Boyer's conclusion that flexible automation (combined with international Keynesianism) offers better prospects of sustained growth than flexible specialization follows directly from his assumption of an inferior trade-off between flexibility and productivity under the latter, an assumption which would not be accepted by theorists of flexible specialization themselves.[39]

The second problem concerns the quality of the data to which the models are applied. As with much quantitative economic history, the statistical evidence is rarely robust enough convincingly to support the econometric tests to which it is subjected by regulation school analyses. Thus, as Boyer acknowledges in his long-term study of French wage formation, the unreliability of nineteenth-century statistics on unemployment and intersectoral wage variations make his conclusions "suggestive rather than definitive," while it is unclear how far the apparent reinforcement of competitive processes of wage formation between the wars is a real phenomenon rather than a result of changes in the method of constructing the industrial production series.[40] Even more seriously, his discussion of real wage movements, as in most such studies, appears to be based on indices of money wage *rates*, although detailed sectoral research typically reveals significant differences between rates and *earnings*, reflecting the influence of fluctuations in hours worked across the business cycle as well as of variations in bargaining power and the incidence of state regulation.[41]

The third problem concerns the conclusions that can be drawn from the confrontation of these models with empirical data. Like all econometric tests, those deployed by regulation theorists can only demonstrate statistical correlations rather than causal relationships between the variables investigated. Thus, for example, it is a central claim of the regulation school that the major cause of the depression of the 1930s in countries such as France and the United States was an imbalance between intensive accumulation and competitive regulation, as demonstrated by the disparity between the rate of growth of productivity and of effective demand during the preceding decade.[42] But the regulation theorists simply infer the transition to intensive accumulation from rapid growth of French productivity during the 1920s, despite considerable evidence that the diffusion of Taylorism and scientific management in French industry remained quite limited throughout this period.[43] And when it comes to the crisis of Fordism, Boyer himself admits that the explanations put forward by the regulation school, at least for the moment, "are both too numerous and insufficiently articulated amongst themselves" (Boyer 1986a: 103–4).

Such problems are common to all forms of economic modeling and econometric analysis, and are in no sense peculiar to regulation theory. But they present special difficulties for the regulation school because of the distinctive nature of its theoretical project. For orthodox economists, following the methodological precepts of Friedman and others, the inability of econometric tests to verify causal hypotheses is unimportant because the value of economic models lies in their predictive power rather than the descriptive realism of their assumptions.[44] For the regulation school, on the other hand, econometric tests are intended to demonstrate the truth of its models as descriptive hypotheses about the real world (Boyer 1989: 1403–5), and their inadequacy for this purpose opens up an unbridgeable disjunction between the theory and its preferred mode of evidentialization. This disjunction, in turn, is simply a further sign of the regulation theorists' continued but unacknowledged attachment to conceptual realism, together with the impossibility of escaping from the deterministic consequences of classical Marxism while retaining key elements of its underlying conceptual framework.

Flexible specialization

For flexible specialization, by contrast, the relationship between theory and evidence is significantly more complex. Unlike regulation theory, the central problem for flexible specialization is not to demonstrate the truth or falsity of its basic concepts, since these are explicitly conceived as ideal-types instead of real forces operating behind the observable phenomena themselves. The appropriate criterion for the assessment of such ideal-types is not their truth value but rather their heuristic productivity: how far does the conceptual framework of flexible specialization illuminate observable processes of industrial change? As we have seen in previous sections, flexible specialization as a general theoretical approach is compatible with a broad spectrum of possible forms of productive organization – including the continued predominance of mass production. But a hypothesis whose validity does depend on empirical evidence is whether – as much of the literature argues – current manufacturing practice is moving in the direction of flexible specialization as a specific model of productive organization, taking account both of the plurality of institutional forms within which it may be pursued and of the possibilities of hybridization.

What sort of evidence might permit us to test this hypothesis? Both macro and micro levels of analysis are in principle relevant. At the macro level, one would ideally like to establish statistical indicators of cross-sectional variations and changes over time in the distribution of different forms of production across industrial sectors, national economies and the international economy as a whole. Thus one would need large-scale data about such issues as product diversification (number of distinct products manufactured, rate of introduction of new models, average batch size); productive flexibility (costs of product changeover, nature of equipment used, minimum efficient scale of operation); work force versatility (skill composition, job content, training); interfirm relationships (extent and nature of subcontracting, reliance on collective services); and geographical agglomeration of economic activity.

But there are good reasons, both practical and theoretical, why reliable macro data of this type are likely to be difficult if not impossible to obtain. The first arises from the nature of the available industrial statistics. Like all official statistics, the classification systems used by industrial censuses in different countries reflect particular sets of theoretical assumptions and administrative practices. Thus as Storper and Harrison rightly point out, national statistical accounts cannot be used to analyze the operation of real input–output systems or industrial sectors "since they tend to classify whole firms or establishments according to their 'principal' activities" (Storper and Harrison 1990: 6). And even within these limitations, as Luria has shown, there are fundamental difficulties in measuring intertemporal variations in product diversity because of widespread inconsistency in classification, frequent code changes and suppression of data to protect proprietary information (Luria 1989: 16). The more rapid the rate of product diversification, moreover, the more serious these difficulties become, since earlier classifications become obsolete more quickly and incommensurability of data from different periods increases.

For many of the other empirical questions thrown up by the flexible specialization hypothesis such as product batch sizes, work force versatility or patterns of subcontracting, little large-scale data is available because official statistics have not been compiled with these issues in mind.[45] But even if industrial census classifications were rewritten with an eye to flexible specialization, significant conceptual problems would still remain in the interpretation of such evidence. The key issue is the context and strategy-dependence of each element of a productive system within the flexible specialization approach. Each industrial branch or sector, for example, has its own specific market and technological characteristics against which any particular indicator must be assessed. Thus a given product batch size will have a different significance in, say, clothing, steel or automobiles, and there is no obvious way to aggregate such data across the economy as a whole. At a deeper level, moreover, the same component or practice may have a different significance depending on its place in the broader strategy of the individual firm: thus as a substantial body of research has demonstrated the same equipment such as numerically controlled machine tools or flexible manufacturing systems can be used in contrasting ways in different national and industrial settings.[46] Similarly, even the most determined mass producer may be obliged to manufacture some speciality lines in small batches, while the best-selling lines of a successful flexible specialist may likewise be turned out in significant volumes.

For all these reasons, such macro-level indicators can only provide a suggestive guide to broad trends in industrial reorganization rather than a definitive test of the flexible specialization hypothesis. The preferred form of evidentialization for flexible specialization is instead the analytical case study conducted at the micro-level of particular firms, regional economies or industrial sectors. Only detailed case studies permit the close attention to context and strategy which is the hallmark of a flexible specialization approach; and only this method makes possible the comparative analysis of relationships between forms of production and institutional frameworks which is central to its theoretical architecture.[47]

Two major problems arise from this strategy of evidentialization through case studies: interpretation and representativeness. As we have already noted, most firms or regional economies characteristically combine elements from both flexible and mass production rather than embodying pure examples of either model. How then can one assess the precise balance between the two models in any given case from the standpoint of the flexible specialization hypothesis? In principle, the solution might appear to lie in a search for objective indicators of flexible specialization such as those discussed at the macro level which could more successfully be applied to less heterogeneous micro data. Considerable mileage can undoubtedly be obtained through this route in documenting the spread of flexible specialization, as a number of suggestive studies have indicated.[48] In practice, however, the conceptual problems raised by such indicators remain the same: their precise meaning in any case cannot be determined without reference to the strategies of the actors concerned. Hence the case study method necessarily entails an ineradicable element of subjective interpretation, in which there is considerable scope for legitimate disagreement among different observers. But if there is an unavoidable degree of indeterminacy about the interpretation of case study evidence, this does not mean that there is no valid basis for discriminating among competing views, contrary to what current fashions in literary theory might appear to suggest.[49] Competing interpretations, like competing theories more generally, can properly be ranked in terms of their plausibility in accounting for agreed features of a common body of evidence according to internally consistent criteria.

These considerations can be illustrated more fully through a brief examination of current debates about the nature of Japanese manufacturing practices. How far can the success of Japanese firms be properly interpreted as evidence of the diffusion of flexible specialization, and how far instead as evidence of the development of increased flexibility within mass production itself? At one level, as Sayer points out, this question may be considered largely semantic, since there is considerable agreement among apparently conflicting interpretations about key features of the Japanese system: the rapid pace of model renewal and new product development; the productive flexibility obtained through organizational innovations such as just-in-time component supply, quick die-changes or mixed-model assembly lines; the prevalance of job rotation, teamworking and other forms of functional flexibility among large sections of the labor force; and the importance of "relational subcontracting" between large and small firms.[50] At another level, however, this question is crucially important, given the centrality of Japanese manufacturing to the characterization of current trends in competitive strategy and productive organization in the international economy.

Some differences between the two interpretations are based on a mis-specification of the opposed view: thus it is no objection to the flexible specialization hypothesis properly understood to emphasize the continued role of larger firms in Japan, nor the high overall volume of different products manufactured within the same firm or plant. Others arise from the results of new empirical research, such as David Friedman's (1988) demonstration of the limited role of MITI and the importance of small-firm industrial districts in the development of key Japanese export industries such as machine tools. But many differences of interpretation arise from ambiguities in Japanese industry itself. There can be little doubt that most Japanese innovations in sectors such as automobiles or consumer electronics originated in domestic firms' adaptation of mass production methods to local conditions during their postwar drive to catch up with the West.[51] And important features of more recent Japanese practice can still be legitimately interpreted in this light, from the limited range of variation on certain models and the long production runs of key components through the continued importance of hard automation and the relatively narrow skills required for many jobs to the dominance of large firms over their subcontractors. But there are also signs that many Japanese manufacturers are pushing these innovations in a more radical dimension in order to trade on their competitive advantage in catering for fragmented markets and volatile demand. Thus leading firms in these sectors appear to be dramatically increasing the pace of product innovation, expanding the range of distinct models which can be manufactured with a given combination of workers and machinery, and devolving responsibility not only for component supply but also for final assembly and product development to suppliers whom they often encourage to work for other manufacturers as well. Such strategies may not lead Japanese manufacturers to converge on a pure model of flexible specialization, but like many Western experiments at industrial reorganization, they have already progressed too far to fit comfortably into an alternative conception of neo-Fordism or flexible mass production.[52]

But even if it were agreed that particular cases could legitimately be interpreted as examples of flexible specialization, a significant problem of representativeness would still remain. As Sabel himself remarks, "for example is not a proof" (Sabel 1989a: 23), and no quantity of case studies, however convincing, could demonstrate the validity of the broader flexible specialization hypothesis. In the absence of a comprehensive macro-level map of the relative importance of competing models of productive organization, this difficulty may appear insuperable. But the Japanese case discussed above suggests an alternative strategy – common to most approaches to the analysis of industrial change – of focusing on those national economies, regions, and firms that have proved most successful in the current phase of international competition. Beyond Japan itself, for example, flexible specialization analyses have concentrated on regions such as Emilia-Romagna and Baden-Württemburg, whose technological dynamism, export competitiveness, and importance to the national economy are relatively well documented.[53] Other analyses, conversely, have concentrated on countries such as Britain in which flexible specialization has been weakly developed, highlighting its role in explaining their poor performance in manufacturing competition during the 1980s.[54] In either case, finally, the claim is not that international competition imposes a single form of productive organization on economic actors, given the plurality of institutional frameworks and the possibilities of hybridization, but rather that tendencies can be observed towards the displacement of mass production by flexible specialization as the dominant technological paradigm of the late twentieth century.

POLICY IMPLICATIONS

Our discussion of policy must begin with a number of caveats that make clear our approach to this issue. In discussing the policy implications of the three broad perspectives outlined above it is necessary to consider which policies actually follow from these ways of conceptualizing forms of manufacturing organization and their wider sociopolitical conditions. It would be all too easy to slip into discussing the general views of the authors, confusing the policies that can strictly be justified by their approach to manufacturing with their political opinions and wider social concerns. Hence we need to consider the specific types and levels of policy intervention that have some direct connection with the explanatory and analytic frameworks in question.

Thus we must avoid overgeneralizing policy into politics. But we must also avoid considering policy implications too specifically, as if the sole relevant domain of policy is "industrial policy" in the narrow sense.[55] It by no means follows that an approach to manufacturing will, by reason of its own conceptual apparatus, foreground industrial policy in the sense of targeted interventions in particular industries. Broader policies for education, training, industrial finance, and macroeconomic management may be considered the most effective ways of securing the conditions for advanced forms of manufacturing organization.

"Post-Fordism"

Of the various "post-Fordist" approaches little need be said. There are obvious political implications of, for example, *Marxism Today*'s arguments but few policy implications. Given the inattention of *Marxism Today* to specifying and analyzing the economic relationships involved in "post-Fordism," this is hardly surprising.

Freeman and Perez's work, by contrast, does have definite policy implications. Both authors make comments on many policy areas, including education and training, but here we are concerned with the strict implications of their basic thesis about the role of technical change. Given the primary role they assign to the diffusion of the new core technologies that form the basis of a new "techno-economic paradigm," then the key areas of intervention are technology policy and R&D policy. The aim of policy should be to identify the emerging technologies and adopt programs that favor their diffusion through the whole economy. Freeman (1987, 1988) comments favorably on the role of Japan's Ministry of International Trade and Industry (MITI) in this context. One could characterize his view as a variant of the strategy of "picking winners," which we discuss further and criticize on page 109 below. In his case, the primary object of such a policy of selective state intervention is not support for potentially successful firms (which is what the phrase often means in US debates) but support for technologies whose generalization is capable of producing growth across the whole economy.

Regulation theory

The regulation school's approach to policy is developed in relation to its general theory of capitalism as riven by periodic and specific systematic crises that are far more than the normal cyclical crises recognized by conventional economics. Regulationists claim that the present period is one of crisis and transition in the structures sustaining capitalism on a worldwide scale. The regulationist approach to policy can best be seen as a combination of a neo-Keynesian expansionary program at the international level on the demand side and a neo-Marxist conception of the restructuring of the supply side, involving a new settlement between capital and labor in order to stabilize current economic volatility and class conflict. The expansionary policies are relatively clear, if both relatively conventional and very difficult to implement on either a global or a national scale. The policies that further the emergence of a new mode of regulation are less clear, and to the extent that they are developed in the political programs of regulationist authors they have little to do with the original neo-Marxist terms in which the problem was set up.

Regulation theory must thus seek policies that both secure a new international regime of accumulation and a new mode of regulation, with the various particular institutional settlements being thrashed out as the result of social and political struggles in the national capitalist economies. Regulationist policy must operate across a number of different levels and be relatively coherent between these levels. These levels are: the international economy, the major economic blocs like the European Community (EC), and the relation of the major social forces within the national economies. The regulationists' policy domain is thus far more extended than those of flexible specialization and its policy requirements are inherently far more ambitious.

The regulationists argue that the current period of crisis within capitalism creates both major conflicts between the main components of the world economy, between the First and Third Worlds, between the major industrial powers, and between the major social interests within the different national capitals. The previous international regime of accumulation based on American hegemony has broken down. The hitherto successful combination of Keynesianism sustaining demand and Fordism generating the productivity gains on the supply side to make an expansionary policy possible is no longer stable or sustainable. Reaching for classic Keynesian or Fordist solutions will not do. We are in a transitional period characterized by instability and crisis in the regime of accumulation and the mode of regulation. There are different routes out of this crisis.

The regulationists' value standpoint is to resolve the crisis on the terms most favorable to the left – to the peripheral economies of the Third World, and to labor in rich and poor countries alike. This resolution, however, will be within the limits set by the continuation of global capitalism. The regulationists reject the old classical Marxist breakdown theory of capitalism as a global system. In policy, as in theory, the regulationist approach claims to represent a "middle way"; in this case between the illusions of internationalist revolutionary socialism and the insularity of reformist social democracy in the national economies of the advanced countries. Regulation theory remains concerned with capitalism as a global system and it seeks policies for the international economy that will promote stability and growth.

Lipietz (1987, 1989) is clearest on this international program. He argues passionately that conventional policies will only exacerbate the plight of the Third World, driven into accelerated crises by the burden of debt. Ultimately an international expansionary policy that permits the First and Third Worlds to grow would be of mutual benefit, and in the interest of the advanced economies. While he is aware of the improbability of "world Keynesianism" (1987: 192), Lipietz argues (1989) for a new international order founded on multilateralism, involving a "new Bretton Woods' agreement. The latter would involve a new international credit money, the cancelation of Third World debt, and the financing of aid to the Third World through the institutions of the new agreement, and a system of open trade – not defensive protectionism. One might agree with most of this: it would be hard to defend First World banks or to argue that development aid should be less than the trickle it is. Our main point of criticism is not just that a new Bretton Woods is unlikely, rather it is that it is a classically Keynesian radical-liberal position. The regulationists' problems are drawn from Marxist theory, and yet their solutions, when they are clear, are fundamentally neo-Keynesian. This is not intended to disparage neo-Keynesianism. On the contrary, it is to insist that such policies can be adopted without reference to the theoretical problems from which the regulationists started.

For the developed countries, and the EC in particular, the regulationists' arguments for revitalizing the economy are less clear and less emphatic. Both Lipietz (1989) and Boyer (1988) argue for the further development of the EC. Lipietz (1989) argues for a new wages pact between labor and capital. This would be based on a trade-off of realistic wages for a measure of control over new technology, security of employment and greater free time. It would also involve a new form of the welfare state involving in particular a guaranteed minimum income and a "third sector" of socially useful subsidized work. Boyer (1988) argues that a variety of scenarios are possible for the restructuring of wage/labor relations and labor markets. But ultimately he appears to consider a form of neo-Fordism incorporating elements of flexibility the most likely. This, combined with neo-Keynesian demand management, a suitable regime of international trade, and restructuring of production, may be sufficient to resolve the crisis. Again, as with policies to stimulate world demand, the difficulty is that the neo-Marxist emphasis on the class struggle leads in practice to policy proposals that, while radical, are in no sense tied to the regulationists' own theoretical problems. Many of Lipietz's ideas are widely scouted by post-Marxists like Andr Gorz or advocates of "associative democracy" based on flexible specialization like John Mathews (1989).

The regulationists seem unclear whether or not the Fordist era in either production organization or wage/labor relations is really at an end. They appear to see the most likely future as one in which elements of flexible specialization and new information technologies and robotics are incorporated into Fordism to produce a neo-Fordist synthesis. Which of the several possible outcomes or scenarios prevail will depend ultimately on the struggles of the key players: state decision makers, capitalist managers, and trade union leaders. Policy in this sense is largely advocacy and shades into a general political program. It is striking how little industrial policy in the narrow sense can be found in regulationist texts and that, in respect of the wider policies for re-regulation, how little there is that is specific to them or dependent on the theoretical problems of regulation. This may well be, to be fair, because figures like Lipietz are effective and pragmatic political advocates. It may also be because the regulationists' starting point in the problems of classical Marxism is quite irrelevant to the actual problems of economic policy.[56]

Our aim here is not to counter the regulationists' actual political ideas, many of which are perfectly reasonable. The point we want to make is that the political positions taken by the regulationists are not strictly policy implications of their underlying theory, or rather, that even if they are compatible with that theory they are not necessary to it and can be sustained on other theoretic bases. Insofar as regulation theory has a distinct policy dimension, it is to give prominence to the need to stabilize the international economy and achieve a more equitable balance between its different components. That, however, is the most difficult level at which to achieve coordinated policies between the various agencies and actors.

Flexible specialization

We may begin our consideration of the policy implications of flexible specialization by noting that it involves a critique of traditional economic and industrial policy instruments. Flexible specialization challenges four widely canvassed ways of promoting the manufacturing sector through specific forms of intervention:

1 classic Keynesian strategies for promoting effective demand in particular national economies, the claim being that state action to promote effective demand will support the manufacturing sector by sustaining the overall level of economic activity and therefore maintaining or accelerating demand for manufactured goods;
2 strategies based on state-directed planning, promoting manufacturing output by coordinating investment policies of firms and concentrating development on advanced technologies and sectors – the three variants of such a policy are socialist command economies directing investment and orchestrating industry through administered prices, indicative planning on the French model, and the state direction of industry through analogues of Japan's MITI;[57]
3 state intervention through deregulation and measures to improve the free working of market mechanisms, active competition policies, and the

promotion of the prerogatives of enterprise management through industrial relations and labor market reform; and

4 a contrary policy of state intervention to promote concentration of ownership and through it to exploit both organizational economies of scale at the level of the firm and production economies of scale at the level of the plant (both also objectives of state planning strategies.

Flexible specialization is not anti-Keynesian in principle or opposed to demand management for general economic-theoretic reasons, such as those implied in monetarist and free-market doctrines. Rather it argues that there are severe practical limits to Keynesianism as a national policy to support manufacturing under current conditions.

The internationalization of trade in manufactures between advanced industrial countries and the liberalization of international trade through GATT and economic blocs like the EC means that domestic manufacturing sectors can only survive if they meet the conditions of international competition. This can only be achieved through appropriate supply-side policies and business strategies. Stimulating domestic demand may merely promote the ingress of foreign manufactured goods and even accelerate deindustrialization if the manufacturing sector is not able to meet international standards of competition. The experience of France under the first Mitterrand government and the consumption and credit expansion-based policies of the Thatcher government from 1983–8 (although the latter is not formally "Keynesian")[58] only serve to confirm this. Moreover, the increasing volatility of international markets and rapidly shifting patterns of demand in the period since the early 1970s favor more specific strategies of response like that of flexible specialization. Keynesianism should not be seen to be obsolete because it is linked in some necessary way with "Fordism." Demand management will not work to secure the home market for long runs of standardized goods by domestic manufacturers under the current conditions of international trade, since foreign "Fordists" might be able to be more competitive. Equally, were Keynesianism possible and flexible specialization widely developed, demand management would not automatically favor "Fordist" strategies, since flexible specialists might be able to exploit production, marketing, and design advantages in a climate of rising demand and growth. Keynesianism is problematic for macroeconomic reasons, not because "Fordism" is obsolete. Flexible specialization is thus not tied to a view of the functional obsolescence of Keynesianism, as some "post-Fordist" ideas clearly are, but to a view of its conjectural and structural inappropriateness in manufacturing economies under current conditions of international trade and competition.

Planning of a centralized and state-directed kind is seen to be inappropriate by flexible specialization theory for a variety of reasons. First, command economies tend to be ineffective, whatever the prevailing form of manufacturing organization. The Soviet model did not prove effective at organizing production on "Fordist" or mass production lines. It supposes that the major economic parameters can be held stable and that shifts in output and accelerated growth of sectors can be preestablished by planning norms. This is difficult in a semiclosed economy where demand is subordinated to programmed supply, it is virtually impossible in an open economy that is forced to match foreign economies, and that is necessarily required to respond to consumer demand. Add to this the volatility of international currency and equity markets, the changing patterns of demand for manufactured goods and the rapidity of product innovation and it becomes difficult to hold any of the major national macroeconomic parameters stable enough for them to be planned. Central planning has failed in command economies, and it is clearly impossible for Western-style economies.

Similar objections also hold for more limited forms of interventionist planning. State direction of investment and the anticipation and promotion of leading technologies presupposes that state officials, with or without the collaboration of business, have the information necessary to "pick winners" and to concentrate national resources on key future technologies. There is little evidence that government bureaucracies are flexible enough and well-informed enough to do this. On the contrary, they will tend to equate economic success with promoting key "over the horizon" technologies; these technologies are frequently not the most commercially successful. Officials will tend to pick major projects capable of state management, and not families of technologies capable of rapid state-of-the-art diffusion. There is a great deal of evidence both that MITI is a "peculiar institution," heavily dependent on the close and informal links between senior civil servants and corporate executives that would be regarded as unacceptable in a Western democracy, and that it has a more mixed record than its most enthusiastic Western advocates allow. Moreover, there is much evidence for both orthodox company-level innovation in Japan and for the success of the local regulation, industrial districts, clusters of small and medium-scale firms, and major company organizational structures and manufacturing strategies which are characteristic of flexible specialization.[59]

Deregulation and the promotion of "free" markets are becoming decidely less fashionable among policymakers now. Flexible specialization is strongly opposed to the free-market model precisely because it emphasizes the importance of social relationships that secure crucial inputs and vital collective services for firms. Such inputs that cannot be guaranteed by the model of sovereign enterprises purchasing the factors of production in open markets include: trained labor, low-cost finance, market and export information, the diffusion of technical information, and ongoing relationships based on trust with subcontractors and partner firms. The high costs to firms of low-trust relationships, the commodification of information, and the absence of ongoing collaboration with labor involve real competitive disadvantages. We have called these factors "dis-economies of competition" and they are most evident in those countries and regimes that have failed to balance cooperation and competition, conflict and coordination (Hirst and Zeitlin 1989b). These "dis-economies of competition" can offset the economies of scale supposedly available to large firms free to choose how best to secure the factors of production in open markets. Some of the most important "factors" are intangible and are also not readily tradeable – trust relationships with other firms and cooperation with labour are two good examples.

One aspect of free-market advocacy is the celebration of small firms. Flexible specialization is not a theory predicated on preferring small firms to large ones. Small firms have no necessary economic attributes, other than some arbitrary cutoff point that defines their size (number of employees, value of capital, etc.).[60] "Large" firms, indeed, multiproduct, multinational companies, can adopt flexible specialization strategies and so can small workshops.

But flexible specialization is clearly opposed to the policy objective of promoting concentration of ownership to secure economies of scale. This policy was widely adopted in the 1960s and 1970s, linked to views of the inherent efficiency of large-scale mass production, in order to promote firms of the size necessary for international competition – "national champions." In the UK, for example, the Labour governments of 1964–70 supported such concentration and created an agency – the Industrial Re-organization Corporation – to promote it. The results varied from ghastly failures (British Leyland) to stagnant survivors (GEC).[61] Such ideas are once again current in the EC, in response to the present merger wave. Policymakers take for granted the need to concentrate capacity in a few Euro-companies in each major industrial sector.[62] However, just as small firms are not inherently more responsive or innovatory in the way that free-market advocates often suppose, so large firms are not inherently more stable or successful, nor are they necessarily able to exploit economies of scale.[63] Size divorced from strategy is no inherent advantage, as the examples of General Motors or US Steel demonstrate. In practice active concentration policy or state indulgence of merger waves has tended to prevail over free-market inspired antitrust and procompetition policies, even in the recent period of widespread belief in the virtues of the market.

There are good reasons to suppose that large firms which have rapidly concentrated by a process of merger and/or acquisitions are not particularly efficient and that they do not possess the means adequately to direct their subsidiaries. Those major multinationals that have sought functionally to decentralize into constellations of subunits, able to cooperate or compete as needful, seem to have judged the matter better and to be one of the complex routes to industrial efficiency outlined by flexible specialization theory.

So far we have considered the negative criticisms made by flexible specialization theorists of other major approaches to policy; we now turn to positive proposals. Flexible specialization in its policy implications is a radical supply-side policy. It starts from the relationships necessary for effective manufacturing strategies and performance, and is concerned with wider macroeconomic and social policies on this basis. It is not a comprehensive economic and political program as such, although flexible specialization concepts can be integrated into such broader advocacy of strategies for reform.

Flexible specialization emphasizes the wide variety of possible institutional forms. There is no flexible specialization equivalent of the "workers' fatherland," a chosen national or regional path that offers a given model to policymakers in other regions or countries. German regions, or Japan or the "Third Italy" offer examples of institutions that sustain flexible specialization, but they cannot be slavishly copied. Strategies have to be adopted that take account of national and regional conditions. Important in this respect are the institutional and political conditions that favor or hinder the patterns of cooperation and coordination which support flexible specialization strategies. Those countries most dominated by liberal competitive politics, by the stress on the sovereignty of the firm, and by antagonistic competition between social interests are least likely to be able to introduce such patterns.[64] The UK offers the most obvious example, and it is further hampered by weakly differentiated and institutionalized regional economies. The broader components of flexible specialization strategies cannot be directly transferred in the way technologies can, precisely because they are not centered on a model of a technology or a form of corporate organization, but on a more complex ensemble of sociopolitical conditions.

Despite the wide variety of institutional routes to flexible specialization they have one major factor in common: the existence of political, normative, and organizational means of creating relationships to foster cooperation and coordination. Flexibly specialized economies at regional or industrial district level are embedded in patterns of social relations that go beyond the market and the formal structures of democratic government. Such embeddedness may arise from preexisting institutional and political legacies that promote solidarity and interest cooperation. These patterns may involve more or less developed forms of "corporatist" intermediation of organized interests. Crucial to the development of such cooperative relationships, either from existing legacies, or by explicit strategies of reform, is an agency capable of exercising social leadership, of providing the conditions for a pact which cements a consensus between firms and between the major interests. This agency is not fixed: it may be a political party, a public body like a regional economic development agency, or a strongly led trade association.

Trust needs to be institutionalized and cooperation presupposes forums in which it can be developed. This is the most important lesson for policymakers seeking to revitalize declining regions or to sustain previously successful industrial districts. This "political" dimension of economic policy is more important than any specific economic doctrine or any particular conception of the leading technology. Politics in this sense is broader than the policies of public agencies; it involves the creation of a regional or sectoral "public sphere" in which firms, labor interests, officials, and politicians can interact and cooperate. Such a "public sphere" is essential if cooperation and competition are to be constructively balanced in the interaction of firms, and if coordination and conflict are to be constructively balanced in the relationship of management and labor. These balances are crucial if economic fluctuations and technical changes are to be handled effectively, if continuity and innovation are to be given their respective roles. Those districts than can change and adapt are most likely to be successful and to weather the effects of volatility in international markets.[65]

The sociopolitical conditions in which manufacturing is embedded form the core of the flexible specialization approach to policy. There are two main routes for developing and institutionalizing flexible specialization: one may be called the strategy of building up, linking firms with collaborative institutions to form and cement industrial districts, and seeking to generalize and link such districts so as to form the dynamic core of a national economy; the other is the strategy of building down through the reorganization of major multinational firms into constellations of semi-autonomous subunits that may cooperate one with another or with other firms in an industrial district. Ultimately, these two strategies can converge. Constellated firms mirror internally the patterns of cooperation and trust that are found in the most developed and cohesive industrial districts. Such firms are also less likely to be invasive or destructive of the industrial districts in which they are located, they find it easier to link horizontally with other firms and cooperative institutions in a district and they are more likely to collaborate than centrally directed and hierarchically organized firms.

Flexible specialization is not just a new management philosophy. It emphasizes those aspects of economic life that cannot be bought on the market and it emphasizes relationships with other firms, public bodies, and labor that cannot easily be accommodated to the management prerogatives of the "sovereign" firm. At the same time flexible specialization is not antimanagement and antibusiness; the purpose of cooperation and coordination is to make firms more productively efficient and more commercially competitive. It involves managements learning lessons and adopting a style of operation not widely taught in Anglo-Saxon business schools.

The core of a flexible specialization policy is thus to create and sustain those institutional patterns that lead firms to cooperate one with another as well as compete. These policies may be informal and firm-centered, or they may be more formal and involve firms collaborating with trade and industry associations, regional governments, and organized labor. In both cases the purpose of collaboration is to secure the inputs firms cannot easily purchase on the market – such as suitably trained labor, low-cost finance, and commercial information. Such collective services enhance the efficiency and competitiveness of firms. Flexible specialization policies thus place considerable emphasis on training, since broadly skilled workers are a core component of such a strategy. Such training cannot be undertaken by firms alone if each firm is competing for labor and most firms adopt the attitude of a free-rider by poaching personnel from those firms that bear the costs.[66] Equally importantly, broad-based training involves cooperation with organized labor to set mutually acceptable standards and employment policies. Flexible specialization is often highly profitable for the firm but it is by no means the case that such firms can bear individually the full costs of R&D, marketing information and new investment at prevailing commercial costs. Cooperative institutions lower and spread these costs; allowing smaller firms to gain economies of scale outside of production by pooling the costs of supporting developed networks of collective services for marketing, consultancy, technological information and so on.

If flexibly specialized firms are profitable they will often be targets for asset-rich major companies that seek to acquire them. Alongside collaborative institutions there is the need for public policy to protect firms from unwelcome acquisitions and to preserve the autonomy of management in successful industrial companies of less than major multinational size. Flexible specialization, therefore, requires something akin to a competition policy, that is, a policy that enables firms to retain autonomy. Competition policy has too often been linked to the dual assumptions that free markets are the most efficient allocative mechanisms and that the more intensely firms compete in open markets the better. Such free-market policies in practice tend towards further concentration. This is because their anti-monopoly aspects cut against the primary objective of increasing the freedom of firms as market actors. Deregulators are in practice more concerned with freeing management from restrictions than with ensuring effective competition. Recognizing that competition must be balanced by cooperation involves giving greater emphasis to protecting firms from predators through company law and policies regulating equity markets.[67]

Flexible specialization strategies thus do not emphasize an active industrial policy in the narrow and traditional sense – that is, targeting key sectors or firms for state aid and intervention. Flexible specialization theory assumes that a wide variety of sectors can be successful components of an advanced economy; varying from high-tech sectors like advanced machine tools or computers to traditional sectors like clothing or furniture.[68] Traditional sectors may utilize advanced technology to gain competitive advantage, and may need public support to introduce and pay for such investment. Public policy should be confined primarily to providing broad-based support for sectoral and regional initiatives, rather than seeking to favor a few major firms in a few key industries. Support for training, investment, the building of collaborative networks, the protection of company autonomy, and so on, does not involve public agencies directly choosing which private firms to favor. *All* firms in a district or sector may benefit from such broad-based policies, if they have the initiative and energy to do so. This policy aspect of flexible specialization is thus more compatible with the requirement for liberal democratic states to be neutral and not to bestow public favours on private agencies by administrative discretion. This is not a free-market or anti-interventionist point, it concerns the forms and objects of a policy of active public support for industry. *Dirigiste* industrial policies favor centralism and large bureaucracies. Far from being a policy of laissez-faire, flexible specialization is an interventionist policy that favors democratic accountability. The close link between centralized state and major business corporations is a real threat to democracy: it is a source of corruption, and, all too often, a means of protecting big but inefficient companies. Westerners have been so seduced by Japanese economic success that many industrial policy advocates see MITI as a model to be copied. Japanese radicals do not, and the unhealthy links between big business, the Liberal Democratic Party, and the state are now a key target of democratic criticism and one of the sources of support for opposition politics.

Flexible specialization emphasizes the effectiveness of regional institutions of economic cooperation. It points to the need to build regional autonomy and to foster the collaboration of industry, labor and public bodies at the regional level. This is important since regional economic policy offers the best way of compensating for the lessened effectiveness of national macro-economic policies. Flexible specialization offers the prospect of radical supply-side policies, vital in a conjuncture where rightist free-market, supply-side policies have manifestly failed at the national level and Keynesian demand management at the national level has severe limits. Moreover, regional intervention is not confined to policies for manufacturing alone, but also involves the health, education, and welfare policies necessary to reinforce them. In many countries the regionalization of intervention and the development of a federal welfare state are proceeding hand in hand (Sabel 1989b). Some countries, like the UK, have continued to maintain a national centralized welfare system in social security, health, and education. In the case of the UK this goes along with highly centralized government, weak institutions of regional economic management, and a declining manufacturing sector. Britain is the major industrial country where flexible specialization strategies have had the least impact, and which faces the greatest problems of deindustrialization and poor international competitiveness. British top executives are strongly commited to retaining exclusive company-level management control, seeing little need to build cooperative relationships with other firms or an ongoing partnership with labor. Britain is thus the acid test of whether flexible specialization can be consciously encouraged by public policy initiatives, since the UK both desperately needs such strategies to become competitive and yet has few of the institutional resources needed to develop them.

5

THE JAPANESE MODEL OF POST-FORDISM

Makoto Itoh

A CHANGE IN CAPITALISM FROM FORDISM TO POST-FORDISM

The Regulation school, which originated in France in the 1970s, has analyzed the post-1973 world depression and the resultant capitalist restructuring as a product of the breakdown of the postwar "Fordist" regime of capital accumulation.[1] According to this school, the long postwar boom in the advanced capitalist countries was generally realized through a Fordist regime of accumulation. In 1920s Henry Ford organized his car factory with a belt conveyer system to increase productivity, paying wages substantially above the social average so as to enable his workers to purchase their products and also to maintain the necessary number of workers for heavy monotonous work. Similarly but in a social scale, the postwar Fordist regime of accumulation realized both increases in labor productivity and roughly proportional increases in real wages. These lay behind the continuously expanding effective demand for consumer durables and other products. Such a regime of accumulation was formed and maintained by a set of institutions and a social consensus concerning the social positions of workers and trade unions as well as the role of the state for Keynesianism and social welfare policies. The social mode of organic regulation including such institutions worked smoothly to foster the Fordist regime of accumulation in the long postwar boom period.

Regulationists, in their concept of Fordism, thus focus especially on the role of effective demand, by absorbing one element of Keynesianism. However, unlike Keynesians, they see the factors that achieved the continuous postwar long boom in a wider range of social relations and institutions relating to capital–labor relations, not confined to governmental macroeconomic policies. In this regard, the analyses of economic coordination and the politics of relations of production in the post-war advanced capitalist countries have recently been pursued as a development of Marxian political economy while flexibly absorbing approaches and views from the Keynesian and the Institutionalist school.

Then further in the Regulationist view, the economic crisis since the 1970s was triggered when the Fordist mode of regulation exhausted its potentials. Capitalism was forced to restructure a new mode of regulation appropriate to a new regime of accumulation. The new regime of accumulation is called post-Fordism. The contents of the post-Fordist regime of accumulation are not yet very clear, and may vary even among advanced capitalist countries. Among them Japan is often regarded as a typical or ideal model for post-Fordism. The so-called Japanese style of labor management is underlined as the basis for the successful restructuring of Japanese capitalism with increased competitive power and flexibility of firms in commodity markets, the labor process, and financial markets. The newly developing microelectronic (ME) informational technologies are fully utilized in various aspects of such restructuring.

A change in capitalism from Fordism to post-Fordism then appears to be a model change from the US type to the Japanese type, particularly in the field of labor management. While US Fordism enforced task fragmentation, functional specialization, mechanization, and assembly line principles, the Japanese model of post-Fordism is based on work teams, job rotation, learning by doing, and flexible production. The model change is seen not only as necessary by the capitalist firms in the US and European countries in order to be competitive with Japanese firms, but also desirable even from the view of workers so as to maintain more stable jobs for multiple skills.[2]

This sort of Regulationist approach is somewhat consonant with Japanologists' general tendency to emphasize that Japan has successfully formed an exceptional, stable, crisis-free, and relatively harmonious economic order on the basis of more flexible social human relations in work places.[3] However, there are certain problems in such an approach and tendency. In particular, the Japanese model of post-Fordism is treated as too exceptional, stable, and excessively idealized. In reality, a historical change from Fordism to post-Fordism must be observed in a more global transition from the postwar long boom regime of capital accumulation to a world economic crisis and a new period of restructuring. Japanese capitalism could not be completely outside of the global economic crisis, and has serious difficulties in common with other capitalist countries. In the process of coping with such economic difficulties, Japanese capitalism is dismantling the stable and harmonious social positions of its workers and reducing the role of postwar traditional Japanese style of labor management. A historical change from Fordism to post-Fordism in the Japanese case is surely worth careful studies in both its specialties and universal ground or common features compared with other capitalist countries. Let us begin with examining the common features and specificities in the Japanese model of Fordism.

THE "FORDIST" PERIOD IN JAPANESE CAPITALISM (1955–73)

The notion of a "Fordist regime" of accumulation applies well to Japanese experience in the postwar boom, but there are some specific features or qualifications. Just after World War II, Japan started from a heavily damaged economy, and experienced successively: chaotic and disastrous inflation, the depression of 1949–50 following restoration of a balanced state budget with a fixed exchange rate 1 US dollar = 360 yen, and a sudden export boom prompted by US special procurements for the Korean War in 1950–53. After such experiences, 1955 is generally taken as a starting point for the stable period of Japan's high economic growth. Symbolically, labor's annual spring offensive (*Shunto*) strategy of a unified movement demanding wage increases began in this year. This represented the emblematic transformation from more militant trade union movements into a unified labor strategy focused almost exclusively on wage rates. The so-called postwar Japanese style of labor management was established by this period. This management style comprises three basic characteristics: customary guarantee of lifelong employment until the age of 55 or 60 for regular employees in big corporations, a seniority escalation system of base wages, and company-based trade unions.

If a characteristic feature of the Fordist regime of accumulation is the internal expansion of the market for capitalist commodities, especially consumer durables produced by improved methods of mass production, this pattern is discernible also in the Japanese process of high economic growth beginning from about 1955 until 1973. Despite continuous rapid increases in labor productivity, the Japanese rate of export dependence (exports/GNP) remained constant at around 10 percent in the years 1956–60, and rose only slightly to 11·7 percent in 1970, as we see in Table 5.1. It was far below the prewar Japanese level of 18·8 percent in 1935 and differed in its constant nature from the upward motion (if not very rapidly) following the crisis of the 1970s. It was much lower than that of most contemporary European capitalist countries. Therefore, Japanese high economic growth with an average annual growth rate of real GDP of 10·1 percent in the years 1955–70 was clearly domestically oriented in its market.

A different impression may arise from the rapid increase in the Japanese share of world manufactured exports, from 5·5 percent in 1957 to 9·9

Table 5.1 Japan's export dependence 1935–87 (exports/GNP)

1935	*1947*	*1956*	*1960*	*1965*	*1970*	*1975*	*1980*	*1984*	*1987*
18·8%	0·8%	10·0%	9·5%	10·0%	11·7%	11·2%	12·5%	13·5%	9·7%

Source: 1935–56: Hitotsubashi University Economic Research Institute, [*Annotated Economic Statistics of Japan*], (Tokyo: Iwanami-shoten, 1961). 1960– : Bank of Japan, *Comparative Economic and Financial Statistics.*

percent in 1970 and to 11·5 percent in 1973. But this increase in exports corresponded with Japan's rapid economic growth, almost twice the average of the advanced capitalist countries.

A factor that enabled Japanese capitalism continuously to expand the domestic market in line with its rapid economic growth and increasing productivity was the annual rise in real wages through *Shunto*. The annual labor offensive thus served as an important gear for Japanese capitalism to realize a "Fordist mode of regulation." We should note, however, that *Shunto* and the lifetime employment system for regular workers with the seniority wage system was not due to the Japanese traditional family type of culture in general, but a result of concessions by Japanese capitalist firms to tame militant workers' struggles, especially protests against dismissals. Such militant workers' struggles in the preceding period of reconstruction after World War II had often threatened Japanese big corporations and their managers. Yearly increases in real wages through *Shunto* and the lifetime employment custom with the seniority wage system also enabled big corporations to maintain the increasing necessary number of experienced male workers, with relatively low wages for the younger workers. They also served to purchase workers' loyalty to companies in their workplaces.

At the same time, the increase in real wages was an inevitable result of expanding employment of regular workers in the process of rapid economic growth, and it was actually accelerated as the labor market tightened toward the end of the period of high economic growth. I am sure that there was no intention among Japanese business circles, bureaucrats, and politicians to raise real wages as an essential source of effective demand for increasingly mass produced consumer durables. The realization of a "Fordist regime" of accumulation in Japan was rather an anarchic result of socioeconomic factors.

In fact the increase in real wages was by no means regularly proportional to the increase in labor productivity in Japan. If stable proportionality between rising productivity and real wages is essential to the concept of "Fordism," then Japanese capitalism did not form a pure "Fordist regime." As shown in Table 5.2, real wages lagged behind the increases in labor productivity in Japanese manufacturing in this period until 1970.

As product wages in Table 5.2 are calculated upon the basis of real GDP in manufacturing defined in terms of the manufacturing GDP deflator, the difference between the motion in real wages and that in product wages relates mainly to the disparity between the consumer price index and the manufacturing GDP deflator. In 1960–5, while consumer prices went up fairly rapidly, the manufacturing GDP deflator did not rise much. Reflecting this disparity, while the growth rate of real wages still remained far below the growth rate of productivity, the growth rate of product wages surpassed it, so as to reduce the net profit share from 41·5 percent in 1960 to 36·1 percent in 1965. The reduction in the net profit share was not much, however, and recovered in the next five years to 40·7 percent in 1970.

Table 5.2 Productivity and wages in Japanese manufacturing (annual percentage increase in each five years)

	1955–60	*1960–65*	*1965–70*	*1970–75*	*1975–80*	*1980–85*
(1) Labor productivity	9·7	6·8	12·5	5·1	9·2	6·9
(2) Nominal wages	6·2	10·1	14·7	18·0	8·3	4·6
(3) Consumer prices	1·6	6·1	5·4	11·5	6·5	2·8
(4) Real wages	4·6	3·7	8·8	5·8	1·7	1·8
(5) Product wages	4·7	8·8	10·9	12·9	8·2	7·3
(6) (4)/(1)	0·47	0·54	0·70	1·14	0·18	0·26
(7) (5)/(1)	0·48	1·29	0·87	2·53	0·89	1·06

Sources: (1), (5), Based upon P. Armstrong and A. Glyn, *Accumulation, Profits, State Spending: Data for Advanced Capitalist Countries 1952–1983,* Oxford Institute of Economics and Statistics, July 1986, updated from OECD, *Historical Statistics 1960–86,* 1988.

(2), (3), (4), are from Japan Productivity Center, [*Practical Labour Statistics*], (Tokyo: Japan Productivity Center, 1988.)

Notes: (1) Based on manufacturing GDP per person employed.
(5) Calculated from productivity multiplied by labor's share (= 1− profit share).
(6), (7) Simple division, not percentage increase.

Excepting 1960–5, real wages and product wages rose more or less parallel and both of them less than the growth of productivity up until 1970. This occurred especially in 1955–60, as both were less than half the growth rate of productivity, so obviously the production of relative surplus value took place so as to raise the net profit share from 26·5 percent in 1955 to 41·5 percent in 1960. Viewed from the demand side, this situation meant that Japanese capitalism needed the expansion of markets for their products besides the growth of wage workers' consumption expenditure.

Another important source of increasing effective demand in Japan was investment in plant and equipment. Actually private investment in plant and equipment increased by as much as 22 percent a year on average in 1956–73, more than twice as fast as the real growth rate of GDP, and occupied 25·1 percent of Japanese GDP in 1973. So long as more and more capital goods were absorbed by such investment, the internal effective demand for goods could increase even without a proportionate increase in real wage rates. In this regard Japanese high economic growth was more of an investment boom to make physical assets of capitalist firms grow faster than the physical flow of consumer goods for workers. The very high rate of investment of Japanese capitalist firms introducing new industrial technologies and consumer durables from the US was financed not only by reinvestment of their own profits but also by "over-borrowing" from banks based on the high rate of savings, approximately 15–20 percent of Japanese personal income. A substantial portion, amounting to 40–50 percent of the necessary fund for investment in plant and equipment, was obtained by such loans from banks and other financial institutions with an almost zero or negative real interest rate of deposit for people's savings. This financial transfer of purchasing power thus added to a substantial deviation from a pure and simple model of "Fordist regime" of accumulation in the Japanese case during the long boom.

Although the prewar pattern of family-owned "zaibatsu" concerns was dissolved through the postwar US occupation policy for democratization of Japanese economic order, groups of corporations often remained closely linked to their earlier counterparts in zaibatsu. The relations among big corporations in each group became non-hierarchical and horizontally co-operative even between banks and industrial corporations. Some new industrial groups were formed outside of old zaibatsu groups, but were obviously influenced and promoted by the existence of old groups. Big industrial corporations could usually "overborrow" from their main banks within the same affiliated groups. Each industrial group would extend their industrial activities competitively by the "one set principle" – establishing a set of companies to cover the whole industrial, trading, and financial sectors, so that each group could basically do business independently. This practice promoted the high pace of investment in plant and equipment in the process of introducing new types of industrial products until the 1960s.

There was a third important domestic source of effective demand in Japan's high economic growth, namely increasing real income in rural agricultural households. The domestic market for agricultural machinery, fertilizer, electric home appliances, houses, cars, etc. was continuously expanding in the rural areas. As the proportion of employed population in primary industry decreased sharply from 41·0 percent in 1955 to 19·4 percent in 1970, the number of agricultural families also decreased, but only from 6 million to 5·3 million. In the same period, while average nominal wages in all industries increased 4·3 times (real wages increased 2·3 times), the average nominal income of agricultural families cultivating a typical Japanese farm of 0·5–1·0 hectare also increased 4·4 times, reaching 1·3 million yen a year. This matched the average annual 1·4 million yen nominal income of all Japanese households. The portion of purely agricultural income in it, however, declined sharply from 66·9 percent to 29·0 percent as more than 70 percent of income of agricultural families came to be earned outside of farming, mainly as wages.

Labor productivity in agriculture simultaneously increased substantially, enabling farm households to sell more labor-power outside the village in the form of seasonal or part-time workers or even as regular workers in nearby factories. Thus, the lifestyle and living conditions in rural areas changed greatly from the prewar situation in which an impoverished agricultural population remained roughly constant during a period of rapid industrialization. The changes show well how Japanese villages served both as a powerful source of cheap wage workers and as an expanding domestic market especially for consumer durables.

Although a considerable portion of the income of rural families became dependent upon wages, which were often earned by main family members temporarily working in the urban areas, the increased rural purchasing power surely served widely to spread the modern, automated lifestyle throughout the country. Modern houses with a range of home electronic equipment replaced old houses with thatched roofs. The government policies effectively helped such changes in rural villages. In particular, protective policies to maintain rice prices for producers, both by subsidy and by import restrictions, worked in steadily leveling up the agricultural income. At the same time, public expenditure to construct highways and to modernize rural roads helped to form an automobile society throughout Japan. While the level of social welfare policies for the urban working people was still relatively low in this period, these economic policies for rural agricultural areas did serve to modernize Japanese rural areas.

Following the relatively favorable period for agricultural areas due to severe food shortages just after World War II, Japanese rural areas experienced a modernization process and joined in a long boom, by extending effective demand for various capitalist products. This is an important feature of the Japanese Fordist period, and it differs from the prewar dual structured regional development between rapidly growing urban cities and stagnant, poor agricultural areas. The regional development was in this regard substantially equalized in the postwar Fordist period much more than in the prewar period.

However, disparity in socioeconomic life between different regions surely remained and continuously served as a basis for the dual structure of Japanese labor market. As we see in Table 5.3, there used to be a wide difference in wages between large and small businesses. The wages of small manufacturing enterprises hiring between 5 and 29 persons was just 54·2 percent, and those of enterprises hiring between 30 and 99 was 65·8 percent of those of big enterprises in 1960.

Table 5.3 Wage disparity between different sizes of manufacturing enterprises

		Enterprise size (employees)		
Year	*Over 500*	*100–499*	*30–99*	*5–29*
1960	100·0	73·6	65·8	54·2
1965	100·0	83·7	78·3	72·6
1970	100·0	83·4	76·2	71·4
1975	100·0	85·7	75·8	70·3
1980	100·0	81·9	70·5	67·6
1985	100·0	79·4	68·7	64·7

Source: Ministry of Labour, *Monthly Statistical Research of Labour.*

The double structure of labor markets related not just to the wage disparity but also to a wide range of differences in labor conditions between regular workers in big corporations and those working in medium and small businesses or irregular workers even in big corporations. Usually without a lifetime employment system, without the seniority wage escalation system, with fewer fringe benefits and with a very much lower rate of organization for trade unions, workers in medium and small businesses were really outside of the so-called Japanese style of labor management. In 1962 the proportion of employees in work places smaller than 300 persons in the whole nonagricultural areas was 53·5 per cent, and it was 80 percent if the public-sector employees were exempted. Japanese big industrial firms could utilize affiliated medium and small corporations quite flexibly under the hierarchical subcontracting system, so long as the latter could elastically adjust the number of employees and could use much cheaper wage costs. At the same time, big parent corporations used to assist the affiliated smaller firms in their subcontracting system to modernize equipment both technologically and financially, often endorsing long-term borrowing from their main banks. Thus, the subcontracting system was effective in making big firms grow easily more flexibly and faster, and also in promoting investment in plant and equipment more broadly than without it.

What about the role of the state of strategic industrial planning on the part of MITI (Ministry of International Trade and Industry)? Although Japanese economic policies were sometimes overrated even as "the most dexterously performed planned economy in the world" or as an ideal Keynesian practice, they did not really constitute a planned economy, and could not be effective without a whole set of more basic favorable conditions for continuous economic growth. Among those basic conditions, there were abundant supplies of relatively cheap labor from domestic agricultural areas and of primary products from abroad, the availability of a series of new industrial technologies imported mainly from the US, and a relatively stable international economic order with the Bretton Woods monetary system under the US economic hegemony. Actually in the long process of high economic growth, Japanese state budgets kept well balanced and did not practise any scale of fiscal deficit policy to level up the macroeconomy. We have to note, however, that the yearly "natural" increase of tax revenue beyond initial budget plans continuously enabled expansion of public expenditures to construct infrastructure such as ports, roads, etc., for industrial development. At the same time, monetary policy to keep interest rates relatively low in the process of mild inflation served to make the interest costs easy to pay for the industrial firms and promoted their investment in plant and equipment by means of overborrowing. Strategic industrial policies on the part of MITI were more powerful in the period of postwar reconstruction and the initial phase of high economic growth, when the allocation of foreign currency (mostly dollar) necessary to pay for imports of industrial machines, raw materials, and technologies served as a means to control priority for investment among various industrial areas and individual firms. This means of industrial policy became ineffective as the shortage of foreign currency was reduced due to increased industrial competitive power. The foreign currency allocation system was abolished in 1960. The import restrictions were another protective industrial policy and were kept longer. For instance, imports of passenger cars were not exempted from the "negative list" for restrictions until as late as 1965. Through the 1960s, however, the import restrictions for manufactured goods became generally reduced and abolished. Tax reductions such as allowances for accelerated depreciation – often combined with favorable conditions for long-term finance, as in the case of the synthetic fiber industry in the early 1950s or the electronics industry since the late 1950s – also worked to a certain extent as a protective policy for strategic industries, especially in the early process of high economic growth.

As a whole, the weight of industrial policies to sustain and control Japanese economic growth was in effect tendentially reduced through this period in the 1950s and 1960s. As for the five-year plans by the government since 1955, they could not be long-lived, and were altered every two to three years. They served just as rough guidelines for the activities of private capitalist firms and could not be a main driving force for economic growth. In fact, real economic growth rate overshot the planned rates in the government forecast in the period of high economic growth, and then suddenly became unable to achieve the government plans in every aspect in the 1970s. Thus, while the contribution of the administrative plans and guidance to high economic growth should not be overestimated, the major cause of economic crisis thereafter must be not so much in mismanagement of economic policy as in the more basic changes in the conditions for capital accumulation.

The process of capital accumulation in the high economic growth period was not at all harmonious and was even destructive to its own basic conditions. For one thing, as extensions of employment in the cities continued in the 1950s and 1960s, depopulation of rural villages began to spread toward the end of this long boom, and caused difficulties in maintaining population levels in increasing numbers of remote villages. Regional economic development became substantially distorted in a way difficult to restore.

Simultaneously, overaccumulation of capital in relation to the limited supply of labor-power appeared for the first time in the history of Japanese capitalism, and made both real wages and product wages rise faster than a rise in productivity at the beginning of 1970s (Table 5.2). The pressure of the labor shortage has reduced the wage disparity to some extent since about the middle of 1960s (Table 5.3). The Fordist regime of accumulation in Japan became thus untenable toward the end of the postwar long boom, as a basic condition in flexibility of the labor market was much eroded and then disappeared. As a corollary, labor discipline in work places tended to loosen even in Japanese capitalism, and the working days lost by labor disputes increased.[4] The difficulty of capital accumulation basically due to labor shortages – together with the rising prices of primary products which finally included the oil shock in the world market as a result of the global overaccumulation of capital in relation to the limited elasticity of supply of such primary products – brought the period of Fordist prosperity to an end. The difficulty of overaccumulation of capital to squeeze the real rate of profit was accompanied by the accelerating inflation in the process of the breakdown of the Bretton Woods international monetary system.[5]

"POST-FORDISM" IN JAPANESE CAPITALISM (1973–)

Japanese capitalism was not at all exceptional in experiencing an acute inflationary crisis in 1973–5. Japan joined other advanced capitalist countries in experiencing a fall of the real rate of profit due to overaccumulation of capital in relation to both the laboring population and the limited flexibility of supply of primary products. The difficulty of capital accumulation was combined with the explosion of the money and credit supply so as to generate an inflationary crisis also in Japan. As the Japanese government implemented an "adjustment inflation" policy and consciously accelerated inflation in order to mitigate damage to Japanese exporting industries by a sharper appreciation of the yen, Japanese inflation became particularly vicious at that time. The annual rate of inflation of wholesale prices reached 16 percent in 1973–5, and it recorded 31·6 percent in 1974. As overaccumulation of capital proceeded with accelerating inflation, speculative stockpiling of various commodities increased and disturbed the Japanese economy by making it difficult for firms to obtain necessary materials at profitable prices. The highest rate of reliance on imported energy among the advanced capitalist countries seemed especially harmful to Japanese capitalism in this process of acute inflationary crisis. Although the Japanese rate of unemployment was still quite low, at 1·9 percent in 1975, Japanese manufacturing and mining production fell rapidly by 20·4 percent from the end of 1973 to the beginning of 1975, and one in three Japanese firms was estimated to be running a deficit at that time.

Nevertheless, Japanese capitalism managed to recover relatively quickly from the first oil shock, and showed certain strength in the face of the following global great depression and the shock repeated by the second oil crisis. Japan's strength in international trade, dramatically reversing the trade balance from a big deficit during the oil crises to a wide surplus, seemed particularly impressive. The financial positions (as shown in the ratio between the financial assets and liabilities) of Japanese big firms also substantially improved and often turned from big borrowers to net lenders or financial big investors even in the case of manufacturing corporations. A relatively low rate of unemployment in Japanese official statistics was seen as further evidence of the exceptional strength of the Japanese economy. It mildly increased to 3·0 percent by 1987.

The so-called Japanese style of management is frequently cited as the basis for the exceptional strength of Japanese capitalism. The lifelong employment system for regular workers, the seniority wage system, and trade union organization based on company are generally seen as the three main characteristics of the Japanese style of labor management. "Look east and learn the Japanese style of management" has become a fashionable slogan throughout the world in the "post-Fordist" period of crisis and restructuring. Capitalist managers hope to find in it ways of obtaining loyal cooperation from workers in changing methods of production and operation by introducing new information technologies. Even among progressives in the West, including some Regulation scholars, the Japanese style of management tends to be idealized as a model to realize a "post-Fordist mode of regulation" in a harmonious, stable economy with less unemployment and some form of a profit-sharing principle for workers.

From the managerial point of perspective, the loyal cooperation of regular workers under the Japanese style of management was surely useful for raising productivity and industrial competitive power by introducing new microelectronic automation technologies. Japanese unions and workers generally did not resist transfer from old jobs or workplaces to new positions within the company. Japanese unions, which were organized on the basis of the company unit and not on the basis of specific skills or kinds of work, generally focused demands on the annual wage increase as long as their members were not dismissed prior to retirement (55–60 years old). These postwar patterns of Japanese management and trade union practice were actually fully utilized for the formation of flexible production lines and business operations as an essential characteristic of the "post-Fordist regime of accumulation."

Flexibility of Japanese capitalist corporations was indeed much increased in various ways by the combination of both spreading microelectronic information technologies and the cooperative attitude of Japanese workers in adjusting themselves to new technologies. A range of practices became common among Japanese capitalist corporations in their attempt to restructure and to "rationalize" through the pressure of economic crisis and depression. Among these practices were flexible production of multiple models of cars, electric equipment, and so forth on the same conveyer belt line, flexible adjustment of supply according to the movement at points of sale (POS), flexible reallocation of workplaces including multinationalization by more business and manufacturing corporations, multiplication of commodity products and operations, flexible and more economical combinations of segmented workers often increasing a proportion of irregular part-timers and workers employed by outside firms, and flexible rotation of workers by giving them multiple tasks in workplaces.

The "success" in executing these practices explains to a considerable degree why Japanese capitalism could intensify industrial competitive power in the world market much faster than other advanced countries in the current great depression. It is clear that not only the technological changes but also institutional and ideological elements played certain roles in organizing the formation of a "post-Fordist regime" of accumulation by Japanese firms in such a way as to increase their managerial flexibility.

A social institution, which served Japanese big firms to increase managerial flexibility besides their own relations with employees, was the subcontracting system. In the process of adjusting to new information technologies, Japanese big firms attempted to restructure the whole set of subcontracting relations with smaller firms by cutting off inefficient subcontractors, and enforcing "rationalized" equipment and costs to remaining subcontractors. The famous Toyota "Kanban" method of management to reduce the costs of stocks of parts and materials cannot be realized and maintained without hard effort on the part of subcontractors to keep just-in-time delivery of their products. The subcontracting system became utilized also for flexibly using irregular workers who work in the parent big firms but are employed by subcontractors. The labor laws were deregulated so as to enable such a managerial practice. The newly intensified subcontracting system serves as a means to increase flexibility for Japanese capitalism to use segmented cheaper workers.

The role of economic policies of the Japanese government or MITI in the transition from declining to growing industrial sectors and from lower value-added to higher value-added products with new microelectronic technologies must not have been great and surely not so much as often cited as keys to the successful Japanese model. Certainly they served as a general guideline expressed typically in the Maekawa Report for individual firms to restructure by introducing more and more information technologies. However, unlike in the initial phase of high economic growth, there has not been an effective policy means by which the government can control priority of investment or positively protect certain industries. Especially in the 1980s, the neoliberalist economic policies worked in general to reduce the role of the state. The working of the market and the efficiency of private capitals tended to be emphasized so as to reactivate them. The attempts to restructure Japanese industries were thus pursued in practice and in the main by private capitalist firms. The role of the state to assist them was rather effective in weakening the social positions of workers and their trade union movements.

At any rate, if recent Japanese capitalism is seen as an ideal "post-Fordist regime" for achieving a harmonious and stable economic life, this is simply not true for most working people. The economic life of general wage workers has become rather unstable and difficult to improve. As we have seen in Table 5.2, real wages have stagnated since the latter half of the 1970s, forming a wide and growing gap with labor productivity (Table 5.2 (6)) after 1975. Unlike in the "Fordist regime" of accumulation, wages now tended to be treated mainly as a cost to be economized, as Japanese industrial restructuring followed a more export-oriented strategy. As a result Japan's rate of export dependence (export/GNP) rose from 11·2 percent in 1975 to 13·8 percent in 1984 (Table 5.1). At the same time, Japanese unions, which had succeeded in raising real wages if not to levels fully proportional with increasing productivity, ceased to be effective even in this area.

However, a big gap between the increase in productivity and the stagnant real wages did not serve to restore the profit share of Japanese capitalist firms until quite recently. The gap served rather as a flexible basis to absorb the increased costs of energy resources and some primary products imported from abroad, and also to maintain export dollar-prices relatively lower even in the process of severe appreciation of the yen from 360 yen to the dollar at the beginning of 1970s to 120–40 yen to the dollar in the latter half of the 1980s. The domestic prices of manufactured products and the manufacturing GDP deflator tended competitively lowered as microelectronic automation technologies spread. As a result, "product wages" measured in terms of the manufacturing GDP deflator went up much faster than the real wages in Table 5.2, so as to make a recovery of profit rate difficult.[6] A characteristic feature of the Japanese post-Fordist regime of accumulation is thus its competitive pricing strategies.

As microelectronic technologies generate factory automation (FA) and office automation (OA), the experiences and skills of regular workers became more and more superfluous in the workplace. Capitalist firms began to employ more irregular cheap workers, many of them housewives as part-timers. While the number of male employed workers increased by 3·7 million, from 24·3 million in 1973 to 28·0 million in 1986, female employed workers increased more by 3·9 million, from 11·9 million to 15·8 million. Among female workers, the number of part-timers increased from 1·7 million to 3·3 million in the same period. Although these changes partly reflect a shift from the primary industry (agriculture) to urban tertiary industry, they reveal a feature of "post-Fordist" transformation with more flexibility in labor management. Especially, unskilled women part-timers became effectively utilized in flexible FA and OA lines in more and more workplaces. As a result of such "rationalization," the rate of organization of Japanese trade unions, traditionally based on regular (male) workers, declined from 33·1 percent in 1973 to 25·9 percent in 1989. The social power of trade unions to raise real wages in the process of increasing productivity had thus to weaken.

As Keynesian expansionist economic policies failed in reactivating economic growth and resulted in a huge amount of state debt with an increasing burden of interest payment, Japanese economic policies turned to a neoconservative direction since 1981, in accord with Thatcherism and Reaganomics. Aiming at reconstruction of a balanced national budget without tax increases, the Japanese government began to reduce the financial support to health and medical services, curb subsidies to private universities and schools, and to trim government subsidied programs in various areas. Such a reduction of the economic role of the state and a farewell to the welfare state as well as to Keynesianism were implemented together with deregulation in various fields, and also with privatization of public enterprises by emphasizing the need for reactivating the fundamental rational workings of a market economy. Although such a policy stance was well suited the post-Fordist regime of accumulation with more flexible competitive strategies of capitalist firms, it certainly added to the economic difficulty of working people.

More and more housewives have been driven into the labor market as suppliers of cheap, often part-time, labor, as real wages of regular male workers stagnated and costs of education, taxes, medical services, and housing increased as a result of neoconservative economic policies. At the same time, the life pattern of housewives working outside the home, many as part-timers, was generalized in Japanese society. The result has been to split and reduce the value of labor-power per person which is necessary to sustain the economic life of a whole family.[7] This effect contributed to the stagnation of real wages for general workers, forming a sort of vicious circle.

Since the Japanese style of labor management was formed in the main for regular workers in big corporations, its function of protecting the economic life of its members inevitably weakened as a growing number of unorganized irregular workers replaced experienced regular workers. The lifetime employment system for regular workers which was an essential element of the Japanese style of labor management is no longer secure. In the process of privatization of the Japan National Railway (JNR), for example, more than one third of the total, 130,000 experienced workers, were fired. Privatization and division of JNR into several private companies, together with privatization of Nippon Telegraph and Telephone Public Corporation (NTT) and Japan Tobacco and Salt Public corporation, was an important part of the administrative reform carried out by the Nakasone Cabinet. Privatization actually constituted a powerful blow to the strongest and most militant trade unions within the Sohyo Union Federation. Sohyo had to decide to dissolve itself in 1989 so as to form a new Federation (*Rengo*) fused with the traditionally more conservative *Domei* (the Japanese Confederation of Labor). Although two smaller national centers for the trade union movement remain, the dissolution of Sohyo symbolizes the weakened social position of the workers movement in Japan through the combined pressures of great depression, microelectronics industrial renovation, and neoconservative policies of the state.

Meanwhile, in the depressed industries like shipbuilding, steelmaking, and petrochemicals, numerous regular workers were dismissed as yards and factories closed. While Japan's official rate of unemployment rose to a historic height of 3·0 percent (unemployment is very strictly defined in Japan and must be at least doubled in order to be compared with the rate of unemployment in other advanced capitalist countries) in 1987, labor markets in Hokkaido and Kyushu deteriorated further with the decline of heavy industries and mining. In these circumstances, middle-aged and elderly unemployed workers face great difficulty in finding appropriate work. Wage differentials segmented by categories such as gender, educational background, regular or irregular positions, etc. are widening in the "Post-Fordist" restructuring of the Japanese economy. As we have seen in Table 5.3, wage disparity between big and middle or small enterprises in Japanese manufacturing, which had once declined in the process of high economic growth, widened again since the latter half of the 1970s.

Thus ironically in the very period when the Japanese style of labor management became idealized abroad as a stable and harmonious ground for both capitalist firms and workers, the essential features of Japanese postwar labor management were collapsing. Flexible managerial strategies in the "post-Fordist mode of regulation" are being incorporated into Japanese firms, including their growing global operations. The same process is increasing instability and income stagnation for the mass of Japanese workers who now face the violent forces of the capitalist market economy largely bereft of protection by trade unions and state welfare policies.

The recent phase of Japanese economic recovery from 1987 depends again more on the expansion of the domestic market. The Japanese rate of export dependency (exports/GNP) has thus been reduced again (Table 5.1). This shift back to a more domestic-oriented recovery was a result of adjustment both to the increased trade frictions with the US and other advanced capitalist countries, and to the effect of appreciation of the yen. However, it was not caused by nor has it resulted in an increase in real wages proportional to productivity, unlike the process of previous high economic growth until the 1960s.

In addition to general replacement demand for consumer durables and houses, a new and important element of domestic demand has arisen from the "new rich" who have gained directly or indirectly by wide appreciation of urban land prices, rent for real estate and stock prices. Inflow of rich foreign business persons, especially into the Tokyo area, added to this element for luxurious consumption demand and for office space. More expensive cars and luxury flats of more than 300 million yen came on the market. The price of land more than doubled in two years until the end of 1987 in the metropolitan area, and the increase in the total estimated land prices in Japan was 248 trillion yen in 1987, almost matching the 276 trillion yen of Net National Income for the year. It is also said that increased housing prices forced people in big cities to give up saving to buy a house and thus to spend more for other consumer durables. Therefore, the recent economic recovery depending more on the domestic demand does not mean a return to the previous Fordist regime of accumulation, but strongly maintains a post-Fordist feature of uneven development with continuous economic difficulty for the majority of working people.

On the other side, the post-Fordist regime of accumulation in Japan has clearly formed uneven development also among different regional areas. The economic depression since 1973 did not push back the redundant working population into the rural agricultural villages unlike the historical experience of previous depressions. As a substitute, the tertiary industries continued to grow on the basis of new microelectronic information technologies in urban areas. The proportion of Japanese working people in primary industry declined from 17·4 percent in 1970 or 12·7 percent in 1975 to 8·3 percent in 1987, while in tertiary industry it increased from 47·4 percent in 1970 or 52·1 percent in 1975 to 58·5 percent in 1987. Thus, depopulation has proceeded further – with devastating effects in the rural villages located far from urban areas. Also, appreciation of the yen, subsidy cuts, and the open-door policy for agricultural products as an integral part of the neoliberalist deregulation policies, which also follow US political demands, continuously hit and endanger Japanese agricultural farming and the rural economy.

Reallocation of manufacturing and business activities to more economic sites of operation upon the basis of new microelectronic information technologies has served to stimulate the local economy to a certain extent, but it did not help those deteriorating rural villages remote from cities or resort places. The difficulties of old manufacturing and mining areas with depressed industries also remained. Therefore, uneven regional development even within the local prefectures became conspicuous in the process of post-Fordist restructuring of the Japanese economy.

At the same time, centralization of business and financial activities, especially those of headquarters into the metropolitan Tokyo area, has been strengthened as an ironical result of the development of information technologies. The technological possibility of forming a more decentralized regional economic order by means of various information media does not seem easy to realize under the current post-Fordist regime of accumulation in Japan. While a generally accessible range of information is rapidly widening, the importance of more specific timely information often of a face-to-face type seems to be growing in the recent unstable world of business, particularly as Japanese economic activity has become more and more internationalized. It is observed that more than 300 headquarters and operating departments are annually moving into the Tokyo area.[8] Even giant corporations in Kansai (around Osaka) have often shifted their headquarters to Tokyo, sometimes even by splitting the headquarters across the two sites. A ratio of average personal income in Osaka prefecture (which is the center of Kansai business activity), compared to that in Tokyo Metropolis, fell from 84·3 percent in 1973 to 78·6 percent in 1985. Such regional centralization of economic activity is a basic cause of the soaring land price problem in the Tokyo area, giving rise to "the new rich," and pushing out general working people into more distant living places from their working places so as to enforce many of them to spend more than three hours a day to commute both ways.

THE NEED FOR ALTERNATIVES

With all its specialities, the Japanese model of post-Fordism represents common tendencies in the advanced capitalist countries. The model was formed as a result of capitalist restructuring through the worldwide great depression. A change in industrial technologies by the growth of new microelectronic information technologies enabled a reduction in size of investment in plant and equipment. Introduction of new methods of production and business operation became realizable by smaller units of flexibly mobile investment. The social position of trade unions has been clearly weakened as an effect of technological changes so as to enable capitalist firms flexibly to employ cheaper irregular workers. Rapid changes in costs, types, and models of commodities – combined with more international marketing – have intensified competition among capitals. Capitalist firms are more interested in their market shares rather than in maintaining monopolistic prices.

To a certain extent, the neoconservative state policies that became dominant in the capitalist world in the 1980s reflect and foster such changes in a capitalist economy. The Keynesian welfare state with various social controls of the free market system was ideologically abandoned. Deregulation, privatization, and cuts of social expenditures were implemented so as to reactivate the flexible and competitive power in a capitalist market economy. The social positions of workers and trade unions have severely been attacked by such a shift in state policies for neoliberalism.[9]

Insofar as the model change from a Fordist to a post-Fordist regime of accumulation in Japan contains these socioeconomic alterations, it is not specific to Japanese capitalism but more or less common to advanced capitalist countries in the past decade. At the same time, the change has to be regarded as a historical reversal backward over a century and is really not confined merely to a model change within a long wave cycle over half a century. Scaling up the units of investment in plant and equipment, strengthening the social positions of workers and trade unions, and increasing economic roles of the state were mutually related together and formed a basic tendency in capitalist societies from the late nineteenth century until the postwar age of Keynesianism or Fordism. The post-Fordist regime of accumulation reversed all these tendencies especially in the 1980s under the neoconservative policies in major capitalist countries.[10]

The alleged strength of the Japanese model of capitalism must be in its relatively smooth and fast adjustment for such a reversed historical tendency. It is clear that the loyal and cooperative attitude of Japanese workers and trade unions specifically facilitated introduction of new micro-electronic types of technologies in workplaces and remodeling Japanese firms suitable for the post-Fordist regime. As I have argued, this remodeling has not at all been a harmonious and even development, but is uneven, and regionally distorts the Japanese economy. The success story of the Japanese model of post-Fordism thus actually belongs mainly to capitalist corporations with their improved financial positions and strengthened competitive power in the world market, but not to the majority of working people, peasants, and weaker people like the aged, the sick or many women.

After a decade of neoliberal post-Fordist policies, the social choice of people began to oppose such an unfair and uneven development. Beginning with a series of local elections, the government party LDP (Liberal Democratic Party) found it hard to win in 1989 and lost the majority in the House of Councillors in the national election of July 1989. LDP lost some seats and JSP (Japan Socialist Party) increased its number of seats again dramatically in the general election for the House of Representatives in February 1990, though the LDP managed to maintain a stable majority in this election. As a result there was a reversed balance in the number of seats between the House of Councilors and the House of Representatives. There were three major political issues fought in these elections; the Recruit scandal, the introduction of consumer tax, and the policy for the freer import of rice and other agricultural products. All these issues relate in one way or another to the neoliberal political attitude in believing in the freer market system which tends to allow moneymaking even by means of political corruption, and to shift the burden of economic depression to the shoulders of the general populace while enabling the new rich easily to become wealthier.

The need for alternatives seems obvious and actually begins to move people's political choice even in the alleged stable and harmonious Japanese society. There are two conspicuous related tides in this recent movement. The rural areas, which were traditionally strong bases for the conservative LDP, have become quite unstable and now often provide serious setbacks to the LDP at various elections. The uneven regional development and worries about future regional development seem to move people's political choices. Simultaneously, the participation of women in political action has become more and more positive, from the grass roots in regional social activities such as citizens movements against nuclear power-generation, the co-op movement, local elections, etc. Although the reorganized trade union movement, the major portion of which is represented by Rengo, is attempting to strengthen its influence in the political landscape, the need for alternatives is thus no longer solely based upon it but begins to work more broadly from below.

Therefore, the alternative socioeconomic strategies for the 1990s cannot be a simple return to the Keynesian macroeconomics or the Fordist regime.

The alternatives must be flexibly proposed through more democratic participation of the general populace in decision making at various social choices, regional issues and multiple levels of social life. We have to see whether trade union movements in Japan can be reconstructed to strengthen workers' control in the workplace, though this seems at the moment the most difficult aspect in the Japanese model of post-Fordism from the view of economic democracy for the people, by the people. Flexibility in the post-Fordist regime of accumulation on the basis of new microelectronic information technologies, however, can possibly be and should be remodeled in such a democratic way for the future. The democratization of socioeconomic order must thus remain an important demand for the pathway in the 1990s for common people no less in Japan than in other contemporary countries.

ACKNOWLEDGMENT

I am grateful for Mark Selden and Allen J. Scott for their thoughtful comments and editorial advice in the process of completing this chapter.

Part III

NEW TECHNOLOGIES AND THE ORGANIZATION OF INDUSTRIAL PRODUCTION

6

THE REVITALIZATION OF MASS PRODUCTION IN THE COMPUTER AGE

Benjamin Coriat

INTRODUCTION

The starting point of this chapter is common to many people working in the field of industrial economics. Most observers agree on the two following basic propositions:

1 as a response to decreasing productivity gains and the new structural instability of the market, the present moment is marked by a set of interrelated organizational and technological innovations at the level of the firm, and
2 this set of innovations has given rise to a new technological trajectory, where the search for more flexible arrangements is the key distinctive feature; along with this new trajectory new principles of economic efficiency are emerging and spreading throughout the whole economy.

From this common starting point, different interpretations of what is happening to the old classical Fordism are emerging. A number of different theses have been sustained. One can identify two basic, opposed, arguments.

1 *The long waves hypothesis* (Freeman 1984, Mensch 1978): after a more or less difficult and long transition, the economy will recover and enter onto a new path of economic growth. According to this view, attention must be concentrated on the process of transition. Efficient R&D and technological policy is required, along with a strengthening of education and training systems.
2 *The flexible specialization hypothesis* (Piore and Sabel 1984): the new technological trajectory means the end of the whole Fordist structure and emergence of a new form of industrial organization where flexible small and medium-sized firms play the dominant role, since they have a greater capacity to react to the new structural instability of the markets.

Many different variants and submodels of these two basic approaches have also been devised.

Along with Robert Boyer, in an earlier essay I attempted to trace a vision of the process (Boyer and Coriat 1989). In brief our thesis is that as regards the norms and conditions of production, the new technological trajectory is much more a revitalization of mass production principles than an alternative and new way of producing goods. The paper then develops the thesis that the arrangements around the "wage relation nexus" (*rapport salarial*) will finally decide the definitive configurations of the general structure of post-Fordism.

In this chapter I would like to elaborate this first vision. To do so, I have chosen to begin with a short but general reconsideration of the elementary principles of efficiency embodied in the typical Fordist "engine of growth." The second part of the chapter is then dedicated to a preliminary attempt to formalize the main changes experimented with during the present days by the classical principles of efficiency.

Furthermore, I introduce some new analytical tools to approach the different components of the so-called "post-Fordism." My starting point is that the transition from Fordism to post-Fordism, in fact, means a transition in the principles of economic efficiency of the whole structure of modern capitalism. It is obvious that during the last twenty years a set of very deep changes has taken place. These changes are specific to each level of the capitalist structure and largely interdependent. So while it is easy to localize and describe the different "local" changes, it is more difficult to obtain a clear and organized representation of the new articulations between the different separate levels. It is, however, the purpose of this chapter to give a first characterization of the specific local changes and of the new general structure which is emerging.

FORDISM AND ITS PRINCIPLES RECONSIDERED: SOURCES AND ORIGINS OF ITS EFFICIENCY

The "theoretical skeleton" of Fordism consists of a series of coherent and interrelated norms as far as the different levels of analysis of the capitalist dynamics (productivity, production, consumption) are concerned.

In the following my purpose is to give a theoretical characterization of the implicit dynamics of the Fordist productive structure and organization. I begin with the simplest level (origins of the productivity gains in the labor process) and proceed by trying to integrate the other more abstract levels (norms of production, of competition, industrial organization, etc.). The point here is to "open up" the fantastic engine of growth of the classic Fordist mode of accumulation, to extract and make visible the different sources and natures of its productivity gains.

The basic principles of the classic Fordist "engine of growth"

We can pose the following "stylized facts" concerning the sources of efficiency of the Fordist model.

- At the level of the labor process productivity gains (or if one prefers, the "norms of productivity") are based on a simple and rough parameter: *division of labor + mechanization and rigid automation.* The type of division of labor (coherent with mechanization and rigid automation) is itself sustained by the old "classical" idea that maximum output is closely related to the maximization of the efficiency of the *individual* worker. Here the idea is that important "learning" gains are based on fragmentation and repetition. This is the reason why we have chosen to term the implicit principle of efficiency at work in this paradigm as (the pursuing of) a "Smith effect,", in reference to A. Smith, who certainly placed maximum emphasis on this particular concept of efficiency. Taylor and Ford systematically exploited this concept and principle of economic efficiency through scientific management and the introduction of the assembly line. In this sense, they contributed to the hegemony of the basic paradigm, first introduced by Smith and the classical economists.
- Consistent with this vision of the sources of productivity gains, the specific *norms of production* of the Fordist mode of accumulation were based on an attempt to capture in a systematic way the benefits of *economies of scale* through long series of standardized products. A "Young effect" is thus one of the characteristics of the origins of the productivity gains pursued by the classical Fordist structure. The Fordist factory is typically a big factory and vertical and/or horizontal integration is the natural complement of this organizational structure of the economy. In my view, this is the reason why the transaction cost analysis pioneered by Coase and reshaped by Williamson has been mainly used until very recently to justify the growing process of internalization of activities in the firm. The benefits of what we can call the "internalization effect" were thus explicitly evoked to explain and justify the continuous process of concentration experienced by the firm during the classical period of hegemony of the Fordist mode of accumulation.
- The *norms of competition* reflect this basic organization: the competition between firms basically involves a long series of standardized products. In the rivalry between firms, the capacity to offer products at lower costs is more important than any other attribute of the products. So clearly we are in a world dominated by oligopolies, and the competitive game is played under the specific rules of oligopolistic competition. Barriers to entry are formed by the economies of scale required to maintain a competitive position in a world of growing firms and markets. Certainly it was Bain (1954 and 1959) who painted the most precise picture of the universe crated by the progressive hegemony of Fordism.
- A well-defined pattern of *industrial organization* emerges from these characteristics: basically the center of power lies in the hands of the big enterprise (the so-called "monopoly capitalism"); around it a series of dominated small or medium-sized enterprises are subjected to

subcontracting procedures. They are thus obliged to assume many of the inherent risks of market changes. Among these firms some are "captured" since the big enterprise tries to reap the benefits of vertical (or horizontal) integration through consolidated links with some of its suppliers.

- Last but not least, if we consider the *whole macro-model* emerging from this set of norms, the huge productivity gains extracted from these arrangements made it possible to transfer a sufficient amount of them to the wages to be able to capture the benefits of mass consumption along with mass production.

To summarize, one can present the specific Fordist engine of growth as a combination of *division of labor and mechanization* (Smith effect) + *economies of scale* (Young effect) + *vertical integration* in a world dominated by *competition through long series* and *oligopolistic markets.*

The disintegration of the Fordist engine of growth

Continuing with stylized facts, we can advance the notion that the disintegration of the efficiency of the Fordist engine of growth has two basic and largely interrelated origins:

- The first one is the so-called "productivity slowdown" which begins (in the US) at the end of the 1960s, and spreads through the whole economy during the 1970s. Many different explanations were given for this crucial phenomenon (see, for instance, the sophisticated studies of *The Brooking Papers* dedicated to this point). Obviously there is no unique and simple explanation. It seems, however, that one of the important causes of this productivity slowdown is the end of the potential benefits embodied into the old Smithian paradigm: the trajectory "division of labor + mechanization" reached obvious technical and social limits (see Coriat 1982 or in the radical American tradition, Bowles, Gordon and Gintis 1983).
- These limits appeared to be stronger in the 1970s, since markets faced dramatic changes, meaning dramatic modifications in the norms of competition. The individual firm was thus faced with a completely new environment. Markets became unstable and largely unpredictable, and product life cycles experienced dramatic decreases. In such conditions new possibilities were opened up and the whole classical Fordist structure – under dramatic pressure – entered into a process of continuing change which imposed an adaptation to the new reality.

THE PRESENT CHANGES: PATHWAYS TOWARDS POST-FORDISM

In practice the adaptation began at the workshop level. A set of technological and organizational innovations took place and began to modify the traditional way of producing goods in order to generate new sources of productivity gains.

The rise of flexibility and the emergence of new "post-Fordist" principles of efficiency

If we concentrate on the enterprise, the content of the so-called new technological trajectory is basically twofold, since the interrelated fields concerned are technology and organization. Further specification is necessary to understand the full implications of the theoretical issues at stake.

Concerning technology the main novelty is the rise of flexible and programmable automation, starting with individual tools (robots, numerically controlled machine tools) and spreading itself through a series of new integrated systems (FMS, CAD/CAM). The benefits given by these new tools and arrangements can be summarized as follows:

1 *Process-flexibility*: the new automation makes it possible to use facilities at full capacity even under uncertain demand; by producing more than one product in the same basic installation the full capacity installed can be worked on, which is impossible in a "rigid" factory. In the latter, any shortage in demand is necessarily followed by certain percentage of under-utilization of capacity, with obvious negative effects on profitability. This advantage given by new flexible automation is termed "Marshall effect" (see Figure 6.1).

Technical Basis	Origin of the Productivity Gains	Empirical Concepts	Theoretical Concepts	Final Designation
(1) Process Flexibility	New Balance Between Fixed and Variable Costs	Decrease Marginal Costs by Using the Full Capacity Installed	"Static" Economy of Scale	"Marshall" Effect
and				
(2) Product Flexibility	Joint Production with Variable Composition of the Outputs	Decrease in Average Costs	Economy of Scope	"Baumol" Effect

Figure 6.1 Economics of multiproduction from the Marshall effect to the Baumol effect

2 *Product-flexibility*: this is another dimension of the same phenomenon. It basically concerns the benefits of the economies of scope associated with flexible technologies. I refer to them as the benefits of the "Baumol effect" in reference to the seminal contribution made to their understanding by Baumol, Panzar and Willig (1982) (see Figure 6.1).

3 *Dynamically* (see Figure 6.2): the cumulative investment in flexible

A. Rigid Plants: A Process of Cumulative Gaps

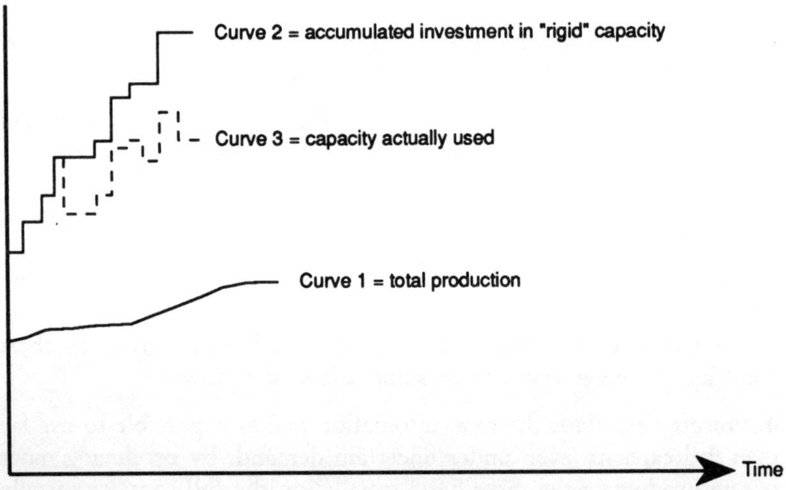

Comments:

- The hypothesis of a slow growth rate of demand is represented by Curve 1.
- The impact on total capacity investment is represented by Curve 2.
- The gap between Curve 1 and 2 relies on the fact that the different "models" yield sales quantities different from the actual production.
- Curve 3 is a representation of actual production.
- The gap between Curves 3 and 2 is a "graphical" expression of under-utilization.

B. Flexible Plants: A Convergent Process

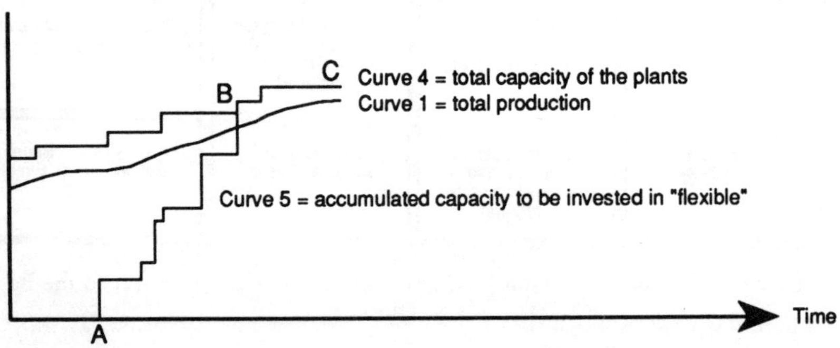

Comments:

- Curve 1 (total production) is the same as in the former case , but;
- Capacity installed to produce the first "model" (A in the figure) can be used (at low additional cost) to produce the new models (B and C) as the form A disappears from the market.
- It is the case that the Curves 4 and 5 (equivalent to Curves 2 and 1) are now basically convergent.

Figure 6.2 Flexible technology: long-term impact of firm capacity strategy

technologies permits optimization of the total amount of capital invested to produce the different products sold on the market at different stages in their respective life-cycles.

Summarizing the theoretical benefits associated with the new electronics-based technologies, we can say that flexible automation gives rise to two fundamentally new phenomena:

a) a "Marshall effect," based on the fact that the capabilities can be used at full capacity even in a changing environment,
b) a "Baumol effect," which refers to the fact that economies of scope can be secured in an ordinary way, using the same basic (flexible) lines of production to produce a set of differentiated goods according to the market needs.

It is essential to note that these new tools and integrated automatic systems are emerging along with new organizational concepts and procedures. It is increasingly clear that the organizational revolution is probably more crucial than the technical one. The growing evidence concerning this organizational revolution is that a new Japanese school is taking the place of the old classical American school. In theoretical terms, the organizational revolution (at the workshop level) is the combination of new principles and techniques involving two dimensions of the classical production management methods: at the level of logistics and of production planning.

By logistics, we refer to the way the flow of semifinished products (raw materials, components, energy, etc.) is coordinated with production operations. One can say that the essence of this new conception consists in a new optimization between elementary circulating and operating times, allowing a high level of economy in lead times.

At the level of planning, either through MRP (Material Requirement Planning) or through Just-in-Time, benefits of different types and origins are gained. Techniques such as Just-in-Time, Kan-Ban, and Quality Circles are the elements of a new paradigm in the field of production management. A set of original sources of productivity gains is thus explored in the new, different layouts and "orgwares" conceived and installed at enterprise level. Figure 6.3 outlines these new developments in detail. (For a detailed analysis of the Japanese school on production management, see Coriat 1991.)

As a consequence of these changes, the classic Fordist structure and organization of production face dramatic shifts. To sum up the situation, the transition from Fordism to post-Fordism can be read as a move toward new principles of efficiency. Fordism is establishing itself on new micro-foundations.

"Post-Fordist" principles of efficiency: the new micro-foundations of mass production

Following the same methodology used in the previous section we can sketch the main theoretical changes at different levels of the old classical structure;

Material Basis	Manifestations	Empirical Measures	Theoretical Bases	Effect
(1) Logistical Revolution	Saving on production time of utilization of equipment and personnel-reduction of transfer times, downtime.	- Productivity of labor - Productivity of capital	- Intensity of labor - Economics of fixed capital and labor productivity	Generalization of the "Smith Effect" (Division of Labor) Liebenstein X-Efficiency Effect
(2) Organizational Evolution (a) Material requirement, planning. (b) Kanban, just in time (with externalization and subcontracting).	- Savings on In-process and stocks - Reduction of In-process (pieces, components, energy) - Reduction of stocks (purchases, warehousing, etc.)	Productivity of Intermediate Inputs	- Savings of circulating capital (I) - Savings of circulating capital (II) - Savings on total circulating capital	COASE Effect (Make or Buy)

Figure 6.3 Economies of organization: content and nature

1 *At the level of the division of labor*: the transition is from the old and rather rough "Smith effect" to a new paradigm based on the idea that the maximization of the output does *not* rely on the performance of the individual worker, but is primarily related to the optimization of operations and processes in a world where the adaptation to unpredictable markets is the new rule of the game (see Ohno 1989 and Shingo 1983). In such conditions, a revolution is underway in the organization of the assembly line and the process of balancing jobs along the production process. In this field, we are passing from allocated time procedures to organization in sharing times, or in other words from rigid standards to flexible and modulable ones (see Figure 6.4). This revolution based on new layouts (linearization and "U" form of productive arrangements) was achieved by Ohno (1989) inside Toyota. In conceptual categories these new methods and procedures belong to the "organizational efficiency" principle first emphasized by Leibenstein (1966) in his seminal essay on "X" Efficiency theory. This is the reason why this concept of productivity may be characterized as the passage from the "Smith effect" to the "Leibenstein effect".

2 *At the level of the production process*: to face the new characteristics of the market (uncertainty, unpredictability) and the changing behavior of the consumer, the transition is from specialized and dedicated layouts to new patterns of production organization, based on economies of scope. The objective here is to pass from the benefits of the "Young effects" to the advantages of the "Baumol effects" (see Figure 6.3). Combining the potential resources of the new organizational concepts (Ohno and the Japanese school) with the new programmable technological tools the productive arrangements are now able to mix the advantages of long series with those of differentiated products in small and medium-sized batches.

3 *At the level of the industry*: a new organization supported by original configurations is emerging. The benefits of vertical integration (i.e. internalization or quasi-internalization through hierarchical subcontracting systems) are no longer – if they ever were – a panacea. Horizontal de-concentration, quasi-disintegration in new subcontracting systems, based on long-term contracts are taking the place of the classical Fordist industrial arrangements. Briefly, we can say that in the place of the Bain principle of efficiency, an Aoki principle (to refer to his vertical versus horizontal anatomy of the respective "A" and "J" forms assumed by the firm) may be found (Aoki 1986).

Thus, some common basic principles of any type of post-Fordist productive structure can be defined. As for the theoretical skeleton of the post-Fordist regime of accumulation at the level of the industrial structure as a whole, the *norms of production* are moving from standardized products to mass-produced but modulable and differentiated ones, while the *norms of*

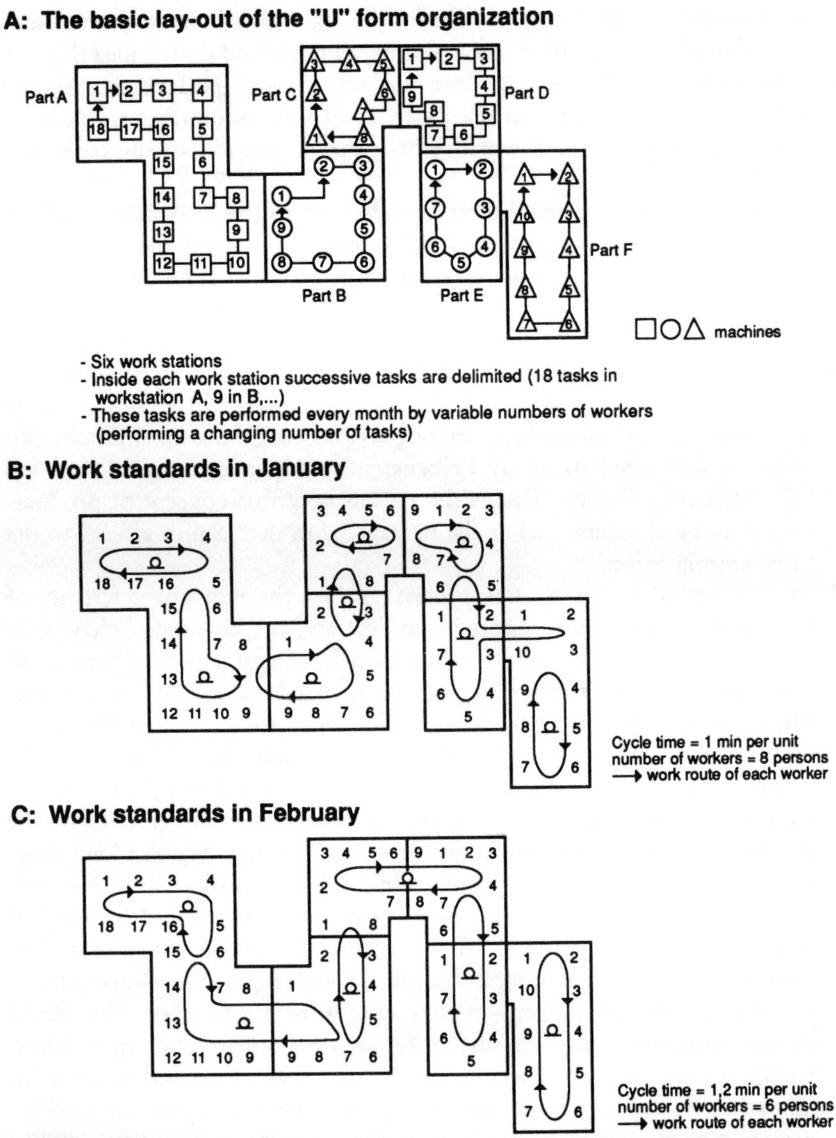

Comments:

If we compare the work organization in January and February, everything differs. The number of workers employed from (8 to 6), the cycle time (from 1' to 1', 2') and the content of the work standards attributed to each worker.

This illustrates what can be called the principle of organization based on "Flexible Work Standard," or based on "Shared Time" to strictly-defined tasks attributed to each worker (as opposed to the old Smithian paradigm).

Figure 6.4 The Japanese revolution in product management: linearization and flexible work standards
Source: Adapted from Monden (1983)

competition are passing from high-volume, low-priced goods to batched and differentiated ones. Here, elements of *monopolistic competition* (in Chamberlinian style) are penetrating the classical oligopolistic forms of rivalry between firms.

How all these new principles can be articulated in a macroeconomic model of accumulation remains a question to be answered. Many different combinations are possible, under different hypotheses of market configurations, product life cycles and so on.

NEW POST-FORDIST CONFIGURATIONS

Two basic configurations (each one including different variants) appear to be emerging to structure the productive process. Following Piore and Sabel (1984) on the one hand, and B. Klein (1986) on the other, these two principles may be characterized as follows.

The flexible specialization principle and its limits

It is not necessary here to insist on the role played in the scientific community (and far beyond) by the very stimulating book published by Piore and Sabel presenting their flexible specialization hypothesis as a possible root for "prosperity."

Since many works have been dedicated to exploring the different implications and meanings of the hypothesis, it seems better to concentrate on what we have learned during the last five or six years, concerning the way the universal need for (more) flexibility has been assumed in different contexts and situations. The following propositions are a preliminary summary of the evidence on both theoretical and empirical levels.

First, it seems clear that flexibility introduced at the level of the individual firm (either through organizational arrangements or through technology) is the basis of a specific competitive advantage. Using the resources of what we have called the "Baumol principle" and exploring in a systematic way the potentialities of economies of scope, the small or medium-sized firm, operating in unstable and changing markets, may be more efficient than the large, rigid one. The theoretical and empirical problem here is that of *the limits* of the strategy of economies of scope with regard to that of economies of scale (see below for a tratment of this crucial question). It is clear that in certain circumstances a strategy based on systematic efforts to secure the benefits offered by the economies of scope can be successful. (See Figure 6.5, where these kinds of successful strategies are illustrated.)

Second, it is a well-established fact that – here again, under certain market conditions and product life cycles – organized networks of small and flexible enterprises are able to compete even on mass production markets with the big classical Fordist factory. Many different forms of "industrial districts"

1. **Flexible Specialization** – Building a competitive advantage on Economies of Scope

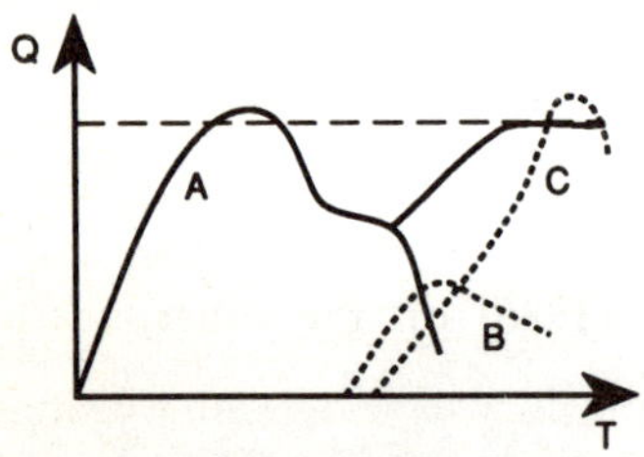

Passive Differentiation

A,B,C are three differentiated products belonging to the same basic family.

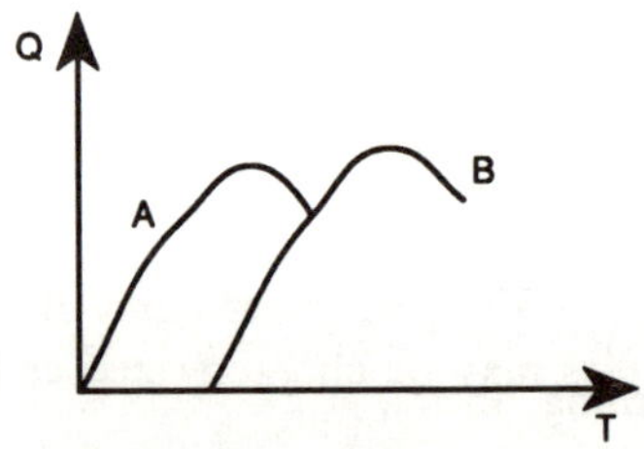

Active Differentiation

When the firm faces a break down of the demand for the product A, it introduces a new product B. This strategy of "permanent innovation" is efficient in markets dominated by products with short life cycles.

2. **Dynamic Flexibility** – Combining Economies of Scale and Economies of Scope

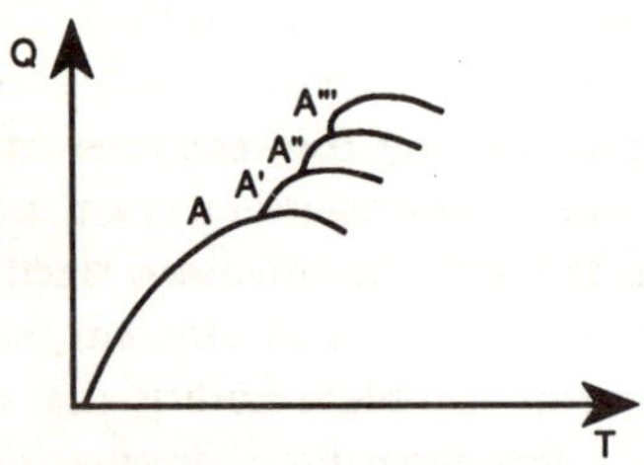

Differentiation by Systematic "Dematuration " policy

The firm extends the life cycle of product A by systematic dematuration policy, introducing A', A" and A'" new characteristics into the product A. Thus, through differentiation, the firm maintains the benefits of economies of scale.

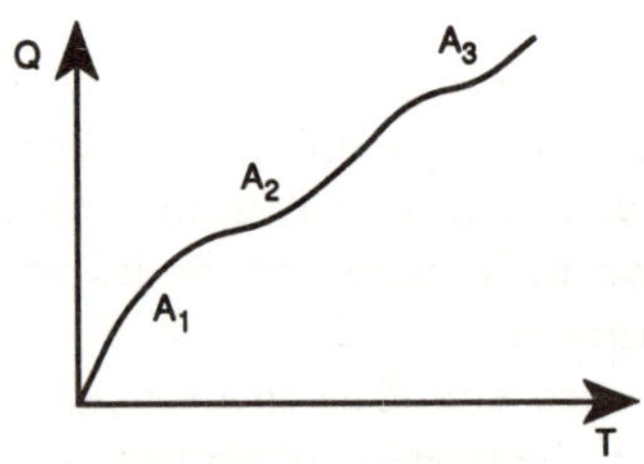

Differentiation by Product Development

When the product A_1 reaches the limit in its penetration into the market, a second generation of the same product (A_2) is introduced by market innovation into the market.

Figure 6.5 Technical flexibility and product life cycles: four basic strategies

have proved their capacity to maintain an advantage even in very competitive environments (see for instance the papers published by the Italian school on industrial districts, in Maruani, Reynaud, and Romani 1989).

Nonetheless there are limits and conditions of validity of these competitive advantages. Two sets of basic conditions are required:

1 The first conditions have to do with the products: the advantages of "flexible specialization" lie solely in manufacturing. The products must be "simple," that is to say at a stage of advanced "maturity" in their life cycle. The product may not include high conception and R&D costs: if the barriers to entry (based on high research costs and investment) are very high, the benefits gained at the level of manufacturing would not be able to cover the costs of continuing R&D and product improvement. This is the reason why, if we consider the products made in the industrial districts we find basically simple and mature products such as textiles, clothing, shoes, and so on. When the product embodies innovations, the innovation is "incremental" (as opposed to "major"), and creates a temporary "niche," since followers can easily imitate the first innovators. The case of the industrial district producing numerically controlled machine tools (analyzed by Capecchi and others) does not invalidate this view. The network of small flexible firms basically consists of subcontracting firms in the sense that the research on numerical techniques and procedures is carried out in the laboratories financed by *big* corporations.
2 The second – and closely related – set of conditions are related to market characteristics: the market, for optimum efficiency of the flexible specialization strategy, must be a segmented one. Thus many different niches can be occupied by small firms operating in distinctive markets. The Italian School – in its best work – insists on the fact that an efficient flexible specialization strategy requires segmented markets and/or rapid changes in consumer behavior.

If these conditions are missing, then large enterprises using the opportunities offered by the new organizational arrangements are certainly in a better position to face market needs, even if small and medium-sized enterprises are much more flexible. In this case, the flexible specialization principle leaves the field open to a strategy based on dynamic flexibility.

Combining the benefits of economies of scale and of scope: the principle of dynamic flexibility

To my knowledge the notion of "dynamic flexibility" was first introduced into the scientific community by B. Klein. Describing the policy followed by certain Japanese firms, he writes:

> As contrasted with static flexibility, dynamic flexibility is not concerned with producing more than one product (e.g., cars and light

trucks) on a single production line – although the Japanese do this too. Rather, it is concerned with designing production lines in a way that they can quickly evolve in response to changes in either the product or production technology. In other words the central preoccupation is to get ideas into action quickly. The main purpose of dynamic flexibility, however, is to make rapid changes in production technology for the purpose of lowering costs and thereby improving productivity.

(Klein 1986)

Following this characterization Cohen and Zysman (1987) insisted on the importance of this principle for any policy designed to revitalize the classic Fordist industrial structure. In a recent work, I have tried to elaborate further on the basis of this idea, and give a conceptual content to the principle of dynamic flexibility as opposed to the principle of flexible specialization (Coriat 1990). Thus dynamic flexibility is defined as being based on an "iterative" process of continuous changes affecting both process and product innovations. More precisely a strategy of dynamic flexibility is followed when the productive arrangements are conceived and managed in such a way that they make it possible to modify the process (at low or zero delay and cost) according to the changes introduced into the products. An alternative solution is to modify the product to benefit from the innovations introduced at production line level. In other words, dynamic flexibility is a strategy designed to reap the reciprocal benefits of product and process innovations. Here the concept of flexible and modulable work standards conceived by Ohno is crucial. More generally a completely new model of work organization, based on strong and continuous investment in human resources, is required here. Flexibility, in this case, is based on the skills of workers to whom broad capabilities to respond to changing circumstances are given.

The main advantage of such a strategy can be shown if we consider its effect on the long-term learning curve. Here one may remember (following the work on the subject by Albernathy 1978, Albernathy and Wayne 1985) that in the long run, the learning curve is *not* continuous. At every important change introduced by the enterprise (concerning either the product or the process) the learning curve experiences upward movements, since the new element introduced destroys the "know-how" and the "experience" accumulated at the preceding stages of the product–process configuration. The strategy of dynamic flexibility can thus be understood as centrally designed to "smooth" these upward movements, and to maintain a continuous upward orientation of the learning curve, even when important changes are introduced. Thus dynamic flexibility supports cumulative gains of efficiency (see Figure 6.6). In a period characterized by an acceleration of this kind (affecting products and/or the processes), it is easy to imagine that this strategy can play a very important role in the competitive game.

A. The case of the Model "T"

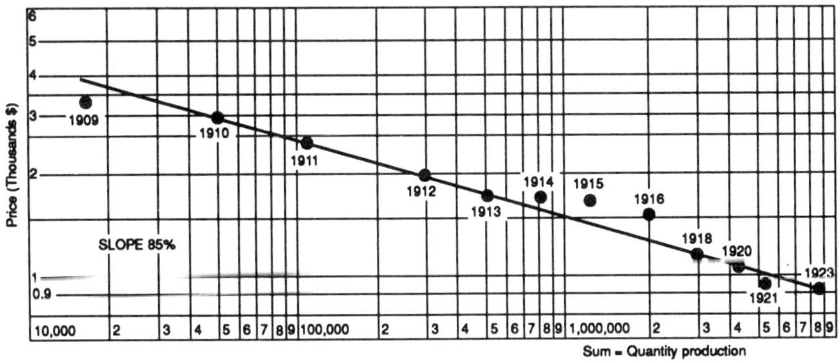

Every time production doubles, the price drops about 25%.

B. Ford = Model "T" to Model "A"

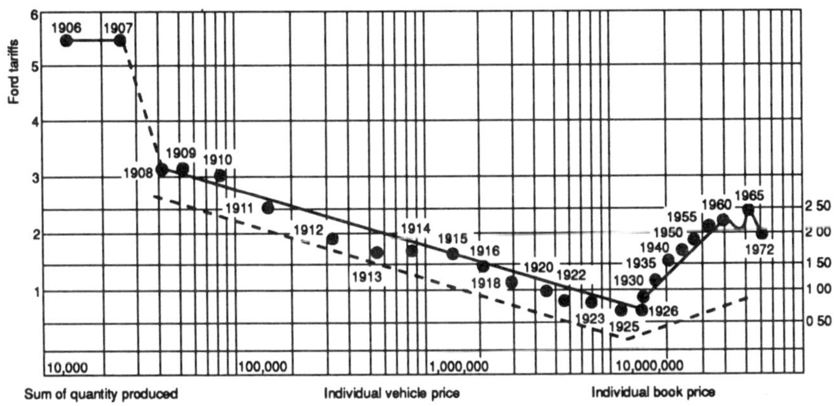

After a steady drop because of Model T's, the prices rise after the introduction of new products.

C. Advantages of productive flexibility (stylized fact)

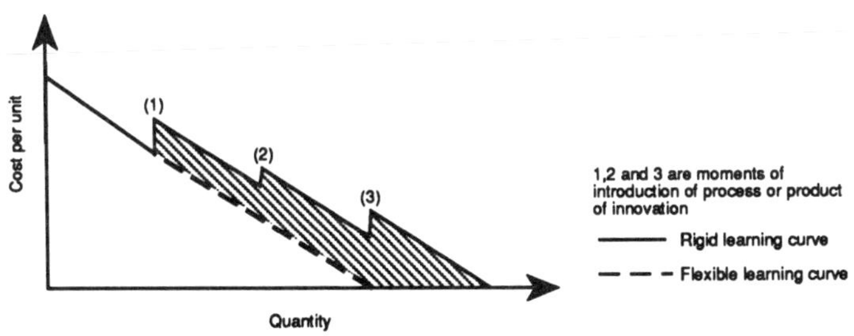

Figure 6.6 Learning and experience curves

But, these strategies also require some "permissive" conditions and reach definite limits. As in the preceding case, these limits are traced by a set of characteristics belonging on one hand to the product life-cycles, and on the other to the dimensions of the markets. As regards the former, the product must be in a phase of "development," having been introduced with success onto the market. Once the product has been introduced and adjusted to consumer tastes, large enterprises can enter into the game, with long-term strategies and policies of experience curves to lower prices, and take large shares of potential markets.

Another condition is that the market for the product must be a *growing* one, or at least a more or less guaranteed mass market (for illustrations one can refer to the market of PCs, compact disks, etc).

In such interrelated conditions of both "product life-cyle" and "extension of the market," it is possible to combine the benefits of economies of scope and economies of scale in successful strategies, without leaving any significant space for medium-sized enterprises using only the benefits of economies of scope.

Of course many different variants of these two basic strategies can be imagined. In Figure 6.5 four basic strategies are illustrated.

THE TRANSITION FROM FORDISM TO POST-FORDISM: CONCLUDING REMARKS

The whole Fordist mode of production is in a course of great modification. It is too soon to try to formulate any reasonable predictions as regards the stable forms and configurations of the future industrial organization considered as a whole.

It seems, however, possible to conclude by focusing on certain crucial points and issues of the present process of transition (see Figure 6.7 which illustrates the new form and principles of the post-Fordist age).

Quality versus quantity: the new content of the norms of production

If we consider the new norms of production, it is increasingly evident that, in most situations, it is no longer sufficient to produce at low cost (through long series, economies of scope, and so on). Differentiation and quality must now enter into the productive process. According to the new tastes of consumers, and their volatility, or to different international standards (both technical and cultural), maintaining a position in the market necessarily means built-in capacity of the sort that allows the productive process to follow the many, largely unpredictable changes. New constraints of quality and differentiation are now combining with the traditional constraints of producing in long series and at low costs. In my view, this is the reason why just-in-time techniques are so important, and so successfully used by the

	Traditional Production Organization	Flexible Production Organization
Without Output Growth – Internal Economies (variable / fixed costs) – Economies of Learning – Economies of Experience	**1) Traditional Production** – Condition: stable and strong demand – Division of Labor: "Smith" style – Economies of Organization: "Taylor" style Type of Firm: small or medium series of standardized products Norms of Competition: by prices	**3) Flexible Production** – Economies of Scope: "Baumol" style – Economies of Organization and X Efficiency "Liebenstein" style Type of Firm: medium series differentiated Norms of Competition: monopolistic
With Output Growth – Internal Economies – Economies of Learning – Economies of Experience	**2) Traditional Mass Production** – Economies of Scale: "Young" style "Smith" style "Ford" style Type of Firm: long series of standardized products Norms of Competition: oligopolistic by prices	**4) Dynamic Flexibility** – "Young" + "Baumol" style – "Liebenstein" style – Iterative economy process / product Type of Firm: long series of evolutive products Norms of Competition: oligopolistic + quality effects

Figure 6.7 Origin of productivity gains and form of competition: a typology

Japanese firms. Far beyond the economies in "circulating capital" (stocks, in-flow products, etc.) the key to success must be attributed to the concept of production through flexible (or modulable) work standards, first introduced into the production lines by Ohno inside the Toyota Motor Company. Along with the new production norm, a new model of work organization based on high investments in human resources is emerging (Coriat 1990).

Price versus nonprice competitiveness: the sophistication of the norms of competition

As far as norms of competition are concerned, this means a parallel modification; nonprice competitiveness as opposed to strictly cost competitiveness is becoming more and more important. This is a direct consequence of the growing role played by the constraints of product differentiation and quality, arising from the structurally unstable markets. But, as I have tried to show, the growing importance of nonprice competitiveness must also be related to the opportunities created by new technical tools and organizational concepts. Since many different strategies can be pursued by firms, new ways are being found to bypass the rough and very rigid requirements of pure cost-competitiveness. Flexible specialization practices, or strategies of dynamic flexibility, open certain opportunities to create profitable niches or segment the market. Thus the firms using these new opportunities can conceive strategies based on principles of nonprice competitiveness. More generally, flexibility revitalizes forms of "Chamberlinian" competition even inside the classical oligopolistic forms of economic rivalry. The whole competitive game is turning out to be much more complex and refined than it used to be.

Is a new post-Fordist "virtuous circle" of growth conceivable?

Based on all these new principles and arrangements, a large variety of models can be imagined. But it is clear that all the virtual possible macro-arrangements are *not* equally stable and efficient. I have elsewhere identified three basic – more or less sustainable – *models of post-Fordism.* (Coriat 1990).

To conclude this chapter, I have chosen to focus on one of those three alternative models. The central point of this model is that it is constructed on the basis of a possible new "virtuous circle," which embodies and assumes all the novelties and characteristics of the post-Fordist age, as regards the set of interrelated norms emerging from the new general environment.

As shown in Figure 6.8 this model rests on two related pillars: quality and product differentiation on one side, quality of the organization and of worker qualifications on the other. One must note that there is a strong reciprocal link between the two pillars, in the sense that:

a) high and continued investment in organization and human resources is the strategic basis established to face the competition through differentiation and quality,
b) high quality and adequate differentiation policy as regards market changes allows firms to capture some *monopolistic rents* on the market, even in very competitive environments.

Thus, qualifications nurture quality and quality nurtures qualifications. This "*qualification–quality*" model is supported by a specific "engine of growth" (see Figure 6.8, and for concrete example of such a model one may refer to some of the German and Swedish strategies; for the German case, see Streeck 1988). According to market conditions, this model uses, either separately or together, the benefits of flexible specialization and of dynamic flexibility strategies. Thus the different new principles of the efficiency associated with the rise of the post-Fordist economy (e.g. the Marshall, Baumol, and Arrow effects) can be pursued in a cumulative way.

In my own vision of the current transformations of the capitalist economy, there is only a very slight probability that this model will become a reality.

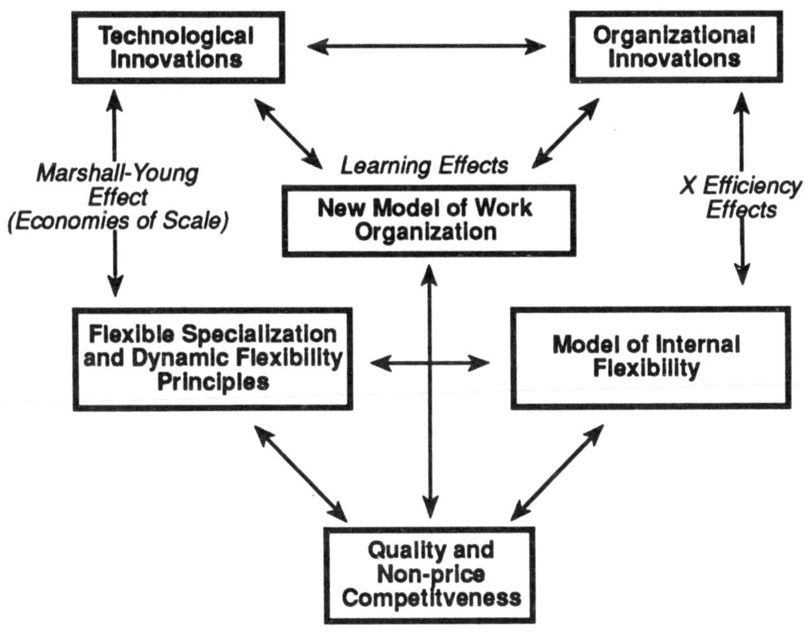

Figure 6.8 The quality skills model

But no-one can contest the right to dream built into the constitution of human beings; and as is obvious, the German car makers such as Mercedes or BMW have found a very profitable way of making dreams come true.

7

TECHNOLOGICAL TRAJECTORIES AND THE CLASSICAL REVIVAL IN ECONOMICS

Michael J. Piore

This chapter is part of a continuing effort to clarify and revise the argument which Charles Sabel and I originally put forward in *The Second Industrial Divide* (1984). It is a response to some of the criticism and interpretative commentary to which the book has been subject. At the same time, it is designed to draw attention to more recent developments in mainstream economics and suggest some of the connections between these mainstream developments and the French regulation school in terms of which our argument was initially conceived. The basic concern is how to understand and conceptualize economic growth and technological change.

TECHNOLOGICAL TRAJECTORY

The key notions upon which I want to focus are, first, that of a *technological trajectory* and, second, the two specific trajectories which in *The Second Industrial Divide* we term *mass production* and *flexible specialization*. The concept of a technological trajectory can be located quite precisely, in both the theory of regulation and the theory of mainstream economics, but the terminology implies a certain interpretation of these theories which many who associate themselves with these schools of thought might not accept.

In mainstream economics, a technological trajectory is the locus of the mystery surrounding economic growth. I use the term "mystery" because it seems to capture the essence of a story, possibly apocryphal, which I have always thought most basic to mainstream growth theory. The story is about Bob Gordon and Simon Kuznets, who were commentators for a panel of papers on economic growth at the annual meetings of the American Economic Association. Simon Kuznets was, of course, the father of national income accounting and, in a sense, invented growth measurement. Bob Gordon is a member of my arrogant generation, and the story took place at the time when we were at our most arrogant age. Bob, as the story goes, spoke first, and began by dismissing all the papers with the comment: "The problem of economic growth was solved by Griliches in his article of 19 . . ." In the silence which ensued while the audience waited for him to continue, Kuznets rose from his chair, walked to the podium, gently put his arm around Bob's shoulders and said "Professor Gordon, you cannot *solve* the problem of economic growth."

In mainstream theory, the mystery of growth is that an industrial economy is able to generate progressively more output over time with the same resources. This effect is measured by *Solow's residual* – that, I imagine is what the Griliches' paper to which Bob referred "explained." In a crude sense, a technological trajectory is yet another attempt to penetrate that mystery. The criticism is that it, and the various specific trajectories identified in *The Second Industrial Divide*, are basically just a new terminology for the same old thing: they may be nice names, but they leave the mystery unresolved. This chapter can be read as an attempt to respond to that criticism.

In one respect the criticism is not well-taken. The terminology involves a lot of hand waving but it invokes at least one substantive proposition. It implies that there is a very specific growth process, so specific that it entrains very definite implications for the institutional structures of the economy. The valid criticism is that Sabel and I assert that relationship without specifying the growth processes in a way in which the institutional structures actually derive from them. The notion that there is a specific growth process – a process so specific that one should be able to deduce the institutional characteristics of the economy from it – however, directly conflicts with a central (albeit implicit) assumption of neoclassical growth theory: that the sources of the residual are so various that they have no strong institutional implications. In the class of models which Solow pioneered, growth is the product of a vast multiplicity of little human actions. That growth is the impenetrable mystery. This view, however, is peculiar to *neoclassical* growth theory and is not universal to mainstream theory as a whole. Indeed, the most recent growth literature in mainstream economics coming out of Chicago, quite explicitly rejects this position. Its ambition is to formulate a theory with strong substantive implications. We turn to the question of how well it has thus far succeeded below.

The school of regulation is not a homogeneous approach to economic analysis either. What Sabel and I found interesting in the regulation approach was the marriage between the neoclassical lessons about an economy as a selfequilibrating system and the Marxist idea of an economy as evolving through history, generating tensions and conflicts, which were ultimately disequilibrating and were finally resolved through the political process. Defined in this way, as a marriage between neoclassical and Marxian approaches, the notion of a technological trajectory definitely belongs to the Marxian dimension: it is that set of forces which propels the economy through history, causing it to outgrow any particular regulatory framework and enter into crisis.

The idea that something quite specific propels the growth process is, of course, not simply Marxian. It was shared by the classical economists, beginning with Adam Smith, and is, as I have already noted, the central tenet of the classical revival at Chicago. In Marx, or at least among Marxians, there are a variety of distinct ideas about what it is that ultimately generates the historical thrust of the system. The major division is between those who see that thrust as rooted in some kind of natural evolution of the technology and those who see it as generated by class conflict.[1] So long as there is only one technological trajectory, however, this distinction does not really matter: in the highly integrated process envisaged by Marx, they both lead to the same set of institutional outcomes. When one postulates, as we have, two trajectories, the outcome is open and the source of these trajectories becomes a central issue. In Marxian terms, a major criticism of *The Second Industrial Divide* is thus the failure to address it. But the flaw in the argument which leads to this reproach is essentially the same as when the argument is viewed from a mainstream perspective: the argument is about the way in which alternative technological trajectories generate particular institutional structures, but we fail to define the trajectories in a way in which the structure can be deduced from them.

Again, however, to say that *The Second Industrial Divide* defines a trajectory by the institutions with which it is associated is to concede too much. A good deal, in fact, is said about the *trajectory* of mass production, independently of its associated institutions. What is said makes the following assertions: first, the trajectory of mass production involves the search for growth (or technological innovation) through the process of the division of labor, as exemplified by Adam Smith in the transition from pin craftsman to pin factory. Second, growth of this kind has a set of distinct technical characteristics. The critical characteristics, in terms of the institutional structures of the economy are fourfold:

a) economies of scale within the firm,
b) increasing returns for the economy as a whole,
c) the specialization of productive resources and,
d) the divorce between conception and execution in production.

The technological trajectory of flexible specialization is not specified this precisely. We did, however, delimit the institutional forms associated with it. The organizational literature at that time, especially but not only in economics, distinguished two basic institutional forms: "hierarchies" exemplified by the large, vertically integrated corporation and "markets," where basic units were isolated small firms communicating through price signals in a market. We associated flexible specialization with a third institutional structure, one of a series of small (in relationship to the hierarchical corporation) units engaged in intense direct communication and embedded in a dense social network. Because there was relatively little research on this organizational form, we drew primarily on work on industrial districts in central Italy and upon old craft communities, generally characterized as "pre-industrial," which in most areas, had disappeared. There now exists a much more substantial body of work on what have come to be termed "network" organizations, and there is an emergent consensus in the management literature that this organizational form is often appropriate not only to organize relationships among small firms but also to structure the relationship among the internal units of large corporate enterprises and between the corporation and its suppliers, customers, and even its external competitors. But the question remains: what is the relationship between this network organization and the characteristics associated with the mass production trajectory? Is it responsive to the absence of all of those characteristics? Or merely some of them? If it does not generate growth through the division of labor, how is growth generated?

Before turning to discuss these questions, it is worth recognizing that what *is* said about mass production does clarify several issues in the current growth literature. First, it clarifies the relationship between flexible specialization and contemporary developments in organizational design and manufacturing technology associated with flexibility, to which flexible specialization is frequently linked. A number of the latter do not affect any of the basic characteristics of mass production. In particular, they maintain the distinction between conception and execution, they continue to rely on resource specialization, albeit in a somewhat novel way, and they seem to involve both substantial economies of scale and increasing returns. This difference led, in an earlier paper, to a distinction between flexible specialization and *flexible mass production* (Piore 1989). A given productive operation in the latter produces a gamut of products and is thus not dedicated to a single make and model as in classic mass production. But the variety is limited and the products specified in advance. In flexible specialization, by contrast, the product line is open-ended. This distinction also involves certain cognitive differences in the approach to the productive process, but the significance of these cannot be deduced from the list of characteristics defining mass production, and requires an expanded definition of flexible specialization, to which we turn below.

Combining mass production and flexible specialization

A second point which does emerge from the list of characteristics defining mass production is the basic difficulty of combining mass production and flexible specialization in a single system. To assert that growth under mass production involves increasing returns is to suggest that *any* alternative trajectory threatens its dynamism.

The essential idea of mass production is captured by Figure 7.1. That idea is that the economy moves down the curve, *ab*, over time; it does so by increasing output, through standardization of products. Smith pointed out that there was another method of production, craft production, which produced specialized goods but with stagnant technology, so that it always had cost *a*. The result: the consumer is faced with a choice, governed by the price differential between $P_c–P_m(1/q)$, and q grows over time.

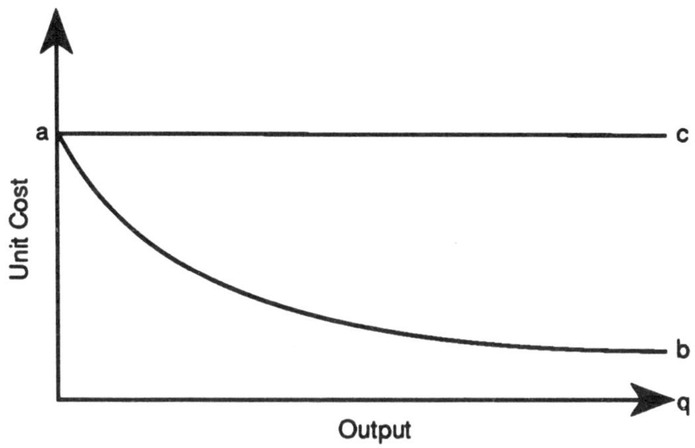

Figure 7.1 Mass production curve

The postulate about the emergence of flexible specialization as an alternative growth trajectory is that, at some point in the late 1970s, P_c begins to fall due to a new dynamic of technological change. As it does so, $P_c–P_m$ declines, and customers are drawn away from mass markets, and q falls. The fall in q, however, itself aggravates the process, raising P_m and further narrowing the price differential.

In analyzing these developments, one can make a further distinction between economies of scale associated with fixed investment in specialized resources and increasing returns associated with the technological development to which the specialization leads. The latter may be irreversible. In other words, the expansion of production may generate "new knowledge." Whether or not this is the case depends on the precise way in which specialization and technological innovation interact, a point to which we return below. If there is irreversibility, or some irreversibility, the technological gains may be preserved in smaller units and at least a part of the effects of a declining q are reversed over time as the scale of new plant and equipment is adjusted to the reduced market size. Then, mass production will makes a resurgence, an effect which we currently seem to be observing.

But our characterization of the mass production trajectory suggests that the resurgence will be temporary; the dynamism of mass production depends on the progressive expansion of q and once the scale is adjusted, the technology will stagnate.

Flexible specialization and the classical revival

A further observation which one can draw from this characterization of mass production relates to the classical revival at Chicago (Lucas 1988). The classical revival, as we suggested earlier, has basically accepted the notion that there is a technological trajectory in the sense that we are using the term, i.e., a growth process that has substantive implications for the economic structure. But the literature assumes that this process is consistent with a competitive economic structure and the concern is with modeling that process as a kind of "general equilibrium" growth. Indeed, the competitive nature of the associated institutional structure is such a fundamental tenet of this literature that it is not even explicitly stated, let alone argued. What makes this curious is that the new Chicago literature draws upon exactly the same classical tradition that we use to define mass production. How does this happen? What is the difference?

In terms of the list of four defining characteristics, the difference centers on the distinction between increasing returns at the level of the economy and economies of scale at the level of the firm. We have defined mass production in a way in which the two characteristics are inextricably linked: you cannot have the former without the latter. And the Chicago literature assumes that the increasing returns for the economy do not imply economies of scale at the level of the firm. In the Chicago revival literature, increasing returns involve *external economies*. And Robert Lucas even coins the term "external human capital economies" to characterize the growth process. The key question then becomes: are the economies associated with the classical growth process external?

The idea that such economies are external comes from Marshall. But Chicago literature does not actually refer to Marshall himself; it tries to make this point instead by reference to Allyn Young (1928). In so doing, it confuses precisely those issues that we are attempting to highlight. The confusion arises because Young uses Adam Smith's pin factory to illustrate his argument. And, in the pin factory example, resources are too specialized to generate "external economies" of scale. Thus, Chicago would need to postulate an alternative trajectory in order to sustain its models. It need not as a matter of logic, of course, join the whole debate: it could simply assert that there is a single trajectory with the properties for which it is looking and that Smith (and Young) were either mistaken about what the pin factory growth process *was* or about its overwhelming importance. But given the extensive literature about network structures which is now developing, this may be a little harder to do than it was in the past.

Chicago might have gone back to Marshall, instead of Allyn Young, but this would not have solved the problem. There is now a very extensive literature trying to interpret exactly what Marshall observed in the light of contemporary experience. One influential branch of the literature argues that what Marshall observed was essentially akin to the dynamic *industrial districts* which emerged in Italy in the 1970s. Indeed, the very term "industrial district" is taken from Marshall's text (Becattini 1979). But the Italian industrial districts are not, as we have noted, markets: the intensity of communications which occurs within them and the social structures in which they are embedded makes it very unlikely that they could ever be compatible with a competitive price system (Piore 1990). Moreover, there is a real question about the "externality" of the economies associated with these kinds of network organizations. The research on both the network corporation and the Italian districts suggest that they survive and prosper only if the economies which are external to particular productive units are internalized as parameters in the decisions of some higher level organization unit, e.g., the corporate headquarters in large organizations; the municipal government, the trade union, and/or a business association in the geographic regions (Piore 1990).

My own reading of the literature is that Marshall did not observe industrial districts in the sense that this term is used in the network organization literature. I think that Allyn Young's choice of the pin factory example was indeed apt: that what Marshall observed might best be termed a *dispersed hierarchy*, i.e. a set of narrowly specialized and hierarchically coordinated productive units. These units operated like the mass production factory of the functionally divided corporation, but the relationships were organized by contracts rather than by internal rules. His observations call into question much of the new theory of the corporation, particularly that of Oliver Williamson and Alfred Chandler who argue that modern business forms are a response to economies of scale. But they are not associated with an independent growth trajectory; the economies of growth are internalized in contracts among these units just as they are in the vertically integrated corporation. Neither reading of Marshall suggests that he was observing a competitive market economy in the sense that we use that term in contemporary economics or that such an economy could sustain growth. To resolve the problem about the relationship between growth and organization form, it thus appears that one must go beyond Young and Marshall to the classical economists themselves.

CLASSICAL GROWTH THEORY

Adam Smith

The two key propositions of classical growth theory originate in Adam Smith:

1 growth is produced by the division of labor, and
2 the division of labor is limited by the extent of the market.

But the classical growth process was actually defined by Adam Smith's *pin factory* example. Or rather, the classical growth process is defined by the example of the transition from the master pin-maker who made the whole pin to the factory in which "one worker pulls the wire, a second worker cuts the wire, a third worker heads the pin, a fourth points the pin, and so forth."

Why should the division of labor enhance productivity? There is a certain ambiguity in Smith here. The ambiguity is important, and we will return to it shortly. But the underlying reason is the specialization of productive resources: specially trained workers and specialized tooling.

The specialization leads directly to Smith's second proposition, that the division of labor is limited by the extent of the market. Because the resources are specialized, they cannot shift to alternative tasks. They become "inflexible" or inelastic. The specialization is thus justified only if labor and capital can be kept employed in the tasks which they are "specialized" to do.

This proposition about the extent of the market is an assertion about increasing returns: as the level of output expands, it becomes possible to further divide, and hence further specialize, productive resources. Must these increasing returns be internalized, i.e. do increasing returns manifest themselves as economies of scale at the level of the firm? Smith did not seem to see a connection: or, at least, he never identified the conflict between growth and competition which such firm level economies of scale imply. But modern human capital theory suggests that in the pin factory example the increasing returns must be internalized: the pin factory resources are not just specialized, they are *specific*; they can be used only in the particular pin-making task. By the time we reach the end point of this growth process in the automobile factory, the resources are specific to the make and model of a particular product. When the model changes, the worker must be retrained and the old equipment scrapped and replaced. Specific investments of this kind will not be made unless the firm can capture the returns. In the pin factory growth process, moreover, the market will *never* support the additional firms which would be required to eliminate the specificity of the investments: in a larger market, it will always pay to divide the tasks still further and make the resources still more specific. Increasing returns, economies of scale and resource specializations are thus inextricably linked. The Chicago-Marshallian growth process, like flexible specialization, requires a different growth mechanism.

Marglin

Clues for an alternative growth mechanism are offered by the debate about the relationship between the division of labor and productivity growth.

Credit for opening, or rather reopening, that debate goes to Stephen Marglin and his 1974 article "What do Bosses Do?" (Marglin 1974). Marglin systematically reviews Smith's rationale for the pin factory and adduces evidence which suggests that each of the pieces of Smith's argument is either empirically false or quantitatively trivial. Marglin then argues that the division of labor is the product not of efficiency but of class conflict. It is designed to enhance the control of the bosses; it does not contribute to growth. Although it would constitute a considerable digression to engage Marglin's argument fully here, the argument is difficult to sustain. Its chief limitation for our purposes is that it offers no alternative growth theory: either growth is an illusion or, as it is for Kuznets, a mystery. The empirical evidence which Marglin offers suggests that the particular *form* of growth may have been chosen initially because it enhanced the power of the "bosses," and the attempt to alter the growth process may be resisted for that reason, but the evidence is consistent with the notion that the growth which results from the process is real enough. And, in fact, if one believes that growth to be real, the postulate of an alternative, but rejected growth trajectory, which we are seeking to sustain here, would only strengthen Marglin's assertions about the role of class conflict: the conflict might determine which of several alternative paths was chosen and/or why, once chosen, the commitment was maintained. To invoke class conflict in this way, one need not make growth itself an illusion. But, Marglin's criticisms of Smith, while cogent, are not helpful in the construction of an alternative theory of growth.

Marx

Much more helpful in this regard are the observations of Karl Marx, not the Marx of the class conflict, from which Marglin drew his inspiration, but the Marx of technological determinism. Marx's ideas about the efficacy of the division of labor and the enhancement of productivity are, at root, very different from Smith's and much closer to modern theories about cognition. Two of Marx's propositions appear particularly promising in terms of developing the notion of an alternative growth trajectory. One is the distinction which he draws between the social division of labor and the detailed division of labor. The second is the relationship which he develops between the detailed division of labor and mechanization. Marx makes a third observation about the rise of the factory system which is also suggestive.

The social division and the detailed division

The distinction between the social division of labor and the detailed division is essentially the distinction between the pin craftsman and the pin header.

Marx suggests that these are two distinct and technologically powerful developments. The pin craftsman – which Smith took as the starting point in the development process – is actually itself a distinct technological development, involving the separation of pin making from the making of other products (like sheep-herding, cow tending, clothmaking, and the like) which were once all produced together in a self-contained household. The separation contributes to productivity because it isolates an activity with a distinct conceptual core, and by focusing on that core and understanding it better, one will produce better pins, possibly in a more effective way. The isolation of pin heading yields no such advantages, because pin heading is not a conceptually distinct operation; it has no meaning independent of pin making (Marx, *Capital*: 390).

Reconceptualization

Indeed, for Marx, the contribution of the detailed division of labor was very different. It enabled the production process to assimilate the logic of the machine. In this sense, it is virtually the opposite of the social division of labor. One separated out the task of pin heading in order to *escape* the conceptual apparatus of the pin and think of the task instead in terms of a rationale associated with machinery. Having done so, one is then able to mechanize the task of pin heading. A by-product of this process of technological development is the narrowly trained worker who does the "pin heading" before it is mechanized. This is an extremely important part of the story which Marx tells about capitalist development but it is not central to the issues here.

Reintegration

A third insight into the process of growth is suggested by Marx's comments on the factory system. For Marx, in contrast to Marglin, the factory system was a real engine of growth. But, Marx cautioned, the growth potential of the factory does not explain its development. The factory itself was created for one series of reasons: better to control a productive process and supervise a labor force which was previously dispersed under the putting out system; to share a common source of power, etc. Once the separate tasks were brought together under one roof and performed in proximity to each other, people were led to perceive of the productive process in new ways and this changed perception was itself a source of innovation. The new innovations were associated: they all facilitated the *integration* of the distinct operations. Thus, the innovations associated with the factory are different from those associated with mechanization: the latter involves looking at each operation *separate* from every other operation; the former involves looking at the operations in *relation* to each other (Marx, 414–5).

How might Marx's views enable us to understand the emergence of flexible specialization? Marx's observations can be read in at least three different ways. One reading is as a story about a dominant technology: the shift from handicraft production to machine production. The natural way to extend this story is to postulate a shift in the dominant technology. And, surely the new technology upon which one would want to focus is information technology, i.e., the computer. A second reading is that mass production grows out of a focus on the detailed division of labor, and that flexible specialization shifts the focus of technological change backward in the cognitive chain toward the social division of labor. A third reading might focus on the distinction between the fragmentation associated with the detailed division of labor and the integration associated with the factory. Innovation in flexible specialization seems more like the latter than the former.

It would be hard to reject any of these as explanations for contemporary patterns of technological change, and, indeed, although not presented in precisely those terms, much of the contemporary literature develops one or the other of these themes. But I would be inclined to abstract from Marx a fourth perspective, not necessarily inconsistent with the other three, but more abstract, more general: that technological innovation is a cognitive process and that one needs a cognitive theory to explain it.

The particular cognitive theory upon which Marx's own explanation of industrial progress is built is something like the following: technology is the embodiment of certain concepts or conceptual frameworks in terms of which we think about transforming resources. Each of those frameworks involves a set of abstract principles which tell us how resources can be organized and deployed. The ultimate foundation for technological progress involves the reorganization of our understandings in new, more revealing and hence, more powerful conceptual frameworks. But, in the meantime, there is a second process which involves applying the conceptual knowledge we already have to perfect existing technologies and organizations.

The social division of labor involves the organization of production into distinct, conceptually coherent groups of operations. This enables us to perfect the underlying technical know-how involved and apply it in more effective ways. Thus, for example, building and farming are conceptually distinct, and the separation of the two enables us to better concentrate on the development of the concepts appropriate to each. Similarly within building, carpentry and masonry involve conceptually distinct materials and, for this reason, we do better in each when we specialize in them as separate endeavors.

But the organization of knowledge in these ways has certain effects. One of these effects is that existing conceptual categories tend to limit the imagination. By directing attention in one way, they block alternative perspectives. In craft pin making, the production process is dominated by the logic of pins on the one hand and the human being on the other. Mechanization involves a different logic. In order to apply that logic, one needs to *separate* pin heading, for example, from the pin. The rationale for the detailed division of labor is thus different from the rationale of the social division. It is to escape one conceptual frame in order to move to another frame.

Finally, any partitioning of the production process into a distinct series of operations creates a problem of reintegration, and this problem arises in both the social division of labor and in the detailed division of labor. There are several different ways in which one might understand the issue of integration (or reintegration). One is that it is simply a mechanical by-product of the process of partitioning: Marx's view of the development of the factory, however, implies that integration is more than this: that it is itself a distinct cognitive process. There are other clues in the literature that this is the case: a study by Persio Arida shows that innovation in automobile manufacturing oscillated historically between phases in which the major improvements involved the perfection of separate operations and phases where the improvements derived from achieving a better integration among operations (Arida 1977). This pattern is evident in a variety of technologies: for a period, IBM focused on perfecting individual machines while Digital's competitive advantage was achieved through *systems*. Audio-stereo technology seems to oscillate between periods where sales promotion focuses on the components individually and periods when the system as a whole is being hyped. The linguist Roman Jakobson calls the focus on the individual element *metaphor* and the focus on the sequencing of elements *metonomy* and, in studies of child speech development and of speech impairment in strokes known as aphasia, traces these different operations to separate and distinct parts of the brain.

But while integration appears to be a distinct cognitive process, there are certain dimensions of it which seem basically mechanical. First, in the detailed division of labor, the elements of the production process lose their original cognitive meaning in terms of the product. Indeed, the escape from the cognitive frame of the product is the *purpose* of the detailed division. This loss of independent cognitive meaning is what we mean by the divorce between conception and execution. It therefore leads to both the inflexibility (or rigidity) of task definition and to a strictly hierarchical organization in which the execution of the individual tasks is subordinated to a higher authority capable of reassimilating them into a cognitive frame in terms of which they have economic meaning.

In the social division of labor, the operations do not necessarily lose their economic meaning. But they might: in a social division of labor, one can make a further distinction between a partitioning of operations that maps directly onto the structure of consumption and a partitioning that does not. Examples of the former are such crafts as baking, tailoring, shoemaking, and, indeed, pin making. But the building trades are an example of a partitioning of operations in production which retains cognitive frames but does not map directly onto the structure of consumption: the carpenter, electrician, plumber, mason, etc. In the first set of crafts, the consumer performs the integration. In the latter case, the integration must be performed prior to consumption in the production process itself. This, I would like to argue, is the definition of flexible specialization. It achieves innovation through the tension between a deepening of understanding within a given cognitive frame and the pull to reintegrate back into a different frame in order to produce a sellable commodity. And it lends itself to a network structure.

This broad story can be reduced to a series of more sharply delimited propositions and summarized as follows:

1 Economic growth is the product of the division of labor.
2 The division of labor involves the *partitioning* of economic activity. One needs to understand both the cognitive and organizational implications of different partitioning principles.
3 The division of labor has two distinct dimensions: *specialization and integration.*
4 Cognitively, specialization has two distinct functions:
   a The *deepening* of knowledge: the development of a deeper conceptual base, e.g., specializing in carpentry as a way of focusing on the properties of wood.
   b Reconceptualization: a change in the conceptual frame itself.
5 Reconceptualization also has two dimensions:
   a Transfer: the movement of an element from one conceptual frame to another, e.g., the mechanization of pin-heading.
   b Invention: the creation of a new conceptual frame.
6 The problem of reintegration is created by a divergence between upstream conceptual categories and downstream conceptual categories, and especially a divergence between the categories in which production is organized and the categories of consumption.

SUMMARY AND CONCLUSION

To summarize and conclude: what I would like to suggest is that mass production involved growth through the detailed division of labor: growth through reconceptualization and within reconceptualization, primarily transfer. As a result of the transfer the conceptual categories in terms of which the elements of production were performed were distinct from the categories of the final product. This created a problem of reintegration, which led in turn to a hierarchical organization form.

Flexible specialization involves growth through the "social division of labor": the *deepening* of knowledge within given conceptual categories. But the partitioning of the productive process is different from the partitioning of the consumption process, and hence, there is a problem of reintegration, similar to the problem in mass production. Network structure facilitates both the deepening and the integration because, to integrate better with other conceptual specialties, the specialists are forced to develop their own specialty more fully. The conceptual level of understanding in this form of growth permits horizontal coordination, thus avoiding hierarchy, but the degree of interaction across specialties is too intense to permit a market.

The market organization of the kind which the Chicago School envisages would seem to be more likely when the conceptual structure of production maps directly onto the conceptual structure of consumption. Why this should ever be the case, however, is unclear.

8

THE STRUCTURE OF INDUSTRIAL PRODUCTION AND THE BOUNDARIES BETWEEN FIRMS AND MARKETS

Giovanni Dosi and Roberta Salvatore

INTRODUCTION

A puzzling feature of contemporary industrial structure is the relative stability of a skewed distribution of firm- and plant-sizes over all the industrial history for which data can be estimated: approximately, a Pareto distribution. In turn this plausibly implies that the relative frequency of activities and transactions occurring within single organizations as compared to those mediated through the markets has remained roughly constant, despite the secular growth of industrial output and despite the secular tendency toward technical specialization among production tasks (i.e. an increasing "division of labor"). Two other phenomena add to this puzzle. First, *sectoral* distributions may well look different from the aggregate one (over "industry" or "manufacturing" as a whole) and seem to be also more volatile over time: the clothing industry is quite different from that producing mainframe computers, in terms of size distributions, degrees of concentration, and probably also in terms of variability of market shares of individual firms.

Second, despite the statistical aggregate stability of the distribution, the relative ranking of single firms is quite volatile: it suffices to compare the *Fortune* list today with that from one of the earliest years.

The interpretation of this evidence involves major theoretical questions: what explains the approximate borders between market and organizations? How do they change over time? How does one account for the coexistence of widely different sizes even within the same productive activity? How does one reconcile the sectoral evidence on diverse patterns of entry, exit, and changes in market shares with rather stable aggregate size distributions?

Any answer to these questions also entails direct consequences for the "political economy" of contemporary economic systems. In particular, it has been suggested by several authors that the changes in technologies and production organization in the last couple of decades are part of a secular transformation in the boundaries between corporate organizations and markets, biased by novel advantages of smaller-size firms. (And, of course, for this interpretation to hold, one must also assess the stability of these advantages and show their long-term effects on industrial structures.)

In this chapter we shall present some ideas and conjectures for an answer to the foregoing questions, drawing from the evidence and theoretical results which have emerged within "evolutionary" analyses of industrial dynamics. In the second section of the chapter, we shall set the background of our work by briefly discussing some interpretations of firms' boundaries and industrial structure from the economic literature. The next section presents a few elements for a theory of firms' boundaries based on the organizational and technological specificities of their "core competences" and on the features of the competitive environments where they operate. In a fourth section we present a simple stochastic model of firm growth, which, however, maintains some of the stylized phenomena of firm- and industry-specificities identified by "evolutionary" industrial economics. We show via simulations that a Pareto-type size distribution is likely to emerge *by aggregation* of sectoral dynamics which may well diverge from it. Next, we study the consequences for industrial structures of a generalized "technological shock" which, supposedly, "devalues" the competitive capabilities of incumbent firms and plausibly favors smaller entrants. The ensuing changes in the size distributions of firms look somewhat like those which have been empirically identified in recent data. However, we shall argue that permanent competitive biases in favor of smaller-size firms must involve also a continuous "evolutionary superiority" of the organizational and technological knowledge embodied in the entrants *vis-à-vis* the incumbents. This, in our view, is most unlikely. Rather, one may plausibly interpret the current transformations in "industrial demography" as the outcome of deep modifications in the patterns of technological learning and market selection, which, nonetheless, do not eliminate the advantages that size, cumulative learning, vertical integration, and horizontal diversification, under certain circumstances, provide.

REGULARITIES IN INDUSTRIAL STRUCTURES AND INDUSTRIAL EVOLUTION

As mentioned at the beginning of this chapter, the evidence on the size frequencies of business firms shows remarkable stability of skewed distributions, resembling rather closely Pareto distributions (cf. Ijiri and Simon 1977 and the bibliography cited there). Of course, if each firm were specialized in one single product, the aggregate dynamics of size distributions would be simply the outcome of

a) changes in output shares of each firm, and
b) product-specific output dynamics (i.e. demand changes).

However, most firms are multiproduct and, thus, the observed aggregate distributions must also account for the degrees to which firms are vertically integrated and horizontally diversified (i.e. they manufacture various products sold in different markets), and also for the fact that firms may merge, divide, sell off parts of themselves, etc.

For the time being, let us stick to the simpler – and far from realistic – hypothesis of single-product firms. A long tradition, going back at least to Gibrat (1931), has shown that the observed skewed distributions can be generated via a stochastic process of *proportional growth* (plus some additional hypotheses which do not always have a clear empirical meaning, as briefly discussed below).[1]

If $X_{i,t}$ is some measure of the size of firm i size at time t (e.g. turnover, employment, etc.) suppose that

$$X_{i,t} \,/\, X_{i,t-1} = \alpha \,.\, \Phi_i(X_{i,t-1}) \,.\, \varepsilon_{t-1} \qquad (1)$$

where α is the growth rate of the market, common to all firms, $\Phi(.)$ is a function expressing the dependence of current upon past size, and ε_{t-1} is a random multiplicative term. Assuming that

$$\Phi(X_{t-1}) = X_{t-1}^{(\beta-1)}$$

for all firms, and taking logarithms, we have

$$log\ X_{it} = log\ \alpha + \beta\ log\ X_{i\ t-1} + log\ \varepsilon_{t-1}. \qquad (2)$$

The proportional growth hypotheses (Gibrat's law) implies, first, that $\beta = 1$ (i.e. the rates of growth do not depend on initial size) and second, that $\log \varepsilon_t \cong N(o, \sigma^2)$ is independently-distributed and in particular is independent of initial size. With some additional assumptions in order to prevent the variance of equation 2 from exploding this equation[2], one actually obtains a skewed limit distribution (a Yule or a Pareto distribution) which appears to fit rather well aggregate empirical data (Ijiri and Simon 1977). "Gibrat's Law" has also been tested in various forms on more disaggregate data, with rather mixed outcomes. In particular, most evidence seems to reject the hypothesis of independence of the variance of growth rates from initial size, and some evidence appears also to restrict the applicability of the "law" to particular size classes.[3] We are not going to discuss here the details of either Gibrat type stochastic models of firm growth or the supporting evidence. A few general remarks will suffice.

Neutral combinations between market and organization?

First, a general hypothesis of independence of growth rates from the size that firms have already achieved seems, at least *prima facie*, at odds with a Marshallian theory of production based on U-shaped cost curves.[4] More generally, the few attempts at accommodating the evidence with equilibrium models of industrial structures seem to us rather far-fetched: they rely, for example, on *ad hoc* auxiliary assumptions on capital adjustment and unobservable distributions of scarce managerial talents (Lucas 1978) or on stochastic processes of convergence to supposedly given optimal techniques under imperfect information and perfect rationality (Jovanovic 1982).

Indeed, the most likely candidates for a theory of production compatible with Gibrat-type models imply constant returns to scale: no persistent scale-biased factor influences growth opportunities (an exception being Kalecki's model with increasing constraints to growth related to financial factors). At the extreme, all possible combinations between organizations and market are neutral equilibria. That is, all possible distributions of industrial structures, ranging from monopoly to perfect competition, can be expected *ex-ante* to show identical efficiency properties.

The "structure" (i.e. the relative invariant size distribution which data reveal) can be theoretically generated by some random process with little or no past memory,[5] *plus* some auxiliary restrictions (such as on birth and death rates). In this case, the random process stands for a multiplicity of minor factors which temporarily make one firm more (less) efficient than the competitors, but are *not* persistent over time. (These considerations, *mutatis mutandis*, apply also to stochastic models based on serial correlations of growth rates with decaying effects over time.) Alternatively, one could also interpret the random process applied to growth rates not as the outcome of a multiplicity of small efficiency-related "shocks," but rather as the random degrees of success of some *extra-economic* variables. In fact, this interpretation can easily be drawn from "radical" theories of the firm (see Marglin (1974) and the discussion in Putterman (1986)). Firms exist essentially because of an asymmetric distribution of power in society. Hierarchical control is primarily finalized to the extraction and the appropriation of an economic surplus. Plausibly, firms grow as a function of the (stochastic) success of their owners in accumulating and concentrating capital in their hands. Indeed, we do not have any difficulty in accepting factors related to power and control among the determinants of the characteristics of firms and industries as we see them, beyond and in addition to efficiency considerations. However, we are rather skeptical of a general "neutrality" hypothesis (i.e. "all forms of industrial structures are identical in terms of efficiency"), either in the Gibrat-type interpretation or in the "radical" one. To trivialize, it is difficult to imagine a Boeing aircraft produced with the same efficiency in an Italian industrial district by thousands of family-run firms, or, conversely, a (now defunct) East German Kombinat producing fashion garments with the same efficiency and creativity as an Italian district. More generally: microeconomic evidence, admittedly quite impressionistic, seems to suggest that in each industrial activity there are some "typical ways" of organizing production. There can be more than one, but not very many. That is, there can be a few combinations between organization and markets that are "neutral" in terms of efficiency, but the set is small and possibly not dense. This also implies that at a disaggregated level one may not expect to find either the structure or the Gibrat-type growth that one sees in the aggregate.

Industry and firm specificity

Second, a few analyses within and outside an "evolutionary" approach (cf. Nelson and Winter 1982, Dosi et al. 1988, and the review in Dosi 1988), have emphasized the sectoral specificities in the process of learning, innovation, and imitation. If, as is likely, there is a direct link between innovation and growth, these specificities of learning ought to have a direct link with the "explanation" of why industrial structures are what they are and how they change. In particular, it has been suggested that various activities present different degrees of *technological opportunity*, that is, a different scope for improvements in products and processes on the grounds of a given knowledge base, and also a different "easiness" in achieving these improvements. (One way of formalizing these differences is in terms of the probability distributions of achieving an *x* percent improvement in some performance characteristics of a product or a process, given some search effort, say a one dollar investment in R&D). Technologies and sectors have also been found to differ in terms of *appropriability* conditions, i.e. the effectiveness through which innovations can be shielded from imitation by competitors. Moreover, according to the features of the process of innovative search and the knowledge upon which it draws, the innovative success of individual firms may show varying degrees of serial correlation, often referred to as the *cumulativeness* of innovative activities. Finally, it has been argued elsewhere that radical changes in the knowledge bases, procedures, and directions of technological change, i.e. a change in the "technological paradigm" (Dosi 1982) also imply significant changes in the relative innovative capabilities and growth potential of firms embodying "old" or "new" competences.

These findings have direct implications on the evolution of industrial structures which can be easily seen by comparison with the stochastic interpretation discussed earlier.

- Differences across industries in technological opportunities (assuming some positive correlation between innovation and growth) imply different means and variances of the stochastic term in equation 1. Moreover, if "opportunities" change over the development of each technology, one should also expect correlated changes in the distribution of the ε_t.
- Appropriability of innovation, cumulativeness of technological advances, and tacitness of some elements of technological knowledge imply some *firm-specificity* in the factors that foster or hinder growth. The findings from the economics of innovation, as well as complementary findings from business studies, suggest that Φ_i (.) functions in equation 1 can hardly be assumed to be identical across firms. Some firms may be persistently "better" or "worse" than others. In turn, that means that

every pattern of evolution of industrial structures must not only be interpreted in terms of a multiplicity of little transitory shocks on individual growth opportunities, but also account for the interaction among firms that may be *persistently different* in their technological and organizational characteristics.

- Differential degrees of appropriability and cumulativeness, and different characteristics of the knowledge underlying innovative activities are also likely to imply a different *entry condition* in each industry. For example, in industries characterized by highly idiosyncratic knowledge, by serial correlation in learning, and by high appropriability of innovations, one may expect the rates of entry to be low. The opposite should hold for industries where innovation can draw on a variety of sources of freely available knowledge, appropriability is low, etc.[6]
- Abandoning the assumption of single-product firms, the variegated sources and procedures of learning and innovation are bound to affect the corporate patterns of horizontal diversification and vertical integration and, thus, also the structure of individual industries and of manufacturing as a whole.

To summarize: the microeconomic findings on the innovation process highlight rather permanent factors accounting for learning specificities of both firms and industries. If, as is the case, innovation is a crucial ingredient of corporate growth, then a satisfactory interpretation of the emergence of the observed industrial structures should account for such processes of differentiated learning and the ensuing competition among heterogeneous agents.

Such an interpretation ought to have two major components: first, an account of the "structures" generated by learning and market selection dynamics in individual industries, and second, an explanation of the interlinked dynamics of firm diversification, vertical integration, specialization, entry, and exit from the various markets. Let us start from the latter.

THE STRUCTURE OF MODERN CORPORATIONS: THEIR CHANGING BOUNDARIES AND THEIR COHERENCE[7]

There are at least two remarkable characteristics of the boundaries of the modern corporation that ought to intrigue theorists:

a) their multiproduct scope, and relatedly
b) the non-random distribution of product porfolios across firms.

That is, the activities of the modern corporation display a considerable degree of relatedness, what we call coherence: thus Shell Oil Company is primarily in the oil business, IBM is in information technologies, Intel is in semiconductors, ICI and Dupont are primarily in chemicals, and Boeing is in areospace. Setting aside for the moment the conglomerate form of business organization, what is remarkable about the majority of firms is their relative "coherence" in their lateral and horizontal business activities.

Admittedly, the boundaries of the firms are sometimes quite fuzzy, with interfirm agreements, joint ventures, and consortia rendering precise delineations of the firms' activities imprecise. Still, even interfirm agreements, joint ventures and consortia exhibit a degree of coherence with other activities in which associated firms are involved. At a very superficial level, the most salient attribute of diversification is that it tends to increase over time, and new product lines bear certain technological and market similarities with the old. Thus firms generally begin with a single product and subsequently become multiproduct rather than the other way around; and new product lines very often, though certainly not always, utilize capabilities common with existing product lines. Indeed, many prescriptions in business economics with respect to diversification strategies are characterized by their emphasis on the desirability for firms to "stick to their knitting," though such prescriptions rarely specify what the knitting is to which firms should stick.

The casual empiricist is also likely to remark upon the fact that rates of diversification vary among firms. Thus Western Union has been in the telegraph business and little else for over a century. Shell Oil has been in petroleum for almost as long, having added petrochemicals only in the postwar years. 3M, on the other hand, has diversified into an extremely broad range of coatings, adhesives, and related products in the last two decades. Corporate diversification is also a phenomenon that can expand and contract. Firms not only add business, they also commonly divest. Indeed, there often appears to be a degree of circularity of the fashion in which new businesses are added and subsequently divested. This is particularly so when corporations diversify through acquisition.[8] On the other hand, corporations often display remarkable similarities with respect to their diversification strategies. Almost everywhere car and truck companies have diversified into buses, e.g. Mercedes, Leyland, Fiat, GM, and Renault all make cars and buses. The challenge is to develop a theory to explain not only coherence, but also similarities and differences in the nature of coherence across firms and industries.

Coherence, we should point out, is different from "specialization". Specialization refers to the performance of particular tasks in particular settings. This is the manner in which Adam Smith implicitly defined the concept, and it is the way we will use it. In the famous pin making example, Smith referred to specialization in terms of particular tasks. Thus a firm may exhibit coherence though it may not necessarily be specialized. We are, accordingly, defining coherence in a multiproduct sense. A firm exhibits coherence when its lines of business are related, in the sense that there are certain characteristics common to each. Coherence is increased with the number of common characteristics present across lines of business and the level to which they are present. A corporation fails to exhibit coherence when common characteristics are allocated across firm's lines of business in no greater frequency and to no greater depth than could be explained by chance.

Neoclassical theories of industrial organization, including the so-called new industrial organization (e.g., Tirole 1988) has little to say about diversification and integration issues. While there is some development of market power and economies of scope explanation for diversification, these theories are without strong organizational implications. Thus one can "explain" diversifications through notions of "full line forcing" or "economies of scope," but such theories rarely explain why a set of contracted relationships among specialist firms could not accomplish the same objectives. In this regard, consider economies of scope. If the production of x_1 and x_2 involves scope economics, then:

$$c(x_1,x_2) < c_1(x_1,0)+c_2(0,x_2) \quad (2)$$

However, in standard treatments, the organizational discussion of the joint cost function $c(x_1,x_2)$ is never explained. The amalgamation implied by the cost function could be accomplished by a variety of organizational mechanisms, including contracts (Teece 1980, 1982). A theory of multiproduct organizations thus does not emerge from the economics of scope literature or from other theories of the firm.

Admittedly, transaction costs considerations contribute significantly to explaining the degrees of *vertical* integration, that is "make or buy" decisions with regards to the firm inputs (Williamson 1975, 1985). However, exchange relations – from which transaction costs may derive – alone, do not appear to be the crucial domain in which to search for the causes of the *horizontal* boundaries of the firm. Drawing from another work of one of us, we advance the fundamental proposition that the boundaries of the corporation need to be understood not only in terms of transaction cost considerations, but also in terms of learning, path dependencies, technological opportunities, selection, and complementary assets (Dosi, Teece, and Winter 1992).

Learning

A fundamental characteristic of economic activity is that it provides the opportunity for learning. Learning is a process by which repetition and experimentation enable tasks to be performed better and faster and new production opportunities to be identified. In the context of the firm, if not more generally, learning has several key characteristics. First, it is generally cumulative. What is learned in one period builds upon what was learned in earlier periods. Individual knowledge gained through learning must, however, be constantly used in order to be preserved.

Corporate learning involves organizational rather than individual skills. While individual skills are of relevance, their value depends upon their employment in particular organizational settings. Learning processes are intrinsically social and collective phenomena. Learning occurs not only through the imitation and emulation of individuals as with teacher–student or master–apprentice, but also because of joint contributions to the understanding of complex problems. Learning requires common codes of communication and coordinated search procedures.

The knowledge generated by such activity resides in organizational "routines." Routines are patterns of interaction which represent successful solutions to particular problems. These patterns of interaction are mainly resident in group behavior, though certain subroutines may be resident in individual behavior. Because of the complexity of such behaviors, knowledge embedded in routines cannot be fully captured in codified form. That is, it has a tacit dimension that often cannot be readily articulated. Hence, it is the routines themselves, and the ability of management to call upon the organization to perform them, that represents an organization's essential capability.

Routines can be of several kinds. "Static" routines embody the capacity to replicate certain previously performed tasks. Needless to say, such routines are never entirely static, because with repetition routines can be constantly improved. "Dynamic" routines are directed at learning and new product/process development. Thus R&D activity proceeds through the employment of routines to ascertain where to probe, how to probe, and how much to probe.

Because routines involve a strong tacit dimension, they may not be easy to imitate. To the extent that it is so, routines contribute to a firm's distinctive competences and capabilities. Such capabilities, by virtue of their evolution in particular environments and organizational contexts, are likely to differentiate firms from each other and provide the basis for differential performance vis-à-vis competitors.

Path dependencies

In standard textbook theories firms have an infinite range of technologies from which they can choose, and markets that they can occupy. Changes in product or factor prices will be responded to instantaneously, with technologies moving in and out according to a value maximization criterion. Oliver Williamson (1975, 1985) has done much to enrich the standard treatment by introducing transaction-specific assets. Such assets have idiosyncratic uses and cannot be redeployed without substantial loss of value. Williamson demonstrates how the need for such assets generates particular requirements with respect to an industry's governance apparatus.

The notion of path dependencies goes even further than notions of irreversibilities implied by transaction cost economics. It recognizes that "history matters." Thus a firm's previous investments and its repertoire of routines (its "history") constrains its further behavior. This follows because learning tends to be *local*. That is, opportunities for learning will be "close in" to previous activities and will thus be transaction- and production-specific (Teece 1988). This is because learning is a process of trial, feedback, and evaluation. If too many parameters are changed simultaneously, the ability of firms to conduct meaningful quasi-natural experiments is attenuated. Thus, if many aspects of a firm's learning environment change simultaneously, the ability to ascertain cause–effect relationships is confounded because cognitive structures will not be formed and rates of learning diminish as a result.

Complementary assets

Related to path dependencies is the existence of complementary assets. Prior product and commercialization activities require and enable firms to build such complementarities. Such capabilities and assets, while necessary for the firm's prior activities, may have other uses as well. Such assets typically lie downstream from product-process development in the value-added chain.

The presence of complementary assets, particularly those that can be enhanced by innovation, helps steer the evolution of new technologies. The reason why Singer Sewing Machine got into the furniture business was related to the fact that it had designed and manufactured sewing machine tables to help facilitate the sale of sewing machines. Thus, firms' path dependencies are not just technological in nature. Paths are generated by technological trajectories and constrained further by the complementary assets the firm develops over time.

Technological opportunities

The concept of path dependencies can be given further meaning through the consideration of an industry's technological opportunities. As mentioned above, how far and how fast a particular area of industrial activity can proceed is due to the technological opportunities that lie before it. Such opportunities are partly a lagged function of foment and diversity in basic science, and partly the outcome of search activities by firms themselves. The depth and width of technological opportunities in the neighborhood of a firm's prior research activities are likely to affect a firm's options with respect to both the amount and level of R&D activity that it can justify.

Convergence of path dependencies

While path dependencies are unique, they do not necessarily run in parallel. They can both diverge or converge. The latter are of special interest, as the confluence of particular evolutionary paths (or "trajectories") can have important organizational ramifications.

Convergence may occur because of serendipity of knowledge or as an evolutionary outcome of a series of unique historical events. Thus, for example, the convergence of computers and telecommunications is the result of the application of digital microelectronics to both fields.

The survivability of a business activity depends on the selection environment and, in particular, the level of competition (both in the product market and the capital market), public policy, and the frequency of technological discontinuities (i.e. changes in the dominant "technological paradigms").

Selection environments

Firms sell their products in markets where customers' average response to relative price changes (within the industry or among different suppliers) and to changes in relative product performance may well differ. Economic notions such as "own-price elasticity of demand" try to capture the former phenomenon. Indeed, the sensitivity and speed of demand adjustment to variations in prices and product-quality can be interpreted in terms of the "tightness" of the selection environments within which firms operate.

Selection may or may not be "tight". Whenever it is not, less efficient firms may be able to survive and possibly even prosper for quite some time. The selection environment for a product may be stronger or weaker than that for a corporation depending on how corporate management treats internal divisions and products that are not performing up to par.

The availability of free cashflows is perhaps the key regulator of selection. If a firm or a business is generating sufficient cash flow to sustain itself and take care of reinvestment needs and opportunities, it will typically be able to survive. If it does not, then it will need additional debt or equity financing. To obtain funding of either kind exposes the firm to capital market discipline that might not otherwise intervene. Hence, the capitalization and the "inherited" capital structure of the firm are likely to affect the strength of selection processes with respect to particular business units.

Core competences

A core competence (Teece 1988) is a set of differentiated technological skills, complementary assets, and routines that provide the basis for a firm's competitive capacities in a particular business. "These people are good at X" summarizes external perceptions as to the nature of these competences. Typically, such competences have an important dimension, making replication by others difficult but not impossible. Indeed, when a core competence exists, its replication proceeds either through deliberate efforts directed at internal expansion or through the exit of key individuals who collectively take certain competences with them, and then raise capital to buy the necessary complementary assets.

Great companies like IBM, Exxon, and Toyota consist of clusters of such competences. Hollow corporations, on the other hand, lack such capacities. Business gets performed under contractual agreements struck with others. When the contracts terminate, the firm's ability to deliver value evaporates. We define a hollow corporation as a business entity that does not have any core competence and uses contractual mechanisms to link particular market requirements with productive capacities. Thus a hollow corporation is a nexus of contracts for design, manufacture, distribution, and service.

Likewise, a conglomerate is a portfolio of autonomous business units held together in a holding company. It is a "mutual fund" of business without core competences at the corporate (holding company) level. As discussed in Dosi, Teece, and Winter (1992), we do not believe that organizations such as those that lack core compences have survival properties unless they are somehow shielded from product and capital market competition.

Proposition from the theory

A firm consists of a cluster of core competences and supporting complementary assets. The degree of coherence displayed among them depends on the interaction between learning, path dependencies and opportunities, inherited complementary assets, and selection. Specifically:

Proposition 1: Specialist firms With rapid learning and tight path dependencies, one will observe "single" product (specialist) firms growing rapidly. Compaq Computer, Sun Microsystems, Intel, Boeing, and Airbus are good examples.

Proposition 2: Vertically integrated firms With slow learning but high path dependencies and specialized assets, one can expect specialist firms with some degree of lateral integration and significant vertical integration (e.g. Exxon, Shell, Dupont).

Proposition 3: Coherent diversification With rapid learning, broad path dependencies due to the presence of generic technologies, and tight selection, one can expect to see coherent diversifiers, like IBM, Hewlett Packard, W. R. Gore, and Raychem.

Proposition 4: Conglomerates With high path dependencies, slow learning, and weak selection, we predict the emergence of conglomerates or other highly diversified companies displaying little intercorporate trading and technological transfer.

Proposition 5: Network firms With rapid learning, colliding path dependencies and tight selection, one can expect to see incumbent firms becoming enveloped in a dense skein of intercorporate relationships involving partial equity holdings or joint learning. Thus, because of the convergence of telecommunications and computers, AT&T is becoming embedded in a thicket of intercorporate relationships. So are biotechnology firms like Genetech and Cetus because the biotechnology research paradigm yields products and processes competitive with pharmaceuticals and chemicals, to name just two examples. If learning in the relevant paths were slower, then there would be a good possibility that firms could diversify internally without the assistance of interfirm agreements.

Proposition 6: Hollow corporations With converging path dependencies, opportunities may arise for entrepreneurs to use contractual mechanisms quickly to assemble diverse capabilities directed at the development and commercialization of a particular product. We refer to such entities as hollow corporations. Unless these capabilities are underpinned by equity links and unless distinctive organizational reactions emerge to "glue" such organizations together, they will not survive, except where the selection environment is weak.

In Dosi, Teece, and Winter (1992), the details and a few interpretative consequences of these propositions are discussed. In this view, the coherence (or lack thereof) of the modern corporation is thus explained by the characteristics of learning (or its absence), path dependencies, and complementarities, mediated by transaction costs and market selection. Pavitt *et al.* (1987), in a work highly complementary to this interpretation, find that the patterns of technological and market diversification can be approximately mapped into a taxonomy of learning and innovation modes.

For the purposes of this work, two major consequences can be drawn. First, the nature of the core competences of each corporation and of the learning regimes within which it operates explains to a good extent its horizontal and vertical boundaries. Second, the observed relative stability of the output shares of the larger firms in manufacturing as a whole implies indeed that *on average* the characteristics of learning and market selection have not dramatically changed. Manufacturing firms – especially large firms – have been able to grow, in the markets in which they were originally established and, even more important, via diversification and vertical integration, at rates, on average, comparable to the overall market growth.

However, it follows from our interpretation that the scope for growth via horizontal diversification is not uniform across firms: on the contrary, it depends, again, on features of the "technological paradigms" which they master. The evidence on the long-term changes in the relative ranking among the largest corporations, witnesses, in our view, precisely their diversity in the opportunities and capabilities of efficient diversification (jointly, of course, with the different rates of growth of the markets in which they operate).

Clearly, each strategy of diversification or vertical integration is as such a decision of entry into an industrial activity. These strategies, together with divestment decisions, entries of new firms and bankruptcies determine the *changing demography* of individual industries. In turn, such demographic changes, jointly with the variables affecting the variations in market shares of each firm, determine the size distribution of business units in each industry.

INDUSTRIAL EVOLUTION AND INDUSTRIAL STRUCTURES

A major claim of evolutionary theories is not only that firms differ in their competences, innovative success, and ultimately, growth opportunities, but also that industries differ in their learning modes and in the ways diverse learning abilities affect competitiveness and market shares. In this respect, the taxonomic exercise by Pavitt (1984) suggests the existence of broad groups of industrial activities differentiated by the sources of innovative opportunities, and the size distribution of the innovating firms. Malerba and Orsenigo (1989) find robust evidence mapping the characteristics of the innovating firms into the features of sectoral *technological regimes* (identified by proxies for technological opportunities, appropriability conditions, the nature of the knowledge base, and the cumulativeness of innovative success). Gort and Klepper (1982) and Gort and Konakayama (1982) identify relatively robust patterns of evolution of the supply structure along the "technological life cycle" of a sample of products. Literature on the "population ecology" of firms has emphasized:

a) the sectorial differences in the rates of birth and death of firms; and
b) the significant effects of the age of firms on their growth (which often turn out to present a negative sign).[9]

Most likely, different learning regimes must also imply intersectoral differences in the variability of market shares of incumbent firms (i.e. what Caves and Porter (1977) call *mobility barriers*) and in the rates of entry and exit of new business units in each industry. Indeed, Audretsch and Acs (1990) find a significant empirical difference in industrial "turbulence" (defined as a synthetic measure of entry, exit, and changes in market shares of incumbents) between those industries characterized by a regime of "cumulative" learning by incumbents and those others where new entrepreneurial activities are more likely to carry innovative advantages.

Clearly, all this variegated empirical evidence *prima facie* conflicts at least with the simplest Gibrat-type accounts of firm growth:

a) firms (and business units) are not identical in their growth opportunities, and
b) the distribution of growth possibilities among different types of firms vary across sectors and across "regimes of innovation."

Then, how does one explain the observed industrial structures? Ideally one should be able theoretically to generate these structures on the grounds of an explicit representation of firm-specific learning process and a selection mechanism which favors (or hinders) the growth of business units characterized by above-average (or below-average) competitiveness. Consider, for example, the dynamics of market shares as from Silverberg, Dosi and Orsenigo (1988):

$$f_i(t) = A[E_i(t) - \bar{E}_i(t)]f_i(t)$$

where $f_i(t)$ is the market share of the i-firm at time t; $E_i(t)=G(\mu_i(t))$ is the competitiveness of firm i – which in turn is a function of some vector of its technological and organizational characteristics $\mu_i(t)$; $E_i(t)$ is the average competitiveness of the industry; and the dots stand for the rates of change. The composition of the vector μ, its changes over time and the $G(.)$ function depend in principle also on the learning regimes of each industry. That is, the learning mechanism specific to each technological paradigm governs the transition probabilities in $\mu_i(t)$. A necessary condition for a Gibrat-type process of growth is that

$$(\mu_i^j)_{t+1} = (\mu_i^j)_t + \varepsilon$$

for every j component of μ, with ε an independently and identically distributed (i.i.d.) random variable. However, one can easily imagine different types of dynamics, ranging between two extreme archetypes. In the first one somewhat resembling an extreme version of the "first" Schumpeter (1934), $(\mu_i^j)_{t+1} = (\mu_i^j)_0$, $\forall t$: no learning occurs and industrial innovation entirely relies on the "creative destruction" associated with the entry of new firms. The other archetype is that whereby

$$P(\varepsilon > 0 | \, (\mu_i^j)_t = \varphi((\mu_i^j)_t):$$

learning is cumulative and the transition probabilities are some positive function of the levels of competitiveness already achieved by each firm, (somewhat resembling the "second" Schumpeter (1942).

Indeed, a major theoretical task ahead is to study the limit distributions in market shares according to different learning and selection processes. Falling short of that, here we present a model and some simulation results that try to capture by some proxies the specificities of learning and selection regimes.

Let us start from the following hypotheses:

a) In cumulative learning regimes the probabilities of exit (that is of "death" of a business unit which may or may not be equivalent to the death of the firm) *decreases with the age of the firm* (metaphorically, the older one gets the more experienced one becomes and more able to cope with unexpected circumstances . . .) The opposite applies whenever discontinuities in the knowledge base make the skills of older firms obsolete.
b) The probabilities of death *decrease with size* if there are some economies of scale in production, R&D, etc. At the opposite extreme, they increase with size if scale mainly induces organizational rigidity and inefficiency.
c) The variance of growth rates *decreases with age* (which captures the idea that older firms are relatively more stable in the competences that they incorporate and the market niches that they occupy).
d) The birth rates are proportional to the number of incumbents in each industry (for some empirical evidence cf. Dunne, Roberts and Samuelson (1988)).

Formally, the survival probability of firm i from time t to $t+1$ in sector k is

$$E[I|a^i_t;s^i_t]=Pr\{e>p^k(a_t;s_t)\} \tag{3}$$

where I is a dichotomous variable (0 or 1); a stands for age and s for size, e is a random variable uniformly distributed in {0;1} and $p^k(a_t;s_t)$ is the survival threshold specific to each sector.

In turn, such a threshold is defined as

$$p^k(a_t;s_t)=\frac{\left[a_t+\dfrac{1}{B(\alpha;\beta)}\;s_t^{(\alpha-1)}(1-s)_t^{(\beta-1)}\right]}{Max\left[a_t+\dfrac{1}{B(\alpha;\beta)}\;s_t^{(\alpha-1)}(1-s)_t^{(\beta-1)}\right]} \tag{4}$$

were $B(\alpha;\beta)$ is a Beta function.[10]

New firms enter according to

$$n_t=(r^k+\theta)V_{t-1} \tag{5}$$

where n is the number of entrants (which we shall assume to belong to the smallest size class), V are the incumbents and θ is a random variable with mean zero.

Regarding firm growth, Salvatore (1990) experiments with the alternative hypotheses of

a) a Gibrat process (i.e. independence of growth rates from initial size);
b) serial correlation; and
c) a one-period Markov process.

Here, we present the results of the latter formulation only (however the major conclusions hold for all three alternative specifications). Each firm – whose size is supposedly measured in terms of employees – grows according to

$$g_{i,t} = g^{\delta}_{i,t-1}\varepsilon_{i,t} \tag{6}$$

where g is the growth rate of firm i and $\varepsilon_{i,t} \cong N(O;\sigma_a)$ is a normally distributed random variable whose variance decreases with the age of the firm.

In this model, the evolutionary specificities of each sector are roughly captured by the influence of size and age on survival probabilities. By trying different combinations of α and β in equation 4 one can generate four "archetypical" sectors, with "sector 1": decreasing returns to scale (and, thus, "evolutionary" advantage of smaller sizes); "sector 2": increasing returns to scale (i.e. higher probability of survival of bigger firms); "sector 3": a U-shaped probability function (small and very big firms have the highest survival probability, which loosely recalls Pavitt's description of "science-based" sectors industries, suggesting the permanent coexistence of big oligopolies and small "Schumpeterian" firms); "sector 4": an inverted U-shaped function (i.e. there is plausibly a minimum efficient scale, beyond which organizational inefficiencies start operating). Moreover, age acts in equation 4 as a shift parameter which moves the survival function upward or downward and should implicitly account for the positive or negative "evolutionary value" of past experience. Hence, one may get 12 typologies of industrial evolution (the four basic relationships between size and probability of survival combined with positive, negative or null effect of age).

We do not have any precise idea on the empirical frequencies of the various "types," but plausibly they tend to capture a good deal of the empirical variety of industries, ranging from cases where both scale and cumulative experience determine a differential competitiveness to the symmetric opposite wherein smallness, flexibility and "novelty" are major evolutionary advantages.

Given 11 size classes, the simulations study the size distributions after 100 runs, starting with a uniform distribution across classes. Interestingly, most distributions so generated depart to a greater or less extent from the Pareto distribution with a density function

$$f(x)=qx^{-(1+q)}$$

and a right-cumulated function

$$F(x)=x^{-q}$$

As an illustration, we present in Figures 8.1 to 8.4 the actual distributions and those obtained with OLS estimates of the logs of equation 7 on the four "archetypical" sectors (assuming no effect of age on survival thresholds).

There is no striking conflict between these results and the little we know about the variety of industrial structures at a very disaggregated level: certainly, it does not contradict the evolutionary proposition that the "boundaries" between organizations and markets are influenced by the industry-specific features of learning and selection.

Further, we suggest that

a) manufacturing-wide Pareto distributions are the *outcome of aggregation* over widely different evolutionary processes and industrial structures, and
b) such aggregate distribution tend to be maintained as long as *on average* the conditions of learning and selection do not dramatically change.

To see the plausibility of the first proposition, let us extract 20 samples of 100 firms each from the final-year distribution of firms in the twelve "types"

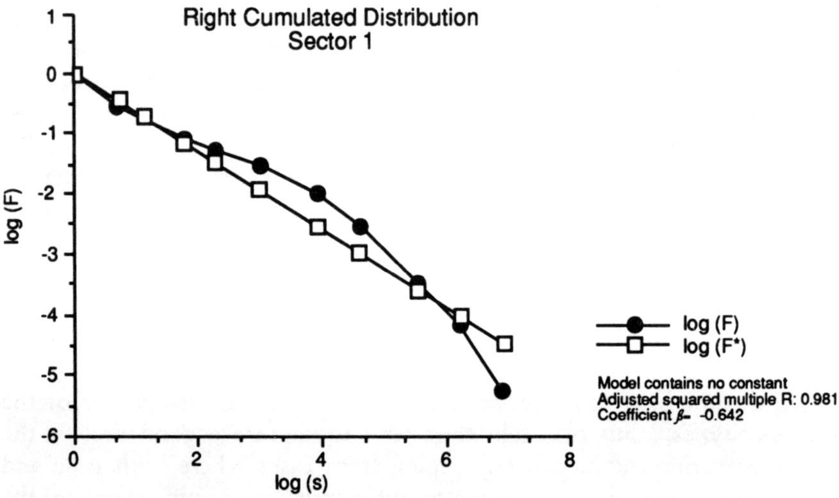

Figure 8.1 Right cumulated distribution (Sector 1)

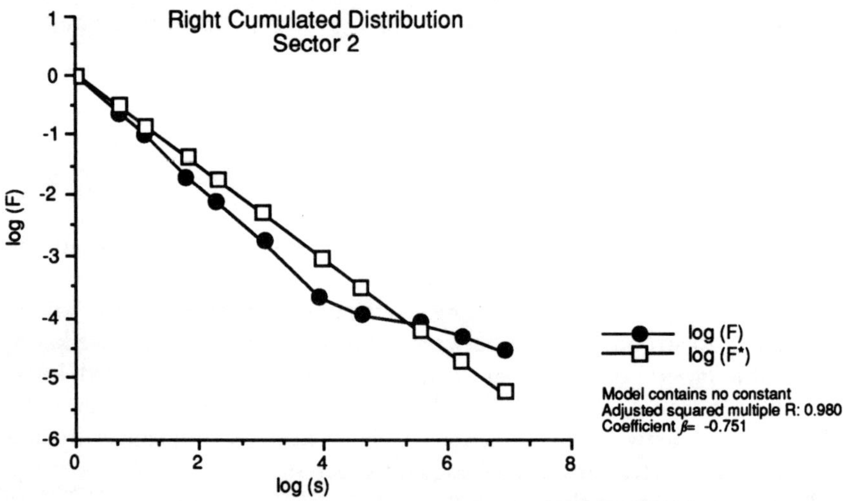

Figure 8.2 Right cumulated distribution (Sector 2)

of industries that the model generates. As can be seen from Fig.5, the fit of a Pareto distribution is quite close, even if it was not so in most of the underlying sectors.

Next, let us ask: what would happen if the general conditions of learning and selection underwent a generalized shock? After all, a few interpretations

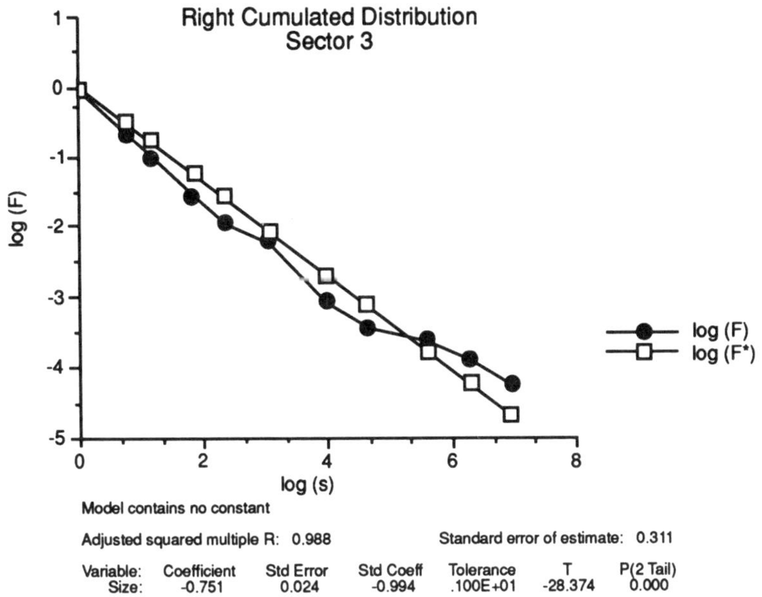

Figure 8.3 Right cumulated distribution (Sector 3)

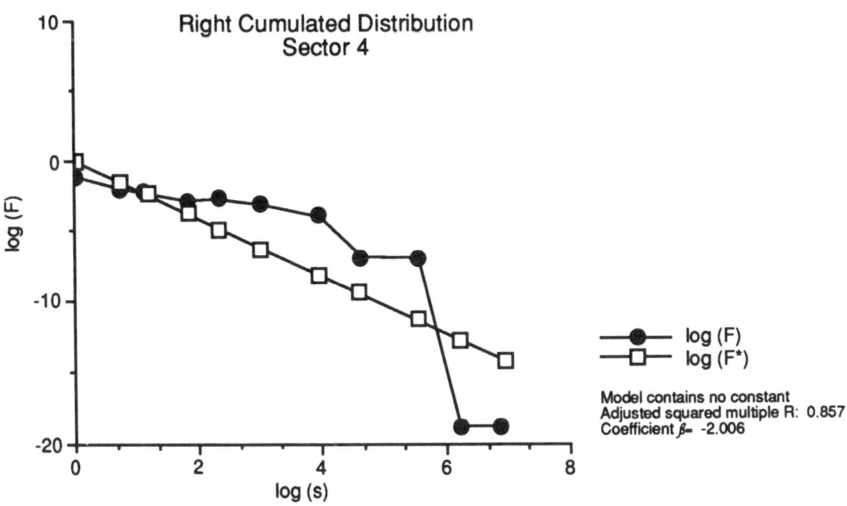

Figure 8.4 Right cumulated distribution (Sector 4)

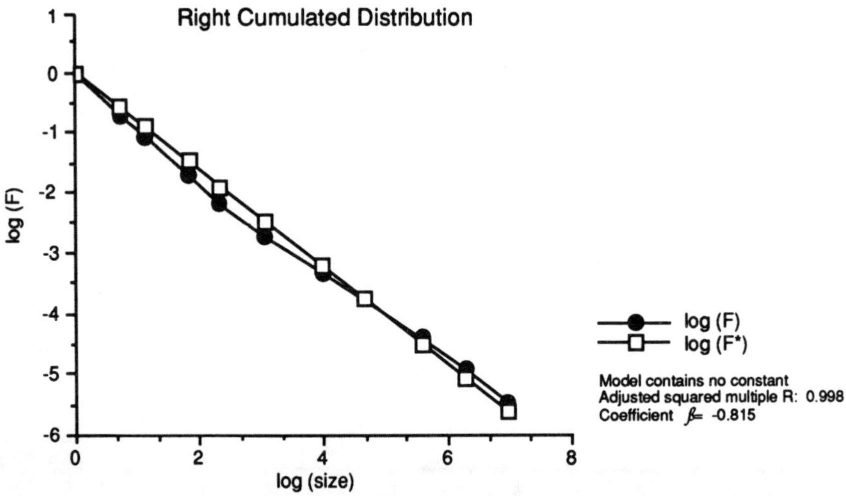

Figure 8.5 Right cumulated distribution

of the current patterns of "restructuring" in Western economies argue precisely that one is observing a major change in the "technoeconomic paradigm," associated with information technologies and, with a "post-Fordist" organization of production. (See for example, Freeman and Perez (1988)). All this – some add – implies a general competitive advantage for *smaller* firms (more rigorously, smaller business units) (Cf. Piore and Sabel 1984).

In this interpretation, one would predict that, first, also aggregate deviations from Pareto distributions, and, second, that these departures have a permanent nature. One actually observes in the 1970s and 1980s significant aggregate deviations from Pareto distributions (for evidence on Italy, see Barca (1985)), indicating a higher-than-average probability of growth of smaller size units. In our view, however, any inference on a competitive bias systematically favoring smaller size units might be deeply misleading.

It has been argued earlier that any observed structure is influenced by both age and size (via their effects on the probability of survival and growth). Suppose that, in fact, the emergence of a new "technoeconomic paradigm" based on microelectronic technologies has partly changed the knowledge base of innovation and widened the domain of unexplored technological opportunities. All this has plausibly enriched the opportunity of entry of new firms, and possibly also challenged the technological and organizational skills of older incumbents. In our model, this economy-wide change in learning regimes associated with the "devaluation" of older skills, may be captured by a negative influence of *age on growth*:

$$g_{i,t} = g_{i,t-1}{}^{\delta(a,i)} . \varepsilon_t$$

Table 8.1 Deviations of actual aggregate size distributions from implied OLS estimate of Pareto distribution

Range	*f(i)*	*f*(i)*	*deviation*
1	43.50	49.79	–12.63
2	17.00	16.65	2.09
3–5	17.84	16.69	6.89
6–9	8.27	6.72	23.09
10–19	7.50	5.05	48.37
20–49	3.61	3.04	24.36
100–249	0.61	0.62	–0.83
250–499	0.23	0.21	11.54
500–999	0.08	0.10	–27.60
1000+	0.01	0.10	–90.60

We have simulated 10 years of such "shock" on growth opportunities and then calculated the deviation of the actual aggregate size distributions from the implied OLS estimate of a Pareto distribution.[11]

Table 8.1 reports the results, which look quite like the evidence reported by Barca (1985) on Italian Census data: the biggest business units grow less than average while small to medium units grow above average. However, in our interpretation, this is the outcome of technological novelty and *not of size* as such. New firms are generally smaller, but their advantage does *not* reside in their "smallness', but in their "newness" (they are carriers of technological and organizational innovations allowed by the new "paradigm").

Under what circumstances can this bias be permanent? If our approach is correct, a permanent bias in favor of relatively "new" (and, *as a consequence*, smaller) firms would require

a) the general disappearance of economies of scale in production, marketing, R&D, etc., and,
b) the total lack of path-dependency and cumulativeness in learning.

We consider both circumstances as highly unlikely. Rather, we suggest, the process of exploration of a new "technoeconomic paradigm" is eventually leading to new "trajectories" of path-dependent learning. Older organizations and new ones that have become big as a result of their innovative success, most likely, are progressively internalizing the competences associated with the new paradigms. As a consequence, one should expect the growth process to return to a form similar to that described by equation 6.; and – as simulations not presented here show – the *aggregate* industrial structure to approach again a Pareto distribution.

SOME CONCLUSIONS

We have argued in this work that the nature of technological and organizational learning, jointly with the process of market selection, underlie the explanation of the observed boundaries between the activities that are internalized within corporate organizations and those that are mediated through the market. The characteristics of *learning and selection* account for the approximate scope for diversification by multiproduct firms. These regimes of learning and selection, specific to each industrial activity, determine also the variety of industrial structures that one observes. The relative stability over time of a Pareto-type distribution of firms (and business units) – it has been tentatively suggested on the grounds of model simulations – may simply result by the aggregation over sectoral structures showing significantly different boundaries between organizations and markets.

Further, we have argued, economy-wide discontinuities in the sources of innovation are likely to "shock" the size distribution of firms and business units by enlarging the "window of opportunity" of new entrants. However, long-term changes in the relative efficiency and competitiveness of small and younger firms *vis-à-vis* bigger and older ones would imply also a long-term fall on the importance of innovative experience and cumulative learning. Indeed, we are rather skeptical on any proposition suggesting that contemporary environments have become more "Darwinian" than in the past (that is, "learning" counts less and random mutation counts more; organizations decrease their abilities of coping with turbulent environments as compared with daring individuals . . .). Rather, we interpret the evidence on the industrial restructuring of the 1970s and 1980s as witnessing profound changes in the sources of corporate competitiveness, which – progressively – *older and younger* organizations are learning to master.

ACKNOWLEDGMENTS

Support of the research by the Italian Ministry of Research ("Progetti 40%") is gratefully acknowledged. Part of this chapter is drawn from Dosi, Teece, and Winter (1992).

Part IV

THE TERRITORIAL FOUNDATIONS OF PRODUCTION SYSTEMS

9

TRUST, COMMUNITY, AND COOPERATION

Toward a theory of industrial districts[1]

Edward H. Lorenz

There is now a vast literature on dynamic industrial districts or regional agglomerations of firms. Initially regarded as a development confined to the Third Italy, industrial districts are now acknowledged to be widely distributed in Europe, the United States, and Japan. The technological dynamism of industrial districts has contributed to a declining belief in the superiority of mass production techniques and to a burgeoning interest in the region as a unit of production.[2]

The case study literature has succeeded in cataloging the most prominent characteristics of industrial districts and has managed to identify the key factor in their technological dynamism: a particular balance between cooperation and competition among the firms within them (Dore 1986: 153–204; Friedman 1988: chap. 5; Piore 1990: 4; Piore and Sable 1984; Raveyre and Saglio 1984: 64–7; Sabel, 1989: 45–52; Sengenberger and Loveman 1987).

Cooperation among producers in industrial districts has two principal aspects. It takes the form of the provision of collective goods,[3] notably training or education and research and development, but also medical care and unemployment insurance. Collective goods generally are provided through the auspices of some local institution: business association, trade union, or possibly municipal or regional government.

Cooperation also takes the form of adherence by producers to a set of *norms* of competition.[4] This entails respecting the property rights distributions implied in contractual commitments,[5] and adhering to certain norms of *reciprocity*: sharing of technical information; subcontracting out to one's less successful competitors; and refraining from wage competition and labor poaching (Brusco 1982; Dore 1987: 171–9; Raveyre and Saglio 1984: 165; Sabel and Zeitlin 1985: 146–9). Reciprocity is key to the technological dynamism of the districts because it reduces the risk of investing in new products or processes, while discouraging the alternative competitive strategy of competing through wage competition.[6]

While the link between cooperation among producers and technological dynamism is well understood, there is as yet no theory explaining cooperation within industrial districts, a fact which probably explains the limited success of policies designed to create industrial districts (Piore 1990: 9–11).[7]

This chapter is offered as a contribution to a theory of industrial districts. I argue that an adequate explanation for cooperation among producers must systematically explore the relation between cooperation and two observed features of dynamic industrial districts: their being embedded within "communities," and the high level of *trust* among producers within them.[8]

The chapter is structured as follows. In the first section I develop the notion of "community," drawing on Michael Taylor's (1982) theory of how community maintains social order[9] in stateless societies. In the second section I argue that the methods of social control used in communities to maintain social order account for, at least in part, the adherence to norms of competition in industrial districts. I also argue that differences in the degree of flexibility and technological dynamism among industrial districts (or in the same district over time) can be accounted for by differing levels of trust among the actors. The third section uses the analysis of trust and social control in communities to address the policy issue of the intentional creation of industrial districts.

I

Taylor (1982: 25–33) defines community as a small and stable group of individuals who:

a) hold beliefs and values in common;
b) whose relations with each other are direct as opposed to being mediated by the state or some other bureaucratic institution;
c) whose relations are many-sided as opposed to specialized (there is no clear distinction between economics and politics); and
d) who practice reciprocity.

These characteristics can be more or less present and correspondingly we may talk of a group constituting a community to a greater or lesser extent. For example, a group of individuals forming a "business community" constitute less of a community than such intentional communities as Israeli *kibbutzim*, since relations between the actors are more specialized and the range of values and beliefs held in common are likely to be narrower. Further, contemporary communities, however wide the range of commonly held beliefs and values, are embedded in societies with states and some relations are mediated by bureaucratic institutions. Correspondingly, contemporary communities are less communal in Taylor's sense than, say, the historical example of primitive tribal groups in stateless societies.

The principal types of social controls used to maintain social order in communities are: the threat of "self-help" retaliation (feud, vendetta, etc.); the threat of withdrawal of reciprocity; and the use of the sanctions of approval and disapproval. These methods of social control are decentralized in that there is no political specialization and all may participate in their use (Taylor 1982: 80–1, 90–1).[10]

In general, these social control methods will be more developed the more nearly the community approximates Taylor's definition. For example, neither the threat of withdrawal of reciprocity nor the "force of public opinion" are likely to be effective methods of securing cooperation in large open societies, since repeated encounters among the same individuals are less likely and information about others is less likely to be common knowledge. Similarly, the threat of withdrawal of reciprocity is less likely to be effective if individuals can rely on state provision of collective services, while the threat of retaliation is likely to be diluted in communities subject to the laws of a centralized state and its police force.

Taylor argues that "community" provides a decentralized "solution" to the collective action problem of assuring security of property rights.[11] The threat of retaliation, the withdrawal of reciprocity, and the use of the sanctions of approval and disapproval may be thought of as selective incentives in Olson's (1965) sense. They alter an individual's expectations of the behavior of others to secure his or her cooperation, rather than altering preferences among outcomes. However, the decentralized provision of selective incentives characteristic of communities assumes a second order collective action problem. Since sanctioning the behavior of others is costly, why not "free ride" on the sanctioning efforts of others (Elster 1989: 40–1; Oliver 1980)?

Ultimately, there are two explanations for cooperation in communities. It is the preferred choice in the context of an iterated Prisoners' Dilemma, where rational actors cooperate, not because of selective incentives but because the anticipated benefits of future mutual cooperation are valued higher than the one-time rewards of defecting while others cooperate (Axelrod 1984). This "solution" assumes accurate information about the attitudes and beliefs of others and that this information is "common knowledge" (Binmore and Dasgupta 1986). Doubts concerning the *trustworthiness* of others are likely to result in a breakdown of the "tacit" agreement to cooperate. Since the character of communities as small and stable groups would tend to generate the necessary information, it is plausible to argue that reciprocity may emerge for reasons of long term self-interest independently of selective incentives (Taylor 1987: 52–3).

Cooperation also may be motivated by social norms which have been internalized through education and socialization and which cannot be reduced to optimizing rationality. Norms may operate at the first link in the chain of actions, but also at the second: It is a norm to sanction those who do not sanction a violation of a social norm.[12] In this case, the "problem" of collective action in effect disappears, since, due to the internalization of social norms, cooperation is the *preferred* outcome.

In eclectic fashion, I propose to argue that both motivations for cooperation may operate. We may prefer to cooperate not only because we believe it to be in our long-term self-interest, but also because we believe it is the right thing to do.[13] However, the reliance of communities on such "selective incentives" as expressions of approval and disapproval to secure cooperation implies that long-term self-interest alone is an imperfect guarantor of the social order. For some individuals the future may not be important enough to dissuade them from acting in ways that threaten person or property. More tragically, it is possible that all prefer cooperation for reasons of long term self-interest, but cooperation nonetheless flounders on lack of information and *distrust* about the intentions of others. Believing (incorrectly) that others prefer to defect, we defect, only to have our behavior sanctioned.[14]

Finally, before considering the implications of this discussion for industrial districts, in communities it is difficult to distinguish the methods of social control from what is being controlled because of the many-sided nature of relations. The violation of a norm in one sphere of activity has ramifications in others and affects the entire social order. The prospect of this discourages would-be violators of social norms (Taylor 1982: 75–6).

II

One implication of the above discussion is that industrial districts should be seen as an aspect of the many-sided nature of social relations in communities, rather than as something autonomous embedded within communities.[15] The relations between the actors forming the industrial district may be more or less specialized and correspondingly they may constitute more or less of a community. For example, the actors forming the high-technology district of Silicon Valley constitute less of a community than the actors forming the French industrial district of Oyonnax investigated by Raveyre and Saglio (1984). Since the Oyonnax community is more stable, the relations between actors less specialized, and the range of beliefs and values held in common wider, it is likely that the methods of social control characteristic of communities will be more fully developed in Oyonnax than in Silicon Valley. I return to this general issue in the third section addressing policy.

Of the decentralized methods of social control found in communities, two are used in industrial districts to assure conformity to norms of competition: the threat of withdrawal of reciprocity and expressions of disapproval and approval (Raveyre and Saglio 1984: 164–7; Sabel and Zeitlin 1985: 153–4).[16] The absence of the threat of "self-help" retaliation can probably best be accounted for by the location of industrial districts in societies subject to the laws of a centralized state.[17]

The case study literature shows that these decentralized methods are complemented by more centralized methods. For example, in Lyon during the nineteenth century the *conseil de prud'hommes* mediated disputes between weavers and merchants, while at the turn of the century the Saint Etienne municipal government regulated the length of the working day for ribbon workers. The municipal governments of Solingen and Remscheld in Germany regulated safety conditions in the metalworking industry (Sabel and Zeitlin 1985: 150).

The use of centralized methods might be explained in terms of the progressive intrusion of the state into formerly self-regulating communities resulting in the atrophy of decentralized methods (see Chapter 11). Decentralized methods work best in small and stable communities where the actors share a wide range of values and relations are many-sided. The use of centralized methods might reflect the relative openness of nineteenth- and twentieth-century industrial districts to outsiders who because of their limited economic dependence on insiders and their lack of common values would be relatively impervious to the threat of withdrawal of reciprocity and the "force of public opinion."

Being able to judge the trustworthiness of others is important in industrial districts, not only because of the possibility that outsiders may be impervious to the methods of social control used in communities, but also because of the inevitable scope for opportunism offered by the necessary incompleteness of contracts (verbal or written) in an uncertain world.[18] Unanticipated contingencies, associated with technical change or demand shifts, provide opportunities to interpret contract terms in ways that shift the distribution of returns to the favor of one side.[19] Neither third-party assistance in dispute resolution nor the force of reputation effects will necessarily assure contract compliance, simply because it may be impossible for the third party (or community) to distinguish between real and bogus claims. Community organization reduces the extent to which we need to worry about the trustworthiness of others with whom we have little or no personal experience, but it cannot dispose of our concerns altogether. Ultimately our decision to trust others will depend on our judgement of their honesty based on their behavior in repeated encounters.

For these reasons, it can be argued that trust and cooperation remain problematic in industrial districts. The case study literature shows that unanticipated contingencies can trigger behavior which in a mutually reinforcing manner undermines high trust relations. Herrigel (1990: 403–8), for example, has documented the breakdown of trust and cooperation in German industrial districts during the interwar years. Sabel and Zeitlin (1985: 158–9) refer to a similar breakdown of trust among producers in the Sheffield cutlery industry during the late nineteenth century, the Birmingham metalworking shops after 1890, and the Saint Etienne silk ribbon industry during the interwar period.

In each of these cases, the breakdown of trust was associated with a widespread violation of established norms of reciprocity contributing to a loss of flexibility and technological dynamism. This occurred in two ways. In the case of Saint Etienne and the German industrial districts discussed by Herrigel, prolonged economic recession led to a loss of faith in the potential advantages of mutual cooperation. In Saint Etienne, for example, faced with a decline in demand for ribbons after World War I, producers disputed over who should bear the cost of apprenticeship training, and investment in new products and processes languished (Sabel and Zeitlin 1985: 159).

In the case of Birmingham and Sheffield, the process was the reverse, with the perception of increased opportunities associated with expanding national markets encouraging producers to break their established links with the community. At the same time, labor in Sheffield, fearing loss of place within the district, sought to secure their property rights through enforcing substantive rules fixing the division of labor.

The examples of Birmingham and Sheffield illustrate a tension within dynamic industrial districts. On the one hand, as with all communities, leveling mechanisms operate within industrial districts to restrict the growth of inequality.[20] Examples include unemployment insurance and the redistribution of demand and provision of credit to support less successful competitors in final product markets. Too much inequality invites distrust by undermining the perceptions of mutual dependency that help sustain cooperation in communities.[21] On the other hand, too much leveling discourages risky investments in new products and processes. What is required is an appropriate balance between rewarding entrepreneurship and limiting the growth of inequality.

The technologically dynamic character of industrial districts depends on the beliefs of the actors in two respects: the belief that others can be trusted to abide by established norms of competition; and the belief that, while all deserve a place within the community, entrepreneurship should be rewarded by social advancement. The following discussion of policy, focusing on the United States, addresses the question of beliefs and how they change.

III

If the argument of this paper is correct, then the dynamism of industrial districts depends not solely on the creation of institutions to provide collective goods such as research and training, but also on the beliefs of the actors: the belief that, while entrepreneurship should be rewarded, all members of the community should be guaranteed a place through adherence to norms of reciprocity. The question of the intentional creation of industrial districts reduces in part, then, to the question of changing beliefs.

One approach that can be readily rejected is that suggested by the intentional creation of such communities as the nineteenth-century utopian American community of Oneida.[22] In Oneida, the forms of cooperation and reciprocity that sustained economic innovation emerged as by-products of commonly shared religious beliefs and experiences. The difficulty with this for policy is that the attempt to establish beliefs, which must be held for other than rational reasons, in order rationally to pursue some goal (economic efficiency) is likely to be self-defeating (Elster 1983: 51–2; Hawthorne 1988).

A more promising approach is suggested by the examples of such high technology districts as Silicon Valley and Boston's Route 128, despite the fact that they were not intentionally created. The commonly held beliefs of the actors forming these communities are relatively narrow and largely understandable in terms of rational economic pursuits.

Arguably the two principal beliefs underlying the dynamism of Silicon Valley are a belief in entrepreneurship, and a belief that technical knowledge is common property and should be shared, subject to certain ethical standards. As Rogers and Larsen note, the exchange of information is key to the ability of the network of firms composing Silicon Valley to manage the uncertainties of developing new products and productive technologies:[23]

> Information-exchange is a dominant, distinguishing characteristic of Silicon Valley. Because innovation entails coping with a high degree of uncertainty, such innovation is particularly dependent on information. Information must be given in order for it to be obtained. The nature of the technical information exchange process in the microelectronics industry demands a high degree of reciprocity among the participants.
>
> (Rogers and Larsen 1984: 79–81)

Information exchange in Silicon Valley is achieved in part through an exceptionally high degree of mobility of engineers and managers among firms:

> Silicon Valley was, by the mid-1970s, distinguished by its unusually high levels of job-hopping and new firm formation . . . it became both common and acceptable for engineers and managers to regularly shift between firms. This consistent reshuffling process strengthened the region's social and technical infrastructure: as they chose to leave established companies, engineers took with them the skills, know-how and experience acquired at their previous job . . .
>
> (Saxenian 1989: 37)

Of these two beliefs, having sufficient quantities of the first (entrepreneurship) to generate technological dynamism among a network of firms is less problematic than the second (information exchange). Given a pool of entrepreneurial talent in a region, a self-selection process is likely to operate to assure that some of it is present in the firms composing the planned district. Some entrepreneurship is required simply to take the initiative to start up a firm.

The reciprocal practices of information sharing among firms in Silicon Valley are not the norm in the United States and no doubt would violate the codes of secrecy operated by most American firms. The common belief in Silicon Valley that technical information is collective property seems to be a by-product of the close affiliation of the firms with a major research university and the fact that engineering professors took the lead in spinning off new high-technology firms.[24]

> The war thus created the technological infrastructure which supported Stanford's first electronics spin-offs during the 1940s and 1950s. The contribution of Stanford's engineering dean, Frederick Terman, lay less in the linkages he forged between the university and local industry, as commonly assumed, than in his active support for aspiring entrepreneurs . . . Terman initiated a tradition of cooperation which his protégés in turn replicated in their relationships with other emerging Silicon Valley enterprises . . .
>
> One of Stanford's most significant, but overlooked, contributions to the development of Silicon Valley was in fostering inter-firm cooperation.
>
> (Saxenian 1989: 32–3)

Barring the intentional creation of new industrial districts by major research institutes, how might the distinctive values about sharing technical information that sustain innovation be promulgated in the United States?

A promising approach to this question is suggested by Samuel Popkin's (1988) notion of the *political entrepreneur*. A political entrepreneur may be defined as an innovator who solves collective action problems not by offering selective incentives, but by persuasion and changing beliefs, beliefs about the value of the collective good and expectations about the behavior of others (Popkin 1988: 16–17). Popkin develops this idea in the context of an analysis of the political mobilization of peasant communities during the Vietnamese revolution. Popkin shows how four national movements (the Catholic Church, the Cao Dai, the Hoa Hao, and the Communist Party) won support for their aims by having their cadres work to solve smaller-scale collective action problems at the village level, including the creation of insurance and welfare systems and the provision of educational opportunities. They accomplished this largely by persuading individuals that the contributions of others were dependent on their contributions, and by acting as monitors to increase the amount of information about the behavior of others.

The analogous case for industrial communities would have the policy-maker as political entrepreneur persuading producers, without the use of selective incentives, that they would benefit from independently providing tangible collective goods such as apprenticeship training or collaborative research and development. The policymaker would also act as monitor. Persuading producers actively to participate in the effort would be key. Active participation, if successful, would increase the amount of communication among producers and would contribute to creating the trust among them that is a prerequisite for sharing technical information. As the adage goes, "nothing succeeds like success."

Of course, there is no reason why reciprocity need be limited to the exchange of information. The actors might be persuaded to practice other forms of reciprocity that are characteristic of European industrial districts such as subcontracting work to less successful competitors and refraining from labor poaching and wage competition.[25]

Reciprocal information sharing as practiced in Silicon Valley is not the norm in the United States and an important question is: how robust are such communities, given their openness to outsiders who may not share their beliefs? The actors in communities such as Silicon Valley lack common religious beliefs or ethnic ties that would allow individuals to distinguish between likely cooperators and defectors. On this point, the recent work of Axelrod (1984) on the iterated Prisoners' Dilemma is relevant. Axelrod has shown that a group of individuals practicing a "nice" strategy cannot be "invaded" by an individual or cluster practicing a "mean" strategy as long as the nice strategy is *maximally discriminating*. A maximally discriminating strategy will eventually cooperate even if the other has never cooperated yet, and once it cooperates will never cooperate again with someone who defects on the first and all subsequent moves, but will always cooperate with another using the same strategy it uses (Axelrod 1984: 62–9). Given the lack of religious or ethnic ties in planned industrial districts, the policymaker as political entrepreneur could play a key monitoring role by providing the information to distinguish between "nice guys" and "meanies."[26]

A final point I want to raise about the maintenance of intentionally created industrial districts in the United States, concerns the balance of rational self-interest and social norms motivating cooperation. The policymaker is likely to operate through an appeal to the self-interest of producers: persuading them of the economic value of collective goods and of the importance of their contributions for eliciting the contributions of others. Given that the desired forms of reciprocity are the exception in the United States, an appeal to producers' social norms or values is unlikely to be successful.

For this reason, such informal forms of cooperation as information exchange and refraining from labor poaching are likely to be particularly susceptible to a breakdown due to changes in the environment that alter a producer's perception of the value of cooperation. For example, a period of slack demand, increasing a firm's fear of competitive failure, is likely to promote information hoarding, as the successes of others are seen as limiting one's own chance of success.[27] Although persuasion may play a role, the likely remedy for this is the institutionalization of leveling mechanisms, such as the provision of credit during downturns and unemployment insurance, that assure all actors some place within the community. Thus, to balance somewhat the heavy emphasis on beliefs in my discussion of policy, the distinctive resources of the policymaker as political entrepreneur are most likely to be effective if complemented by formal institutions that assure some redistribution of resources to the less successful members of the community.

10

A REEXAMINATION OF THE ITALIAN MODEL OF FLEXIBLE PRODUCTION FROM A COMPARATIVE POINT OF VIEW[1]

Bruno A. Courault and Claudine Romani

At a time when new theoretical models to explain the relative contraction of national – and growth of international – markets are being developed, analysis of the "Marshallian industrial districts"[2] has been used by Italian researchers to explain the relative process of certain regions in coming to grips with these new conditions. The dynamics of employment in many regions of that country and the ability of these regions to withstand national and international competition is seen as the result of the flexibility of the production systems of their industrial districts. This flexibility is said to be the reward for a coordination of production between the social base and the smooth regulation of economic, social, and political elements in a local context. The industrial district promotes flexibility due to the coordination of productive structure and its social roots.

We intend to reconsider the generality of such a model, and to understand the conditions under it that might be extended to other countries. Research on the Annonay and Oyonnax industrial systems in France pioneered the consideration of the general principles of the Italian model.[3] A host of other studies has since been carried out by Courault, Rérat and Weisz in France;[4] Romani and Bettio,[5] Brusco, Garofoli, Solinas and Villa[6] in Italy; comparing French and Italian industrial zones. This work has stimulated, even among the Italian researchers, questions concerning the possible extension and durability of the above-mentioned model, i.e. "To what extent does the Italian model in itself constitute a specific productive form?". We must ask whether the originality of the Italian model is a result of its unique local foundations, or whether it describes a generalized economic logic of productive decentralization.

In this chapter we intend to address this question in order to consider the extent to which the industrial district model may be reproduced in other areas.

OVERVIEW OF THE ITALIAN INDUSTRIAL MODEL

A network of small and medium-sized companies

The originality of the industrial district model, as designed by its founders, lies in the central role played by the notion of local space: the district is a place where products are produced and put in circulation, labor is managed, and there exists institutional regulation to ensure the stability and survival of the district. Different authors take different views on the articulation of these three functions and on the influence of the social and economic mechanisms that shape local space (Brusco 1982, Solinas 1982, Bagnasco 1985, Becattini *et al.* 1987, Ganne *et al.* 1988, Sabel 1988, Regini and Sabel 1989, Saglio 1989).[7] Here we shall consider specific aspects of the internal workings of this model as summarized by Piore and Sabel, who call the industrial district systems of "flexible specialization," and set up this idea of flexibility in their book (1984).[8]

We may begin with the organizational logic of the "production decentralization" model, which was formalized by Brusco in the early 1980s, and identifies the production system as an integrated whole, composed of a myriad of specialized units, all of which are linked to the market in a way that maximizes the responsiveness of the productive system to market trends (Brusco 1982). The system is taken to be the canonical form of production flexibility.[9] Decentralization entails at one and the same time specialization of units and diversification of the production system as a whole. The application of this model to manufacturing industries such as clothing, knitwear, silk, footwear, mechanics, machine tools, electronics, metalworking, and even steel production, underscores the attempt of analysts to generalize the model, i.e. to suggest that all locally based industrial specializations are in fact Marshallian industrial districts.[10]

It is widely argued that the production flexibility resulting from this apparatus is linked to labor management. Solinas, for example, argues that the economic structure of the districts rests on a distinctive social base, part of which is its system of labor–management relations (Solinas 1982). This implies that patterns of social relations that are specific to particular places, and therefore to the history of each district, play a central role in this industrial district. Social relations specific to each village, clan, or family, and the specific, local cultural foundations that cement them, underpin the economic organization of each district: they act as a system for the exchange of economic information that in its turn feeds into and promotes the circulation of products throughout the entire district.

The district therefore appears as a paradox: a blend of archaic and modern forms of social life, the age-old, guild-like producers' organizations and traditional social networks combined with the most sophisticated technologies and marketing strategies. Far from vanishing with the extension of modern industries, the clan and extended family have strengthened their foothold and received a new lease of life through the recourse to new technologies and global markets. It appears that the growth of international markets, rather than triggering the classical phenomenon of industrial concentration, has simply rendered more dynamic a system of production rooted in archaic forms of social organization. Relations of proximity and of the family are important in supporting exchange networks which constitute the social basis of production.[11] The work force plays a pivotal role. The foremost figure is neither the white-collar nor the blue-collar worker but rather the independent or "autonomous" worker. The latter is not the same as the archaic category of the "artisan." Unlike the artisan, the autonomous worker occupies many different roles, such as professional worker or home-worker, often switching from one season to the next, and never achieving the fixity or stability of the artisan's former situation (see Brusco and Sabel 1982, Solinas 1982). The existence of this type of worker provokes much interest because his existence is inconceivable without specific local supports, on one hand, and because it extends craftsmanship to an industrial scale on the other.

A local area open to the international market

Another paradoxical aspect of the industrial district model is that it is based on fragmented and localized producers who are nonetheless heavily oriented toward export markets.[12] In the shoe industry, for example, the ability of Italian producers to succeed in international markets appears as a direct consequence of the locally fragmented sytem of production organized as an industrial district. Here, firms are not linked by a tree-like hierarchy, in which only those firms at the top of the tree gain exclusive access to the end market, while the others are reduced to subcontracting; rather, in the footwear industry, many different producers have access to the end market, but this position is never secure.[13]

This is difficult to understand, for the structure of production as a whole appears to be too fragmented and incoherent to produce such a strong position enjoyed on international markets. The strength of Italian producers on the international market lies in their creative flair. This means that in Italy every shoe company, whatever its current market position and status, gears itself up to present a collection each season. If the collection is well-received, the firm gains access to the end market; if not, it is relegated to subcontracting work for that season. The fragmented structure can thus consistently capture international markets, in spite of the power of the major buyers. With their creative acumen, Italian producers define the qualities of supply, and this secures them a position "upstream" of the market itself. In contrast to Italian producers, French and English firms face integrated distribution systems, and this distances wholesalers from production.[14] Italian producers push industrial specialization as far as it will go, without making any investment downstream or in the commercial sector. Italian production flexibility is therefore based on the way distribution follows the productive system and not the other way around (see Courault and Rérat 1988, Courault 1990).

The Italian case thus provides reason for skepticism with respect to theories of production which assert that competition necessarily stimulates industrial concentration.[15] Here the opposite is the case: the end market is in many cases represented by foreign buyers who set up offices within the specialized districts themselves in order to follow the producers' design trends (Courault, Rérat 1988).[16]

A GEOGRAPHICAL EXTENSION OF THE MODEL

We may now ask whether the originality of the Italian district model is the result of the history of Italian local economies and to which extent it can be applied to other geographical and historical spaces. For example, are the traditional forms of local industrial systems and local economies that currently exist in the industrial regions of France similar to those of the Italian districts? To reply to this question we shall draw on two comparative studies. The first is a comparison of French and Italian districts specializing in shoe production (Courault *et al.* 1987; Brusco *et al.* 1987). The second examines the branch agreements passed in the textile and clothing industry in the same two countries (Romani 1988).[17]

In the Italian case, economic development appears to be rooted in specific local rules. In the cases of the knitting industry in Carpi, silk production in Como, and textiles in Prato; the rules that govern labor management are defined at the level of the district (see pages 212–14). In contrast the rules established in the French shoe industry districts have less to do with local foundations than in Italy, as they have been developed in line with the national system of economic regulation.[18]

It must be stressed that in spite of these different local rules for labor markets, production methods are very similar in the districts of the two countries: the production system consists of a myriad of localized units benefiting from the proximity of the work force, neighborhood relations, and reliance on interfirm relations. Moreover, in both cases, the local area constitutes a zone where a localized work force can be mobilized as needed.

The principal difference concerns the underlying forms of organization and governance of the decentralization of production. In Italy, decentralization relies upon a significant mobility of workers between different very small units, and these units are sometimes *end producers* and other times *subcontractors*. In France, on the other hand, decentralization takes the form of geographical deconcentration of units of a given firm. In the Italian case, the mobility of workers between companies in the district assures the flexibility of the units and the system as a whole. A large number of independent workers can thus at any moment be mobilized to provide the missing link in the chain of production. The French situation is very different; as there exist fewer units of production marked by a greater degree of stability in the structural organization, labor mobility involves the workers of a given company, and even this is limited as it is unusual for them to change units of production. Thus *interfirm mobility* of workers plays the central role in the flexibility of the Italian productive system, whereas in France *internal labor mobility* is dominant, as is the case of companies in the district of Cholet (Courault et Rérat 1987; Lorthiois et Rérat 1987).

The Italian district can thus be said to be a local construction in the sense that the industrial system is a product of local social relations (see Bagnasco, Trigilia 1985). In Italian cases professional craft skills and local political regulation are essential, whereas in the French cases the economic logic of the firm – which is not local – is secondary to economic factors (Ganne *et al.* 1988; Ganne 1990).

The specificity of the Italian economy resides in the way that social and occupational mobility underlies the formation of an entrepreneurial class. Becattini (1977, 1990) has developed an analysis concerning the central role that this class constitutes, but he has been challenged by Bagnasco (1987). Becattini insisted upon the role of entrepreneurs acting as a class in the development of industrial activity, whereas Bagnasco underlines the roles of interclass social consensus, based on shared values and habits, in reproducing the social environment for the generation of new activities on the model of the traditional ones. In this view, the importance of the ways the workers learn and enlarge their collective skills is stressed (Courault-Rérat 1989). Though management of labor is essential in both the French and Italian cases, in the latter access to high skill is fully adapted to external flexibility, while internal flexibility is central to the functioning of the French areas.

THE STABILITY OF INDUSTRIAL DISTRICTS

We would like to examine the way the district model articulates local rules and external economic constraints. One can explain the formation of industrial districts in the shoe industry in the different cases we have studied in France and in Italy as the outcome of a series of local rules which concretize in that sector and territory, a specific path of historical development. In each district we can identify a moment when the choice of a product policy became strongly identified with local culture. The rules of each district thus reflect this historic period when the supremacy of a particular product was established and when the locality adopted practices different from those in place elsewhere. In agreement with Bagnasco (1985), Saglio (1989), and Thévenot (1989), we would say that the local specialization in each district reflects *the social construction* of a particular product market. Industrial specialization lasts only as long as the local rules can assure the adaptation of the local productive organization to the state of competition in those markets.[19]

In the case of four different French shoe industry regions (Marseille, Fougères, Romans, Cholet) we find that local rules exist to harmonize local working practices with the needs of production in light of the state of competition. In Marseille and Fougères, high skill levels were required for the production of high-quality or handmade products. Product changes induced the decline of those regions because they led to the implantation of new greenfield sites of production untouched by these traditional working practices: it is in this way that we can understand the emergence of new regions such as the Cholet, which developed later on the basis of new technology and lower quality products. The Italian district of Barletta has recently emerged as an alternative to the Marches, which had itself replaced Vigevano. In a different way Romans, in France, in contrast to Vigevano, saw its productive potential decline as a result of specializing in articles for the luxury market. In both countries there is a tradition of a succession of new districts joining together with the old, and then, after a certain period, supplanting them. Expansion of districts means modernization, crisis, and sometimes decline, where the existing work rules cannot be adapted to the production of new products.

THE DIFFERENT INTERPRETATIONS OF THE DISTRICT MODEL

We have thus far considered the shoe industry in terms of models underpinned by an economic logic interwoven with local social and political factors. It is now necessary to enlarge our field of inquiry to see if these models are specific and unique to their times and places or whether they have general applications. The question has already been widely debated with respect to in the work of Piore and Sabel (1984), who claim that the model of flexible specialization could be seen as a possible historical successor to mass production.

Yet it seems hasty to conclude, as do Piore and Sabel, that the Italian districts represent a possibly general model of industrialization. The cases examined here suggest the importance of different local contexts in the emergence and consolidation of the Italian system of industrial districts, and this notion is reaffirmed in the work of Bagnasco (1985), Capecchi (1987), or Brusco (1982) on the origins and conditions of the emergence of the Italian district model (see Brusco 1990; Capecchi 1990).

The opposition between large companies and small companies once appeared to us to provide a solid point of departure for the study of productive systems, for large-firm systems and small-firm systems follow completely different logics of development (Saglio 1987). In one case, there is a concentration of production capacity as internal economies of scale come into play. In contrast to this, the developmental tendencies of small-scale units of production appear to be less governed by economic laws; hence they are less generalizable. It would be a mistake to reduce them either to the Italian model of "industrial districts," or to the equivalent French model studied by Ganne (1983) and Raveyre and Saglio (1984), and Courault-Rérat-Weisz (1987). Such a reduction assimilates the dynamics of small and medium-sized companies in general to those of particular *local industrial systems*. Our analysis stresses the factors which make small-scale productive units depend on their local bases (social and professional resources, economic initiatives to develop and reinforce the power of local business, and other political factors). Much of the theoretical framework for this line of analysis goes back to Marshall, according to whom the market and competition do not rule out important interactions between solidarity and competition, consensus and the market, in complexes of small- and medium-sized firms.[20]

In this light, the recent hypothesis of Regini and Sabel (1989) of a *convergence* between small and large firms is of particular interest. According to them, small and medium-sized firms have developed in compliance with the organizational model of large companies but on a different scale, for example using forms of alliance or coalition; conversely, large companies have decentralized, either in the form of a network, or through the classic use of small subcontractors, many of whom were initially dependent and who later developed greater independence and separation. This hypothesis thus asserts the existence of global dynamics of production systems. In our view, however, the districts are not the result of convergence, but represent a different intermediate category which bypasses the classic dualism between large firms, on the one hand, and small and medium-sized firms on the other. We would counter Regini and Sabel, by claiming that this intermediate form is efficient in a context of sharpened competition and is potentially long-lasting, but not a replacement for the other two.

We interpret this intermediate form alternatively as the emergence of a productive system incorporating all the small and medium-sized companies, or as a new type of firm, derived from large firms and seeking to recuperate its autonomy by large decentralization of production, with respect to efficiency through the adoption of flexible specialization.

Regini and Sabel seek to legitimize the district model as a norm and thus as a point of reference for development in other areas. This position runs counter to our interpretations of these areas, which situates them within specific local historical and economic contexts. For example, the fashion-clothing system in Cholet (France) has the appearance of a 1960s type creation ex-nihilo; but it is really the result of a transformation of a large number of existing subcontractors. The footwear system in Apulia (Italy) is similar but dates from the 1980s. The latter system combines the productive centralization of a number of large units with small subcontracting units who draw on a large number of independent workers (Brusco, Garofoli, Solinas, Villa 1987). The footwear system in the Marches region (Italy) came into being and expanded from a tiny starting point (small production for the local market), whereas the Vigevano district fell into decline because of the overly specialized products of local firms. The same occurred in the Cholet (France) footwear system, confronted with the Fougères system which has now disappeared from the map. Other local industrial districts (Carpi or Prato, Oyonnax or Roanne) all present this feature of decentralized units of production linked together to serve the final market; it radically differentiates them from deliberate attempts to install districts such as the "technopoles" and other networks established by particular groups (see Greffe 1988; Domergues 1988).

According to the Italian authors we have quoted, districts emerge on the basis of local processes within a defined local historical context (Capecchi 1987; Trigilia 1986). Only two authors, Bagnasco (1978) and Fuà (1985), argue that preexisting social and cultural foundations of districts alone play a major role in the development of the district. This conflicts with what is suggested by the analyses of Brusco (1982) and Capecchi (1987), who place less emphasis on social factors and insist on the role of economic logics, i.e. the influence of product specialization in the regulation of the productive system. Our argument is that economic foundations have a powerful structuring effect on the formation of districts, but that those foundations are themselves inseparable from social forces – especially in the labor market – in the rise and regulation of the districts.

LOCAL SELF REGULATION: THE DISTRICT AS A REGULATED COMMUNITY

We need now to understand how the district operates: the constitution and negotiation of territorial rule systems,[21] in our view, play a vital role in the survival and reproduction of industrial districts. These systems of norms go a long way to determining the "life cycles" of the districts, their conditions of growth, and their survival or disappearance. The district can be thought of as an industrial space in which implicit or explicit norms develop, most often stabilized in the form of agreements between local partners. Though such agreements also owe much to other norms in force outside the district, and internal modes of regulation can never therefore be wholly dissociated from practices that remain external to the district, it is nonetheless the case that the local negotiation of these norms is of central importance. These rules reinforce the overall coherence of the local productive system in its territorial context, as Saglio (1989) has often emphasized.

In Italy, local negotiation plays a threefold role. First, it seeks to establish local conventions which usually remain implicit. These agreements attempt to stabilize relations on the market in a form that we might describe as "relational": this applies both to the norms of strategic solidarity between business and to relations between companies and political actors within the region. For example, we might call attention to various efforts to establish alliances or "partnerships" with the aid of local governments. These efforts have taken place in two areas: the creation of service centers for firms, especially in the industrial districts of Castel Goffredo, Capri, and Como; and efforts to aid professional training, both initial and continued, via the establishment of training centers such as the technical industrial institute of Biella in Lombardy.

Second, this negotiation establishes institutional means of intervening in the labor market at a local level. It provides for local adaptation of national institutions, which are then managed for local outcomes. We would draw attention to the regulation of overmanning through the intermediary of the Cassa Integrazione de Guadagni (CIG) following a decision made within the framework of local agreements (for example, that of Prato, ratified on 22 July 1987). This institution is a powerful tool of social regulation combining industrial and social factors (Romani 1987). The CIG mixes local and branch-level negotiations within a framework which combines the logic of sociopolitical and professional rationality. The institutional dimension thus has an extremely original regulatory function – of which there is no equivalent anywhere in the world – in that as it creates a link between occupational labor markets and local labor markets that are found nowhere else, except in Germany (see Silvestre, 1986). This is a powerful example of the way that external institutional mechanisms can reshape local affairs. The rules of the CIG envisaged that only firms that were undergoing restructuring would be authorized to invoke the "state of crisis" in order to have recourse to the resources of the CIG. The agreement had, as its main purpose, to allow small firms to suspend activity of a portion of their personnel for a period of at least six months. These firms negotiated as a bloc (of 250), to obtain access to the resources of the CIG, affecting 8,600 workers. Such a negotiating process required that the individual enterprises be organized at the territorial level. Only in this way could they obtain the authorization to use resources of the CIG. But by negotiating at the territorial level, they established the precedent that the firm is not the exclusive agent in such a negotiating process: other territorial partners were now legitimated. There are many other examples of the local negotiation of such institutional devices as employment training contracts (Romani 1988). But local negotiation takes place around other issues, too. Analysis of the content of a sample of district agreements negotiated in Italy has shown, among other things, that one of the poles around which local negotiating practices are organized is the definition of the various different components of salary (Romani 1988).

Third, the local production of norms creates a hybrid and autonomous level of negotiation located at the point of contact between the company, the sector, and the region.[22] It redefines the areas of competence of the social actors involved in the negotiation process. For example, in Italy, local negotiation is not the exclusive province of "traditional" social bargaining partners such as the trade unions or employers associations. On the contrary, it involves the participation of a range of agencies in local civil society – local councillors, local associations – whose presence draws all those involved in the negotiation into a process of active cooperation to create rules. In comparison with previous practices, this kind of regulation of professional relations leads not only to a diversification of those involved in collective action but also to a modification of the strategies adopted by these parties. This broader-based articulation of social actors, apparently inherent in local negotiation, gives the local community a "consensual foundation" and reinforces its legitimacy. This has led to a socially consolidated management of economic restructuring, including the production and organizational strategies of firms. The success of those processes goes a long way to explaining the ongoing success of certain local districts, and also the disappearance of others: for example, the failure of the old French regions of Romans and Fougères in producing shoes reflects the rigidity of professional agreements which had been locally adopted and had blocked the possibility for technological innovation (see Courault, Rérat and Weisz 1987).

The power of local actors can also be seen in local negotiation on issues of labor management, employment and layoff policies, and skill and wage setting.

Thus local regulation in the production of norms is a vital part of the broader movement toward decentralization of industrial relations in the 1980s (Romani, 1987). In Italy, the need to render labor more flexible has in effect led to a progressive abandonment of standardized and rigid rules. The redistribution of areas of competence at different territorial scales has therefore given a new push to the local level. This has led to a diversification of labor market conventions and a reduction in national level.

The specifically Italian dynamics thus seem to lie at a crossroads between decentralized regulation, which includes the local, market, and domestic dimensions, and regulation negotiated at a centralized level of the national institutional framework, which is less political in nature and has a greater social content, as in the example of CIG.

DISTRICTS SITUATED IN METROPOLITAN AREAS

The classical model of the district (i.e. in Tuscany) developed in the presence of the extended family, which in turn accompanied the historical form of agricultural sharecropping prevalent there (mezzadria). The dense family networks form the conditions for a small-scale entrepreneurial class and promote solidarity. If the Italian district appears to be well suited to development in this type of rural environment, we may ask whether equivalent forms of localism could emerge in metropolitan and industrialized areas. Metropolitan districts stand out from the classical district model in that they cannot draw on the same socioeconomic conditions. This is the case, for example, in urban Spanish industrial districts.[23] Industrial zones in the Barcelona area have the classical structures of a district: a high degree of specialization in its production units (in this case in small-scale mechanics and metallurgy); a high concentration of small and medium-sized companies[24] in a defined geographical area; and a tight subcontracting network and dense interdependence between companies. Although this district manifests the main characteristics of the Italian districts, its social foundations are in fact quite different. First, it came into being quite recently (the early 1960s) in an area without industrial traditions. Second, its formation followed an exogenous principle of local development: it grew from the massive importation of labor (immigration from the south of Spain, mainly Andalusia) and not by drawing on local labor reserves. Industrial development here was financed by importing capital (from large foreign companies) and not from any original accumulation of capital by a locally based social class. Third, the local small business class in the district is not of rural origin. Rather, this class crystalized around the skilled Catalan worker of the 1930s and 1950s who, after years of factory work, set up his own workshop. It grew by drawing in skilled or highly skilled workers who had migrated from the south, and who had given up working for a large company (Seat) in order to establish workshops that would then do subcontracting work for the company. The small business class has also been nourished by the tradition, strong in Catalonia, of "individual companies," created by someone in charge of the initiative and by his *socio* (associate), frequently a member of the family if the founder is of the south, but more usually a factory colleague if the founder is Catalan. Finally, the relations of solidarity on which the cohesion of the district is founded rely less on family links than in Italy and more on "factory solidarities" and "brotherhood" (*gremios*).

Those suggest the heterogeneity of the conditions under which the districts are formed and in particular their most recent forms. This diversity in the creation and organization of industrial districts renders any general hypothesis regarding the reproduction and spread of the original model extremely hazardous.

11

INDUSTRIAL DEVELOPMENT AND LOCAL INDUSTRIAL SYSTEMS IN POSTWAR FRANCE

Political economy of a transformation

Bernard Ganne

INTRODUCTION

Starting from observations in postwar Italy (cf. Bagnasco's Third Italy, 1977) there has been a rediscovery of the importance of the territory in general and localities in particular in the organization of production. The local dimension, long overshadowed by concern with the macroeconomy, has reemerged as an object of concern. This is due in part to the rediscovery of the variety of economic forms that constitute an alternative to the mass production system that had come to dominate the economy in many countries (Sabel and Zeitlin 1985). It is also due to attempts to anticipate developments by observing a congruence between these older forms of industrial organization and the new ones based on flexible specialization (Piore and Sabel 1984). In light of the performance of the Italian districts, and as if to compensate for decades of neglect, the local area is making a comeback as a new paradigm of industrial development.

A new hegemonic paradigm?

Not long ago, structuralism was hegemonic in French economics and in sociology. For the structuralists, the local is nothing more than the projection onto a small screen of major macroeconomic forces. Ironically, it is often those who were the most ardent supporters of the school of development economics, which favored self-perpetuating industrialization ("*industrie industrialisante*"), who have become the most ardent defenders of territorial and local modes of development (Courlet and Judet 1986a, 1986b). This type of about-face appears to be leading to the development of a new hegemony which, in my view, is as damaging as the old one.

First, the Italian example was too rapidly turned into a model which uncritically champions endogenous growth. The success stories are celebrated, but the real lessons of the districts are not being learned (Michelsons and Baptiste 1989; Courault and Romani 1990; Ganne, 1989d).

Second, the current focus on the few zones in France that are most similar to the Italian districts (Cholet, the Jura, etc.), masks the fact that these zones are exceptions. In France there are few phenomena truly analogous to those observed in Italian industrial districts.

By stressing the direct similarities between types of local development in the two countries, one misses the real question: in which political context do these local dynamics occur? Surely they have different meanings in the centralized French system than in the decentralized Italian one. Undue fascination with the success of Italian industrial districts thus forecloses precise assessments of the importance and place of local industrial systems in general.

A study of the French case responds to these difficulties by permitting the district phenomenon to be put into a theoretical perspective, which underscores its link with the political system. Until the postwar period, local industrial systems occupied a relatively important place within the French productive system. But then many of them disappeared just at the moment when the Italian districts began to prosper and proliferate.

The first part of this chapter contrasts the evolution of French and Italian districts, showing that an effective mobilization of endogenous potentialities is insufficient to guarantee the survival of localized industrial systems. Not only did the social, economic, and political integration of the communities fail to guarantee their survival, but it even contributed to their decline by helping to close the local economy off from the outside world. It was from the exterior, by means of state intervention, that these involuted systems were reopened.

The second part of this chapter draws out the lessons of the French case by showing to what extent the very constitution of local industrial systems depends on political factors.

Methodologically speaking, it would nonetheless be a mistake to think in terms of a local/global opposition, of endogenous versus exogenous forces, or of local cultural forces versus economic and political forces. The point, rather, is to understand the different ways in which the construction of the local is authorized by the political in France and in Italy: the comparison illuminates not only the internal structure of districts, but the type of relation established between the different systems and their environments. Thus, rather than an element-by-element comparison, each system must be taken as a whole, of which its politics forms an important part.

PLACE AND EVOLUTION OF LOCAL INDUSTRIAL SYSTEMS IN FRANCE: SOME CONTRASTS WITH THE ITALIAN DISTRICTS[1]

In the interwar and early postwar years, France had a large variety of local industrial systems, all characterized by the importance of small family firms.

Often these were well integrated with each other. This system disappeared during the 1960s, at the very time when a similar form of economic organization began to gain importance in Italy. Why?

From the 1950s onward, French industry – including the Paris basin, all the regions of the north, the northeast and the southeast – was marked by extensive restructuring (Browaeys and Chatelain 1984). Even so, the variety of industrial systems – and the systems of small firms in particular – remained important. These included the *système Motte* in the north, consisting of federations of family firms; the aristocratic system of paper production in Annonay (Ardèche); cutlery in Thiers; the textile centers of the Vosges; the small metallurgy and watchmaking industries of Franche-Comté; and finally the silk industry in Lyon (Sabel and Zeitlin 1985). These systems joined communities of persons and populations of firms via networks of artisans and merchants in historically bounded industrial districts (Becattini 1987).

Paradoxically, it was perhaps these links – or rather the concern to preserve them as they were – that contributed to the disappearance of the local industrial systems. The rigid maintenance of these systems put the defense of the status quo before the problem of adaptability, thus blocking the economic adaptation. A *dirigiste* modernizing current, created at the end of the 1930s, was able to make its mark effectively after the war by organizing economic concentration, (cf. the X-crise group, Bonin 1988: 11). This was aided by the fact that collaboration with Vichy had politically discredited the *patronat*.

Thus, contrary to what is observed in the Italian case, local integration of the industrial systems does not always guarantee openness or adaptiveness. For French postwar planners, these systems constituted just so many obstacles to be removed. It is thus the massive disappearance of former local industrial systems that characterizes the postwar period in France, with the exception of certain specific cases to which we now turn.

Diffused industrialization and localized industrial systems: limited developments

It was only in certain zones that there developed in France phenomena similar to the Italian industrial districts. In the west these zones include the Choletais in Anjou, and the Vendée. In the east they include the Savoys, the Monts du Lyonnais, and the Jura. Given what is, after all, in France, the recent interest in problems of districts and diffused industrialization, one might suspect that this meager harvest reflects only the deficit of research on the theme. And the inventory certainly remains to be completed. But it is also true that outside of these principal zones, industrialization has not spread in France as dramatically as in the north-east-center of Italy and the 70 area-systems listed by Garofoli (Garofoli 1981, 1985). The reason for this is perhaps that in France such zones are exceptional, isolated culturally, politically, socially, and even economically.

As Bagnasco and Trigilia's observations regarding the white zone of Bassano and the red zone of the Val d'Elsa (Bagnasco and Trigilia 1984, 1988) suggest, diffuse industrialization and localized industrial systems have developed in France in ideologically specific zones, places which cultivate a certain political, social, and religious autonomy from the rest of the country.

In the Vendée and Cholet, we thus find wholly "white" zones, formerly royalist and very Catholic, marked by an ancient opposition to the Republican state; and these characteristics are equally marked in the Monts du Lyonnais. In the Jura and areas such as Oyonnax, it is rather the old socialist and libertarian tradition of the nineteenth century, and also the Utopians, that seem to have played a crucial role, and these traditions are reflected in a type of development based on craft firms.

In recent years, all these zones have had grown rapidly. Examples are the shoe industry in Cholet and plastics in Oyonnax. The organization of flexibility in these regions is reflected in the way they have succeeded in adapting to crises, even using them to facilitate renewal (changes in the productive system in the case of the shoe industry, adoption of new techniques in plastics, etc.).

These zones in a sense cultivated their autonomy, favoring internal complementarities and looking first of all within themselves for solutions to their problems without, however, closing in upon themselves as did local industrial systems.

In contrast to the preceding examples, where the local order hindered openness to the market, here openness to the exterior, and the adaptativeness which this implies, constitute the best guarantee of the longevity of the local system.

Networks of small firms: more and more delocalized developments?

Much recent analysis has shown the superior adaptiveness and capacity for offensive flexibility of small firms. Numerous studies have also shown the importance in France of this type of development in the creation of firms, the transformation of the industrial tissue, and the development of technopoles.

Yet even though these new dynamics may sometimes take place on a local basis, they seem in fact to depend more and more on the formation of professionalized networks, which are not restricted to the local level. Recent studies of French *technopoles* has shown, for example, how little articulation there is between the strictly local milieu and the small established firms: local firms are more competitive than they are cooperative (Chanaron, Perrin, and Ruffieux 1988). A number of studies of old industrial regions have also uncovered the radical degradation of the horizontal solidarity system (de Banville 1984). Our comparisons – Annonay in France and the Val d'Elsa in Italy – have also shown the different dynamics of these two countries (Ganne *et al.* 1988). We observed in Italy a reterritorialization of activity at a wider regional scale, while in France we found increased mobilization of external resources in the form of professional networks: the change in the social model of the firm seemed thus to be accompanied by a transformation of what has been called the territory-zone into a territory-network (Leborgne and Lipietz 1988).

In view of these latter developments it would seem that the texture of the industrial tissue is becoming less and less "local", in the Italian sense. This hardly suggests development of industrial districts on the Italian model.

Thus we observe in the French case that:

1 whatever their strength in terms of local structuring, local industrial systems may disappear; and that they have in large part already disappeared;
2 in recent times the reemergence of industrial districts in the Italian sense has not in fact occurred except in isolated and circumscribed cases; and
3 such recompositions of small firm systems as do exist seem to follow a logic where the strictly local aspect now plays a relatively restricted role.

This suggests striking contrasts, point by point, to almost everything that has been observed in Italy. The Italian districts had their phase of full development at the very same time that France, under the effect of a massive movement of concentration, witnessed the disappearance of old local industrial systems. And it was also the case that – at the very time when, following the crisis, a certain reterritorialization of activities seemed to be taking place in Italy – in France there seemed to be a deterritorialization of the small firm networks.

A POLITICAL ECONOMY OF LOCAL INDUSTRIAL SYSTEMS: AN INTERPRETATION OF THE FRENCH CASE

A relative closure of local industrial systems, a preponderance of state action over that of the main economic actors, delocalizations resulting from local policies: localized dynamics observed in France seem in this way to obey completely different rules from those usually considered to hold in the Italian case. Could the explanation be that the dynamics of the local industrial systems in France depend on other factors?

Economic Malthusianism and local industrial systems

We noted above that localized industrial systems were quite important in the interwar period in France. To understand the particular position and evolution of local industrial systems in France – and their relative closure – recall that the period was marked by a Malthusian economic strategy backed up by the central state.

The greater persistence in France than elsewhere of localized systems (in the 1930s, France had a much less concentrated industrial structure than England or Germany) is due more to the fact that the prevailing Malthusian economic strategy reinforced these systems than to their internal, autonomous dynamics.

This economic Malthusianism consisted of a policy of promoting accords and cartels, and resulted in involution and closure of the districts[2].

As A. Sauvy wrote:

> The [Malthusian] spirit permeated all of the legislation of the crisis, and even the pre-crisis. Conservatives and socialists tried to combat the crisis by acting on the symptoms: scarcity, to drive up prices, underproductivity, to increase employment, protection and closure of professions – these remedies accentuated, at the national level, the very problems they were intended to combat. No [other country] was affected to the same degree as France by such a serious problem.
> (Sauvy 1984: 413).

In 1919 reconstruction was too rapid and ambitious. Many firms simply rebuilt their prewar factories, with no effort at modernization. The new was already outdated (Bonin 1988: 50). As late as 1931, in spite of some concentration in a few branches and increased vertical integration, French industry was still much more fragmented than its neighbours: the capitalization of Saint Gobain and Alsthom, the two largest companies of the time, was only a fifteenth that of the British ICI or the German I.G. Farben. The electrical industry was divided among many firms: no less than two thirds of the industry's workers were employed in firms with fewer than 100 workers, with 40 percent in shops with fewer than 10 employees (Kuisel 1984: 158).

Essentially, French industrial development was based on networks of small firms, with management remaining individualistic, rooted in family capitalism, and firms serving regional markets. Firms were not geared to conquering larger markets, preferring instead occasional associations to fusions (Bonin 1988: 65). They were wary with regard to outside finance, which was pejoratively regarded as the domain of speculators, merchants and "pirates" (Sauvy 1984: 104).

In this period the main fear was overproduction. Although management resisted the *dirigiste* practices the state had adopted during the war, after the armistice it soon renewed the demand that the state regulate competition. This included restriction of external competition through protectionism – customs duties doubled on average in 1931 – but also, and above all, restriction of internal competition through the authorization of understandings and cartels (Bonin 1988: 92). The colonial empire remained for its part a privileged domain reserved to French industry.

In this way the understandings multiplied. Subtle distinctions allowed legitimated gentlemen's agreements (thanks to a law of 1926) and even created incentives to reach such agreements. They were now not only permitted but encouraged. In 1935 a bill designed to require participation in such understandings failed only narrowly (Sauvy 1984: 114; Bonin 1988: 94). It was believed at the time that fixing prices and setting production quotas was the best way to balance production and consumption. Overproduction was to be avoided by limiting competition, which lowered prices to the level of marginal costs, and ruined firms (Sauvy 1984: 111–12). The use of understandings expanded considerably in this period; according to Sauvy, on the eve of World War II there were between 1,000 and 3,000 agreements (Sauvy 1984: 117, note 1).

Certain trades and industries could thus be organized as closed associations, the most extreme case being the shoemakers and shoe-repairers: after the forty or so instances of direct regulation of the shoe trades by the state, the Le Poullen law of 22 March 1936 required authorization of the Ministry of Commerce two years in advance of the opening of any factory or workshop, as well as any enlargement or relocation of an existing factory. On 7 April of the same year another measure instituted the shoe-repairers' monopoly: fear of an incursion by Bata and its modern methods motivated a five-year prohibition on the opening of any workshop or repair booth (Sauvy 1984: 115).

This "republic of the holes in the soles," as it was called, may seem humorous now, but it reflects well the general opinion of the time: F. Ditte, writing in *Le Figaro* in 1931, for example took pride in "our *timid, prosperous* economy,"[3] and compared it to the "presumptuous, decadent" economies of the Anglo-Saxon races (Sauvy 1984: 103). The state intervened in more and more details. The result of such policies was that the real degree of competition approached zero in the France of the 1940s (Caron 1981: 365).

This policy guaranteed the status quo, but was incapable of dealing with the concentration of firms and the necessary transformation of the industrial structure relative to 1900–30 period. It resulted in a freezing of investment whose main effect was a decline of almost 20 percentage points in national income between 1929 and 1938 (Bonin 1984: 94). Less rich and less dynamic, this is how France appeared just before World War II (Bonin 1986: 97).

It is against this background that one should look at the persistence of localized industrial systems in France just after World War II. These systems had little openness to the exterior and were organized to defend the status quo, while the locale itself was more segmented than unified due to the fact that each branch attempted to preserve its link with the state and not with other local producers.

It is not, as in the Italian case, the autonomy of the zones and their structural independence with regard to outside authority that is most important here. Rather it was precisely through their articulation to the national political structure that the local actors seem to have been defined. The "Malthusian state" used the preservation and renewal of local systems as the vehicle for the gradual evolution it sought, with local actors expecting the state in return to guarantee the status quo. But the circle was closing in upon itself. It was thus not in spite of nor in the absence of the state that the different local industrial systems persisted in France before the war, but thanks to the state and in close contact with it. The "Notable-system" was key to local power in a centralized political economy (Grémion 1976).

Annonay is a case in point. I began studying the zone with the idea of finding an industrial ensemble of local small family firms that were closely interconnected. I supposed that the interconnection was threatened by the concentration of capitalism. I was thus astonished to find a highly segmented economic and social space, divided between the vocational community of the leather goods industry and that of the paper industry. These two communities had essentially no overlap (Ganne, 1983a). They rubbed shoulders in a restricted space, with the small firms of the leather sector functioning among themselves as a separate industrial district, while each of the different paper mills remained rooted in a kind of "estate-based, aristocratic" management applied to industry. There were few links between the two industries. The leather-working sector was marked by a specific system of cooperation/competition that developed work force mobility, playing on complementarities on the level of orders as well as that of techniques. There was also a system of reciprocal money drafts, under the benevolent eye of the local bank, which created a system of mutual insurance to cover unforeseen circumstances.

Otherwise, however, the local universe was socially, economically, and politically highly fragmented among different industries, each group relying more on its own activity than on the local system for the solutions to its problems. The fields of intervention of the local economic and political institutions were carefully delineated with the social and equipment side being the concern of the local public bodies, and the economic that of the Chamber of Commerce. Interference between fields of authority was not tolerated.

In short, this was a local system that was highly fragmented between both the economic and the political and the different branches of industry. The key to this segmentation was the type of vertical relation that each body cultivated with regard to the central power. Through negotiations with the state, each industry sought guarantees of its market and the reinforcement of the status quo, as in the policy of understandings and cartels. This fragmentation was also supervised, on the political level, by the notable system. The notable's function was to guarantee and promote diversity and to broker relations with the external world and the central political power. This was the source of both his strength and his weakness: at base, his power was both guarantor of and hostage to the status quo (Worms 1966; Ganne 1985). As the interface between the community and society in general, the power of the notable thus depends directly on the centralized character of the state. We are a long way, here, from the notion of autonomous local power as it is understood in connection with the Italian districts, not to speak of the American communities. In the brokering game that it plays between central power and local actors, the system of notables is at the opposite pole from the American ideal of community. In this ideal, a "community" is a unit of social action capable of mobilizing itself to resolve, without further ceremony, the problems of its survival and development (Grémion 1976: 264); community was to be the antithesis of the state. In contrast, in France, via the notable system in particular, the state remained precisely the frame of reference which defined the action space for local actors. The system of notables impeded the widespread mobilization which would have allowed collectivities to resolve their problems from the inside. It condensed the collective action space. This in turn allowed the notable to position himself as an indispensable intermediary. The price of this system was steep: absence of information, stifling of conflicts, crushing of minorities, formalism of democratic procedures. All of these measures prevented the free play of groups, in order to permit the notable to have his hands free in negotiations with the state (Grémion 1976: 267).

The fragmentation of local societies and the system of notables thus seem to go hand-in-hand, the notable having the function of representing the diversity of interests as those of the locality, a system in which he wields all the power, and therefore perpetuates. The traditional local system thus appears as an integral part of the system of regulation developed at the instigation of the central state in France. In this way we see the danger of equating "local" with community-based or endogenous.

The development of postwar statist interventionism, and the liquidation of former local systems

Under the influence of postwar state planning, the earlier system was challenged and many, perhaps most, of the older local industrial systems disappeared. The central state also promoted the decentralization that was later adopted. Let us take up both points.

In the postwar period, the state attempted to go beyond its role as a mere regulator and become an active agent of economic and social development (Grémion 1976: 471). The system of cartels and understandings was essentially regulatory; in the postwar reconstruction process, it took on an actively directive role, which led it to shake up somewhat the old politico-economic order.

Even before the war, the economic climate had militated for a strengthening of the role of the state. Faced with the incompetence of the heads of family firms and politicians, a technocratic current held that the economy should be run by specialists – technicians, modern managers, and "enlightened" functionaries – in the name of efficiency (Bonin 1988: 114 ss; cf. the importance of currents such as those of the *Nouveaux Cahiers* in the management environment, those of X-crise in the Polytechnique world, etc.). There was a widespread desire that the state should act as prime mover in the modernization process. To do this it had to gain new competences.

The Vichy regime had in fact already taken the first steps in this direction. A number of technocrats at the time regarded the Pétain dictatorship as a historic opportunity to force the conservative, traditional elements of society to accept the upheavals necessary for modernization, which the parliamentary system had bound hand and foot through the electoral system of the Third Republic. The modernizers believed that France was bogged down in a swamp of small family firms; they wanted concentration in the hands of specialists, a sort of *dirigiste* planning philosophy (Bonin 1988: 122).

Though this current failed to take hold at the time, in part due to the war, it continued to mature through schools such as that of Uriage; and it made a forceful comeback after the war. The mixed economy took hold in part because management had shown its limits in the course of the previous period and had discredited itself politically in the war.

We will not go into detail here about all the economic policies of the successive Plans, of the 1960s and 1970s which organized the concentration and reorganization of the French economy. A few points will suffice.

Initially charged only with improving the spatial distribution of activities in France, the policy of Territorial Planning (*Aménagement du Territoire*) was transformed step by step, toward the end of the Gaullist period and then under President Pompidou into a key element of economic policy. It organized industrial groups, created massive industry–port complexes, and developed active intervention at many levels of the system. The local systems formed the pockets of backwardness that economic modernization was meant to restructure. The only permissible reason for not pursuing concentration and growth in establishment size was said to be the size of markets, as noted in 1974 by the authors of an important INSEE report on *The Industrial Transformation of France* (INSEE 1974). Agreement was so widespread at the time that Marxists and state officials could speak the same language, the leftist economist C. Palloix being at one in this regard with J. Ferry, the vice-president of the CNPF, the employers' association (Saglio 1985).

Restructuring had major effects in all the old industrial zones; it led to the disappearance, with little resistance, of the old forms of organization of the local industrial systems.

The disappearance of these old systems was all the more dramatic in that it signaled not only the dissociation of the links that had once existed between the state and the firm, but also a serious challenge to the notable system itself.

In attempting to be more systematic, policy no longer aimed at activity-by-activity control. Its aim instead was to shape the general dynamics of the industrial economy. Specialized functional interventions were replaced by horizontal policies set up progressively by the state, culminating in the 1982–3 decentralization laws.

The success of such policies was not a given, however, since earlier regional reforms, such as that of 1964, had been absorbed and coopted by the local politico-administrative system (Grémion 1976).

But in the concern for planning – general, this time, and no longer just functional and sectoral – a new system of local/national relations was promoted. Responsibility was transferred toward subnational entities. Policies based on new forms of contracts were also used. The more general contracts (zonal contracts, countryside contracts, and then state/regional contracts, etc.) progressively supplanted the specialized, sectoral interventions of the previous system. There were dramatic implications for the notable system. The latter was characterized by vertical, specialized, and privileged relations between the notable and the central apparatus; the new system by means of contracts and management modes that are more "horizontal." They decompartmentalize the local system, so that it may be managed as a unified entity, not as a system of fragments, each of whose interests are brokered *vis-à-vis* the central government by the notable (Ganne 1985).[4]

Thus a real and general transformation of the politico-economic system had gradually occurred in France since the war. This transformation affects the type of relation between and the forms of regulation of localized systems and the central state. A new type of local system, initiated by the 1982–3 decentralization law, recognized a new moral entity in local collectivities: new capacities for decision-making, with control by the central state carried out *a posteriori*, and no longer *a priori*, as previously.

The transition from the republican state as a regulator to the postwar rationalizing state was paralleled by this passage from a society of representation, congruent with the notable system, to a society of organization (Grémion 1976: 471). The idea was that mobilization should go all the way to the local level, where horizontal systems of cooperation would be developed.

This is nonetheless a long way from an endogenous model of local development; local systems do not have the independence which, according to some, is the guarantor of their efficacy. They are, if anything, created by the centralized system of political regulation. This logical relation between center and periphery was explicitly elaborated in the preparatory report of the Ninth Urban Plan, which advocated the establishment of local systems

– based on the mobilization and coordination of endogenous resources – as an antidote to structures that were highly fragmented and uncoordinated.

Towards a standardization of the local?

These changes also explain the relative homogeneity observed in the recent restructuring at the local level. It is indeed according to an identical type of model that contracts governing plans for the development of medium-sized towns and regions are executed. Local differences and specificities are now only taken into account to the extent that they conform to the overall, highly normative framework. There are standardized formulae for evaluating requests for subsidies for investments in local infrastructure and mobilizing local actors. Little space exists for the affirmation of other, less normative, specificities. We observe, in other words, the creation of a "canon of the local."

Decentralization thus functions, paradoxically, as a *unifying frame* of reference. We are a long way, here, from the diversity of models of organization and local specificities that seem to characterize the Italian districts.

This dynamic of unification explains one of the paradoxes that we brought up in the introduction: it is no longer surprising to observe that the reinforcement of the local level is accompanied by greater indifference with regard to the localization of firms and that it encourages development of systems of networks that are more dispersed than localized (Ganne, *et al.* 1988). Decentralization appears to produce paradoxical, perverse effects.

It is thus in its overall context that the problem has to be resituated: the imposition of a single political model of management on the local level, by homogenizing the framework of action at that level, actually encourages economic actors to develop linkages outside their immediate local system. Submitted, as it were, to a single set of norms, the local no longer constitutes a problem to be controlled. The unification produced by decentralization thus promotes mobility of economic activity.

We observe in any case that there is a wide gap between the constitution of the local in contemporary France and Italy. It is important to avoid hasty assimilation of French developments to the Italian case.

CONCLUSION

What conclusions can be drawn about the functioning of the evolution of local industrial systems in France?

The maintenance of old industrial systems in France, far from being the result of local contingent social forces, existing, as it were on the margins of powerful structures, is better understood as an integral part of the centralized French administrative system.

The relatively closed and static nature of the old French industrial districts was congruent with the old form of political brokering at the territorial level which has been called the notable system.

Political authorities sought to break down these "feudal" features and the productivity blockages to which they led in the postwar years. More recently social and cultural homogenization has been pushed by the central state all the way down to the local level. The decentralization law has actually encouraged this trend.

It is to this political economy, which allows us to see how economics and politics become systems that shape each other mutually, that we must refer in order to understand the specificity of the French case, as well as its limits.

The paradox of the French case is precisely that it challenges contemporary stereotypes by insisting that local industrial systems again be placed within the context of the political administrative system. In doing so I have also sought to avoid both the neo-structuralism of the Regulation School (i.e. the current wave of theorization on the subject of post-Fordism) and the neo-culturalist tendencies, which have been revived in the research on the Italian industrial districts (i.e. the increasing importance attributed to local cultures and enterprise cultures).

The risk at present is that the political level is being ignored.

In recent times, a certain fetishization of local development has arisen in response to the success of Italian industrial districts. This "neo-localism" creates several serious forms of confusion:

a) it conflates a number of cases which are in fact very different (Sfax in Tunisia, Cholet in France, Prato in Italy, etc.);[5]
b) it perpetuates, the illusion of local autonomy;
c) it frequently confuses local systems with new forms of industrial organization based on flexible automation, and thus incorrectly confers an image of modernity on these local systems.

The French situation does, nonetheless, seem to provide an opportunity to reflect positively on the variety of local industrial systems. First, local systems can exist without being characterized, as in the Italian case, by local–horizontal integration of economic, social and political dynamics. The French local industrial system is based on local segmentation of these different aspects, each sphere (professional as well as political) favoring vertical relations with the central state.

Second, the local and the endogenous are not the same. If the number of local industrial systems between the wars was larger in France than in other countries, it was not because they were structured in a more deeply endogenous way, or because, as in Italy, they were endowed with a relative economic, social, and political homogeneity that made them more resistant than elsewhere. Rather, it was because these local industrial systems were more heavily regulated and closed in upon themselves, preserving and reproducing, by their mode of articulation with an overall politico-economic system, their fragmented mode of organization.

In the study of the districts and the local industrial systems, the problem is not to oppose the cultural and the economic, the local and the general, or the endogenous and the exogenous, as is implicit in much research, but to understand how differentiated, reciprocal forms of the economic and the cultural which are articulated by politics were mediated through the institutional system.

In the French case, reference to a centralized institutional state structure was indispensable. Whereas in the Italian case districts manage their competition directly, in France a regulating, redistributing state mediates the relationships between industrial systems. The state organizes economic policy, gives out subsidies, and regulates systems of social and professional relations. All of this is, in the final analysis, a question of political habits. It is thus essential to develop ways to contextualize the economic in order to develop a more effective economic sociology.

12

LOCALIZED INDUSTRIAL SYSTEMS IN FRANCE

A particular type of industrial system

Jean Saglio

Have we emerged from the period of growth linked to a system of economic regulation of the Fordist type, and are we now entering a period in which flexible specialization will become more important? In more basic terms: are small firms the future of developed economies? This issue generates heated debate. Questions are asked about the role of small firms in the comparative industrial growth of different countries, about their capacities of adaptation to market fluctuations and technological change, about their emergence in the developed economies and their capacity to respond to the problems of less developed economies. However, the renewed interest shown by researchers in groupings of small firms, and in the dynamism of these groups, does not lead us automatically to conclude that this new object of research is a new social object as such. Even if we suppose that the development of nonconcentrated systems of industries is the future of the economies of developed countries, it should be remembered that this was also a large feature of their past, as it is of their present: if there is a renewal going on, it is perhaps as much in the fact that we are currently concerned with improving our understanding of the working of these firms as in the fact that they are evolving and developing.

It is important to be careful when making lists of the new conclusions that we can draw from recently accumulated observations on such industrial ensembles. Or at least we can examine the ways in which these observations call into question the classical approaches employed in understanding industrialization. In putting forward a differentiated approach to the modes of regulation existing in France today, we will attempt to show that innovation is perhaps not to be found where one expects it, and that the systems that are being set up are perhaps not as new as they appear. On the other hand, it is not unthinkable that the new mode of regulation of the industrial economy, which is being eased into place today, may differ fairly radically from what France has been used to for almost a century now, and which is characterized by the positions and roles of the different actors. So it is necessary to introduce a concept of regulation which is not strictly linked to the notion of state action. Other types of social regulation of industrial development can be found in current French cases, and their analysis has led to the introduction of the notion of industrial systems as a way to understand these new forms of links between societal problems and economic development. Besides altering the means of regulation of production, we are perhaps in the process of abandoning a certain mode of societal usage of the economy. This would also involve supposing that a change is occurring in the very definition of the articulation between French society and its mode of production.

PARADOXES

The fact that numerous studies have been carried out in different countries on this theme of the development of small firms means that we now have access to an empirical corpus which is largely of good quality. The observations that one can derive from the examination of this material lead to the conclusion that there exists here not only an accumulation of knowledge contributing to the enrichment of the usual paradigms, but also a series of paradoxical results leading to a questioning of the pertinence of this paradigm. Most of this questioning arises from some well-known criticisms of the classical approaches. Between firms in the competitive market, it is impossible to develop forms of cooperation and to stabilize places in the production process. Such cooperation, which concerns neither oligopolies nor cartels, modifies the functioning of markets, which are socially regulated.

Workers' behavior patterns are not always conditioned by the need for the impersonal, bureaucratic protection afforded by the large firm. Their involvement in the firm goes beyond simple work force exchanges, and they may be actors in the social networks which link the different firms. Such networks are not only the result of economic exchanges, they are also forms of social community.

In raising some of these questions at this particular point, we do not claim to be drawing up an exhaustive list, but simply to be clarifying those that appear to us to be the most productive, in terms of leading to the construction of a system of comprehension of industrial history.

Places in the production process

The first of these paradoxical observations has to do with the nature of places that small firms, plus the totality of nonconcentrated industries, occupy in the production process. The classical approach insists on the dualism of positions and strategies. Not being in a position to mount effective opposition to large firms, small firms would tend to occupy the places that these others leave vacant in the production process. And thus they would tend to find themselves in dependent positions of subcontracting, in terms of capacity or of specialty: capacity subcontracting when they are supposed to be smoothing out the unevennesses in the production process which the dominant firms cannot or will not deal with; specialty subcontracting when they take on unit or small-batch production which could not be organized in a Taylorist mode of production.

At first sight, a fair number of results seem to uphold this thesis. But at the same time, finer analyses allow us to note that the positions of small firms are not ineluctably threatened by growth and concentration in the industrial sectors. When production is subjected to the dictates of fashion, small firms in the record industry (Hennion and Vignolle 1978), or those of the clothing industry, do succeed in occupying places which are not really threatened by the appetite of more concentrated firms. A doubt remains, however, since it appears that, in such sectors, the system of places is probably more durable than the firms themselves, which seem to turn over at a rapid rate, quite often appearing only for the time needed to make a mark, then disappearing almost as suddenly as they appeared. It may be noted in passing (and we will come back to this question later on) that more in-depth analyses of such sectors seem to show that certain employees, at least, appear to adapt their strategies of employment to this type of instability of firms and institutions.

A second type of case can be observed, where instability and short product lifetime are quite closely linked to the rapidity of technological evolution, or to the necessity of a high degree of adaptibility of the production process. Recruiting more skilled workers, working to order and on projects, the firms of this group are often more stable than the preceding ones, but are not necessarily more concentrated. Considered as specialized subcontractors, they produce by the unit or in small batches, sometimes on the premises of their clients (whose ranks may include large firms), without, however, becoming dependent on them.

A fair number of observers are inclined to think that such institutions are more frequent today than previously. Do we have to see this as an effect of the recent evolution of production and consumption systems which nowadays give over a larger share to less standardized products? And could this then be explained by the interaction of the evolution of the public's tastes and the spread of more flexible production technologies?

And yet at the same time, such situations are not new. The system of fashion, at least in clothing-related industries, pre-dated the industrial revolution. Small firms of toolmakers or qualified technicians are not a recent innovation. Even at the highest pitch of the movement toward industrial concentration, they retained a non-negligible place in the apparatus of production. Thus, rather than a dual system, where these small firms would only occupy slots left temporarily vacant by more concentrated firms, one might just as well envisage that this kind of firm has been able, sometimes collectively, to set in motion strategies offering an alternative to concentration, and, by this means, to parry the threat that larger firms have brought to bear on them.

From access to markets to their mode of construction

It is commonly supposed that only large firms in a monopolistic or oligopolistic situation are capable of dominating and orienting markets and techniques for their own benefit. The examples which we have just raised tend to show that other configurations are possible, in which collectivities of independent firms are nonetheless capable of influencing such markets and techniques.

Of course the two types of situation are not exactly identical. The large monopolistic firm, which dominates a market and which wishes to enforce its management and influence the behavior of its actors, produces arguments of rationalization and prediction which will lead to the required choices (Bauer and Cohen 1981). Conversely, in a system dominated by fashion, each of the actors rationalizes his own position and strategies while discussing a market in which he would individually have no power. On one side, therefore, the management of the market is internalized in, and dominated by, the organization, and, on the other, evolution is linked to relations of power and conflicts over which no single actor is able to exert sufficient influence to turn them to his advantage.

In thus limiting oneself to just the two extremes of the spectrum of possible configurations, it might be tempting to see a validation of the hypothesis whereby one is dealing with two mutually exclusive alternatives. But the large firm is not simply a hierarchical structure where the actors are dominated by the managers. It is also a social organization, and management is not only a problem of technical adequacy, but also a method of mobilization and argumentation designed to convince the other actors and to obtain their participation. So the firm is not only a hierarchy, but also a complex structure of governance.

Observations drawn from analysis of intermediate cases lead to the development of this kind of criticism of a purely dichotomistic approach. Thus oligopolistic situations, like that of the American motor industry in the thirty years following World War II (Tolliday 1988), or the one we observed in France more recently (Saglio 1989), are already intermediate situations. They combine, on the one hand, collective management of quantities and qualities of products, and on the other hand a relative dependence of the system on the solidity of modes of confidence which spring up between the oligopoly's sellers and private buyers. The oligopoly's capacity to impose its solutions thus rests on the strength and stability of the process of enlisting buyers which the firms of the oligopoly succeed in putting into operation.

On the other hand, the markets in which the structure of production is the most atomized among numerous small production units are not inevitably those which are the least controlled by the producers. The fashion system sometimes gives way to more or less formalized collective brands. In this way the different segments of the high-class shoe market are tied to Italian producers who lay down the law. Beyond that, one may also observe real collective monopolies through which a profession composed of small firms achieves acknowledgement of its right to fix the qualities and prices of products (Saglio and Thuderoz 1989). These are not necessarily anachronisms from an old corporatist tradition, but may also be constituted as new responses to very real situations.

If one attempts to take into account this diversity of possible situations and strategies, one cannot easily accept the classic hypothesis according to which the intensity of control of the market is linked to the relative size of the principal producer. The two variables are evidently not so closely related. On the other hand, the establishment and stabilization of a mode of market management appears as the result of complex strategies linking producers and consumers, in which the most probable positions are never certain, and the most improbable are never, however, prohibited. This means that it is not enough to do an analysis of positions in the production system, if one wants to understand the establishment of these modes of management of markets; it is rather a question of considering that these processes are negotiations, and that their products are social constructions of markets (Bagnasco 1985; Saglio 1989).

In the same roundabout way, it can also be understood that the classical approaches which link the size of the firm too closely to the area of the markets in which they participate must equally be called into question. According to such hypotheses, it was considered that artisans participated in neighborhood markets, small firms in regional or national markets, and very large firms with multinational connections in international markets. Already, Fernand Braudel (1979) noted that the relationship was not so simple, and that, from the sixteenth century onwards, the types of relations which became established in these different markets were not identical. Corresponding to the neighborhood markets there was an economy of reciprocal exchanges, corresponding to the markets of the towns and regions a more classical system of market exchanges, and to the multinational firms a system more linked to organization. Bagnasco, following Pizzorno, noted that there could also exist a fourth type of regulation of exchanges, which he named political exchange, and that the different types of exchange are not so clearly demarcated: thus reciprocity may exist between industrial firms participating in regional or international markets.

Although, statistically speaking, the observation that small firms participate less in international markets turns out to be correct, this does not necessarily mean that it is legitimate to conclude that the fact of their size alone renders them incapable of taking part in the different types of exchange. At most one may note that the internationalization of market management by an organization presupposes, in the case of groupings of small firms, the existence of a common collective structure to carry out this management, and which would possess sufficiently powerful forms of incentives and reprisals to compel most firms to respect collective rules.

The specialization of employees

Analyses of the determination of employment rules, and notably the setting of wage levels, have long been concerned, in France especially, above all with large firms. Here, the organized action of the employees, working through disputes and negotiations, came to impose bureaucratic rules which limited the arbitrary role of management, and imposed the taking into account by the management of the values proper to the employee group. On the other hand, it appeared that small firms remained the domain of preference of paternalistic, authoritarian forms of personnel management. One could then pass from such a dichotomy to a dualistic perspective of the labor market, considering that the employees who were the best-off in trade union terms would try above all to get into the most highly regulated sectors of the market, leaving those individuals who had fewer resources to occupy the posts corresponding to the least protected sectors.

More precise analyses have again led to questioning this kind of simple link between forms of firm, forms of work force management, and collective aspirations and strategies of employees. In the first place, it does not appear obvious that the best-off individuals necessarily look for the kinds of statutory protection that the largest firms can offer them. Thus, for example, it was found that many highly skilled workers, who specialized in the manufacture of complex equipment for presses in the metal and plastic-molding industries of the Lyon urban district, did not wish to be employed in the largest firms in the field. On the contrary, these latter were often considered by them as places to begin a career, where one could find a job on graduating from a professional school, but which one should leave after a few years of experience. For the most qualified workers, the small firm offered advantages, in terms of work interest and autonomy, which the large ones could not match. As to the impact of arbitrary employer authority, this was, in their eyes, greatly limited by the possibility, which they constantly envisaged, of changing firms and continuing their career elsewhere (Saglio *et al.* 1984).

Small firms are thus not inevitably the domain of employer arbitrariness and of employee submissiveness. Moreover, large organizations are not the only places where collective action by employees manages to impose forms of work force management which are less unfavorable to the employees than the pure laws of the market would be. In other terms, and to go back to Catherine Paradeise's formulations (Paradeise 1984), bureaucratic organization is not the only mode of closure of the labor market to satisfy the demands of the employees. More collective forms of management, which impose strict rules and allow interfirm mobility, can turn out just as favorably as bureaucratic systems of the firm.

To take account of the possible existence of such systems leads to a reconsideration of the presuppositions implicit in a number of previous analyses. Here again, as in the case of the analysis of market forms previously mentioned, the firm is not exclusively the place of negotiation and enunciation of the rules. The fact that negotiation does not take place in the firm does not necessarily imply that the protection of employees should be reduced to its most simple expression. Between the size of the firm and the form of management of labor markets, relations are more complex than a simple statistical determination. In this respect, the small firm is not necessarily a nightmare of employer arbitrariness, any more than, reciprocally, the large firm is always the seat of impersonal, blind bureaucracy.

Social networks and exchanges

The firm is thus neither the only locus of management of market relations, nor that of industrial relations. The fact that neither of these is managed from the interior of the organization does not, however, authorize the conclusion that they are controlled solely by the impersonal laws of the market. In the one case, as in the other, modes of collective management having to do with formal institutions or informal customs can exist in a lasting way, and can structure the strategies of the different actors.

The existence of such relations, which link the different firms and force them into a certain degree of cooperation, encourages the postulation of the existence of stable relations between them. The available analyses of the functioning of such networks leads, however, to distinguishing this type of instruction from the classical model of the firm. A network of firms is not a quasi-large-firm, since the system of authority and legitimacy which is exercised there cannot be mistaken for the one which obtains in the classical model of the large firm. If, in the latter, one may suppose that legitimacy depends on a form of contract, explicit or implicit, in the case of small firm networks, on the other hand, it is rather the fact of adhesion which is the foundation.

Is it, however, a question of claiming that this type of community is bound together only by adhesion to the same system of values, and that there is thus a certain labor ethic which constitutes the social link that provides the foundation of these networks? One finds that for a certain number of authors, at least in the interrogative form, the sketch of the response is affirmative (Becattini 1989; Raveyre 1988). Such a solution is, however, unsatisfactory, in that this kind of network evidently creates ties between actors whose interests and political orientations do not rigorously match up. The dissatisfaction comes from the fact that it is then necessary to postulate that economic relations are secondary compared to the sharing of a common ideology which strongly binds together the community.

To resolve this dilemma, we have proposed to apply, in such cases, the solution advocated by Blau (1964) in order to understand nontrading structures of exchange. These are then characterized as social exchanges, that is to say that the counterpart of the service is not entirely specified at the moment of the primary service. Thenceforth, the two partners are bound by a system of involvement which, unlike classical market systems, leads to the reinforcement of the totality of their social links.

The uncertainty that results from such wagers is not insupportable for the provider of the first service, in the first place because he can reasonably expect that the counterprovision, unspecified at the time of the first provision, will constitute, for him, a better deal than he could have obtained in a classical market exchange. Thus we have observed cases where the counterprovision consisted of economic and commercial information whose cost for the provider was small, but whose interest could turn out to be quite strategic for his partner (Saglio 1990b).

The second reason why such uncertainties are supportable for the first provider derives from the existence of a larger community than simply the two protagonists in the exchange. In such communities, one can easily identify property that is collective, that is to say available to all the members without their having to participate directly in the cost of its provision, but which is nonetheless reserved to the sole use of the members of the community. This is often the case, between industrial firms in the same group, for economic and technological information. The availability of such information constitutes an advantage for the industrialist. Moreover, if his partner should fail to meet his obligations, he can, by showing that he for his part has indeed satisfied the collective rules, require that these be used for his benefit so as to soften the blow. He thus counts on his other industrial partners in the group to mobilize on his behalf: his colleagues discriminate in his favor in the awarding of subcontracting work and the passing on of commercial information, so as to allow him to get back on his feet. Simultaneously, the existence of such collective property constitutes a constraint for the beneficiary of the first provision, to the extent that his failure to reciprocate entails the risk of his exclusion from the benefits of this collective property.

The non-specialization of relations and resources

The apparent irrationality of certain behavior patterns of actors in such systems of social exchange has often been remarked on, at least with regard to the canons of rationality that can be deduced from the study of large bureaucratic organizations. Within these, actors seek first and above all to obtain the cooperation of the other members of the organization, while minimizing their commitment to these other members. To reach such an equilibrium, management attempts to spell out clearly the contributions and legitimate means of redress of each. Besides this, procedures for establishing relations are also carefully codified. The model of functional analysis then applies perfectly, given the fact that relations are perfectly designated.

Of course it is well known that conflicts of authority are frequent, and that more complex strategies are not completely excluded; but such types of behavior, outside of the norms, can, at least in a primary analysis, be considered as malfunctions of the organization, which the actors will attempt to reduce.

The concrete analysis of the functioning of networks of actors leads quite often to the observation that strategies often diverge from this type of behavior. In taking up the models developed in analyses carried out in the sociology of science (Callon 1986), one observes strategies of engagement in which the actors modify their respective positions, while also modifying, in the same movement, the qualification of the resources at their disposal. At first sight, such strategies of circumvention, playing more on the mobilization of resources than the utilization of legitimate skills, appear paradoxical. They nevertheless quite often allow a reduction of the shortages and gaps that the external observer can pick out in the structure. Thus, the potential entrepreneur has a plethora of ways of procuring capital, involving the mobilization of family resources, requests to local bankers or to state institutions, etc. The argumentation that has to be set forth in order to achieve one's ends will be different in each of these cases, and will most frequently have little to do with classical microeconomic calculation. In the same way, the resources which the worker must mobilize in order to obtain a job very often will have little to do with his skills alone; rather, he will do better if he can mobilize the right network of relations, which will be based on his family, school, professional experience, etc.

Of course it is the case in large firms also that relations between actors are not completely specified by formal organizational requirements. Other kinds of social linkages can be used by the actors, in order to form pressure groups in the strategic debate. These relations, often described as informal, are mobilized by actors when they disagree about the application of formal webs of rules. So informal regulation is used, in large organizations, as a way of contesting the results of formal regulation.

In networks of small firms, the situation is quite different. Formal networks and hierarchical structures do not exist here, and informal networks are often the only way of ensuring cooperation. In systems like this, relations between actors are not specified and qualified by formal rules. To know that in such systems two well-defined actors have a relation to each other is not enough to tell us what will be the nature of the exchanges that they will carry out. In a certain way, in such systems, the division of labor does not imply a specialization of relations.

This absence of specialization implies once more that the argumentations set forth are multiple. In such situations, it is more important for the actor to stimulate the interest of his partner than to convince him of the accuracy of his reasoning or the quality of his skills.

Despite their apparent irrationality, such models of relations are probably more widespread than they appear to be, and they may also serve as an analytic backdrop for the strategies of a number of actors, including those who belong to the most bureaucratic of large organizations. In particular, this schema appears particularly effective in analyzing the concrete commercial strategies used by established salesmen, oligopolistic situations included (Saglio 1989). It is in fact rare that the sales pitches that such salesmen make should be exclusively concerned with relations between the technical qualities and the prices of the products they are offering. More often, with their long-term clients, they attempt to create a whole skein of diverse social relations going from good interpersonal relations to the provision of technical and economic information, so as thus to insinuate the commercial relation into a strategic context which is closer to that of social exchange than to mere trading exchange. In the same way, one observes that, in France at least, subcontracting relations are frequently more stable than might be expected from the application of market laws alone (Lorenz 1989).

Identity preferences and rules of belonging

Such analyses not only lead to the adding of a further category to the catalog of the different types of firm that is customarily made use of, they also encourage us to think that the decision maker, when he selects the actors with whom he is going to establish a stable economic relationship, takes a close interest in the social identity of these producers, and not just in the qualities of the products which they are offering in exchange. In such forms of behavior, it is important for him to share a common identity with his future partners.

In the classical analysis of market relations (Coase 1937; Williamson 1985), this point is treated by omission: the decision maker being the entrepreneur, the only reference group that he uses is the firm. The choice that he finds himself confronted with is between manufacturing in the firm itself and calling on the external market. Already, in Becattini's analysis of industrial districts, the situation is more complex, since a third term can in effect insert itself between these two extreme solutions: the entrepreneur can appeal to the local market, that is, to other firms, but which belong to the same social grouping.

These economic decision makers are in a market where, instead of letting themselves be guided by price structures alone, they manifest orientations of preference directed toward other individuals or institutions with whom they share reference communities. It is indeed a question of preferential orientation, since belonging does not at this point dictate choice in any imperative way. But, at least at identical prices, and sometimes even with a certain excess cost, one observes that the buyer does not behave in a random way, and that he systematically chooses his suppliers among those he takes to be part of the same reference community.

Thus characterized, the phenomenon certainly goes far beyond the industrial districts alone. Is it not to be found, at least in a derivative fashion, mobilized by those advertisers who, in order to sell an object, try to make people believe that the fact of possessing it demonstrates membership of a group with high social status? More directly, it is easily observable that many other reference communities can play the same role of orientating preferences in the market.

In this kind of conception, the industrial district is therefore just one of the forms of economically active social identities. But many other institutions could also be used in this way: family, region, nationality, and ethnic group thus become criteria of economic preference. "*Achetons français*" or "Let's buy British" are the kind of slogans which can have an impact on the behavior of purchasers, including industrial decision makers.

Does the taking into account of such behavior patterns not lead us, more generally, to modify our point of view of the analysis of economic decisions? In the classical and neoclassical analyses, economic exchange is a means for the two partners to increase the value of the economic resources which they possess, and to improve, at least, the efficiency of their systems of resource allocation. The social consequences of the extension of the pure market are well known. It leads to the destructuring of social relations, and their replacement by pure individualism (Weber 1971).

J. Commons' criticism of classical economics emphasizes the point that economic exchanges are not only physical exchanges of materials and services, relations between man and nature, but are also relations between person and person (Commons 1934: 56). In certain cases, it appears that exchange is also a means to manifest one's membership of, and attachment to, the social collectivity. Can one be content to consider this behavior simply as a tactical means of improving the efficiency of one's proper resources? And must we thus indefinitely, on the pretext that we are dealing with the heads of firms, interpret their manifestations of social attachment in economistic terms? Or can we not, on the contrary, sometimes interpret their economic strategies in sociological terms?

It may then be noted that the firm itself can constitute this reference community, but that this is not something that happens inevitably. The classic decision maker is supposed to prefer his firm only in cases where, in view of the production costs (for Coase) or the production and transaction costs (for Williamson), this solution is objectively preferable, that is to say preferable whatever the social identity of the decision maker. By introducing this criterion of the orientation of preference with regard to the reference social identity, one can also explain situations where decision makers choose to reintegrate previously subcontracted production, even when the direct economic advantage is not obvious. In acting thus, they manifest their attachment to a community for whose survival they are willing to invest economically.

At the same time this first example deals only with situations relevant to strategies in the product market. Other types of behavior concerned with strategies in the matter of labor relations can probably be explained using the same analytical tools. For instance, it was for a long time a common habit of observers of French industry to criticize the reluctance of heads of firms to lay off workers in times of crisis, even though it meant weakening their potential at the time when the market was getting back on its feet (Scott and MacArthur 1970; Stoléru 1969). Such a policy of stabilization of the workforce had in part to do with strictly demographic considerations, which Lorenz plainly perceived (Lorenz 1987): the French firm often seeks to exercise a monopoly of employment in its local market and, to retain this monopoly, will make a point of guaranteeing its employees a certain security of employment. On this point, it differs clearly from the English firm, which operates in a more competitive labor market.

We have attempted, in a previous study (Bunel and Saglio 1979), to give a more complete explanation of this phenomenon, showing the extent to which, for French management, the firm is more than a simple possession, or a mere means of accumulating financial resources. In putting forward the concept of the enlarged family circle we wished to bring out the phenomenon of the firm as being also a larger social institution, meaning that it can engender a certain company patriotism which, in part at least, restricts behavior, including that of management. However, it should be borne in mind that such a configuration is not necessarily the case for all existing firms. One can then suggest the hypothesis that the manager's behavior is thereby affected, and that his range of choice is not simply between the firm and the exterior, but can also be influenced by the fact that for him the firm acts – or not – as a reference community.

In stressing such types of behavior, one is deliberately moving away from what constitutes the heart of traditional economic explanations. These, to take up once more Castoriadis' criticism of them, are founded on a functional economic viewpoint, which does not take into account the symbolic functioning of institutions. The idea that symbolism is perfectly "neutral", or – which comes to the same thing – totally "adequate" to the functioning of real processes, is unacceptable and, in truth, devoid of sense (Castoriadis 1975: 168).

In their real behavior, economic agents, including managers, thus perform a certain number of acts which, in all good economic logic, must be considered as errors. Sociological analysis cannot accept such a judgment. So it is up to us, and using the same process in each case, to explain types of economic behavior that can be considered as correct, that is to say in keeping with economic logic, as well as incorrect types of behavior, that is to say behavior which is contrary to such logic to the point where it can lead to the disappearance – pure and simple – of firms.

But to do this it is not enough to consider that the limits of the actor's rationality are due to the imperfection of his information, or to that of relations between actors. We must also take into account those cases, which are more frequent than is generally supposed, where the system of values proper to the individual, and incorporated into certain institutions, becomes contradictory to economic rationality, and yet imposes itself on the actor.

The localized industrial system is then no more than one of the potential institutions or configurations towards which the decision maker will exercise this preferential orientation. In other situations, the reference community will be the firm, or a part of the firm. We know the efforts that are made by the management of large French firms to create the kind of spirit within their work forces which will orient their behavior in a way that is favorable to the firm. Might we not say that the enterprise culture they talk about is analogous to the specific favorable culture of the industrial district?

INDUSTRIAL SYSTEMS AS TOOLS OF ANALYSIS

Hierarchical organization is not the only alternative to the market

In the analysis of a large organization, whether of the administrative or the industrial type, it is most often possible to consider that the social relations defined by the limits of the institution are largely self-contained. The boundaries of the organization thus become the limits of the field of analysis, and the exterior can be apprehended as a context which is, certainly, not immutable, but whose variations have an influence only at the margins on the functioning of the organization, which must however adapt itself to such variations.

This kind of reasoning is not appropriate in the case of smaller work units. Here, on the contrary, as we saw previously for some precise points, the behavior patterns of individuals, and the systems of relations which become established between them, are a lot more strongly influenced by what happens externally. Thus, the large firm can define a primacy of the internal labor market, in the double sense that individuals' careers are acted out in the interior of the organization, and that the rules of functioning which fix working relations and individuals' expectations, and which allow for the management of possible conflicts, are negotiated and enunciated through the play of the actors within the institution. The smaller organizations are not in the same situation. Even if the employees show a certain tendency to remain in the firm, one may not deduce from this that what is involved is an internalization of the labor market, in the sense that the locus of the enunciation of rules, and of the resolution of conflicts, remains external to the firm.

It is the same thing for commercial and economic relations with external markets. On account of the weight it carries, the large firm can impose its own constraints and demands on its suppliers; and, in situations of monopoly or oligopoly, it can even do so with regard to its clients. The smaller organization, on the contrary, sees itself forced to follow currently admitted procedures, accepted customs, and constraints imposed from the exterior.

Nonetheless, as we have already seen, the alternative to the externalization of the management of constraints in the organization is not always the pure market, upon which the firm would have no means of acting, and in which the different actors would be considered as atomized, having no possibility of intervening to modify the rules of functioning.

The classical solution brought to this analytic problem concerning the functioning of labor markets consists of utilizing a concept of industrial relations systems derived from the work of J. T. Dunlop, 1958).

Very schematically presented, for the purpose of what interests us here, this concept corresponds to a double idea: on the one hand that labor relations are not regulated by the market, which is governed by the law of supply and demand, but by webs of rules which orient the behavior and attitudes of the different actors, and, on the other hand, that this system of relations is itself a product of the actors' games and their strategic interactions. In the economy of Dunlop's concept of industrial relations systems, the economic, technical, and juridico-political environment is analyzed as a series of contexts outside the industrial relations system. On the other hand, the introduction of the notion of the industrial system corresponds to an extension of this concept to other types of rules than the simple rules and customs which concern labor relations.

The notion of the industrial system

The notion of the industrial system has the objective of defining the relations that individual firms carry on with each other. The fiction that these take place in an environment which is undifferentiated and governed by the pure laws of the market alone is not tenable. On the one hand the establishing of a market system, even an imperfect one, necessitates the cooperation of the different actors, or the intervention of an organizer who can impose himself on the other actors; on the other hand, the environment is never totally undifferentiated, and phenomena of customs, of quality norms, etc., eventually give a certain stability to relations which, otherwise, would remain undetermined.

A concept of this kind derives from the notion of industrial relations systems. These are in fact governed by webs of rules, elaborated in conflicts between actors, but nonetheless, in return, governing these same conflicts. The totality of the conditions of employment, of remuneration, and of the organization of work, are in this way referred to as the web of rules elaborated in interactions and conflicts between employers, employees and the state.

In this kind of interpretation, two types of distortion must, however, be avoided. In the first place, the stated rules are not necessarily imperative, and so are not always easily observable by the external observation of behavior, since the rule does not dictate the actor's behavior, but rather it defines what is legitimate in terms of behavior or expectation. The existence of the rule thus means that the real divergences which, in their strategic options, the actors decide to operate, can be interpreted in an unequivocal way: thus, the announcement of a minimum wage for each profession, in the French industrial relations system, is not a determination of real wages; most of the time, real wages cannot be simply derived from them. But the mere fact that the rule exists allows the actors to take up positions, knowing in an unequivocal way the gap between real and minimum wages.

In the second place, one must also take account of the fact that the fixing of the rule is not the same thing as a labor market transaction. Reflection on this point has been clarified by the debate, in the English school of industrial relations, on the nature of collective bargaining (Fox 1975; Flanders 1968). Is this an analog of individual bargaining, where the employee tries to improve the terms of the contract of sale of his labor power to the employer, or is it an activity of some other nature? In fact, it appears that, in most cases, the union is not a monopolistic seller which negotiates a collective contract by haggling with the employers. Collective bargaining is an activity that is situated on a second level (that of rules). It is an activity of enunciation, and not of transaction properly speaking.

Likewise, the definition of institutions that allows the coordination of the actions performed by the various actors, and the organization of their cooperation and conflicts, like the definition of the actors in itself, is a product of this historical activity of collective bargaining. In the actors' games, no position, system of alliances or institution is absolutely predetermined and immutable. On the contrary, all of these are liable to be transformed as a function of the game itself, to the point where it appears illusory to look for a definition of institutions such as firms, which is not itself situated in the framework of a precise system of actors and rules. Or, in other terms, it may be considered that the comparability of firms is only possible in the interior of a given set of rules.

The concept of the industrial system derives from this conception of the articulation between the actors' games and the set of rules. In the functionalist economy of Dunlop's conception, the social subsystem constituted by the system of industrial relations is locked into a set of economic, legal, and technological contexts, where reversibility and substitutability are not really possible.

Congruence of the rules, and a priori nonqualification of resources

The notion of the industrial system consists in breaking this strict, predefined distinction between context and system. In their interactions, as in their strategies, actors are constantly inclined to play off one type of element, and one type of rules, against another. At most, the system establishes rules of equivalence which ensure congruence between these different subsets of rules.

The example already brought up, that of toolmaking firms in Lyon, provides an illustration of this point. It has been seen that, in this kind of group of firms, the mobility of individuals between the different firms is one of the means of career management open to skilled workers. This mobility operates within a set of rules which allow the different actors to give a signification to the departure of individuals. In this way, anyone can tell whether such a departure corresponds to a normal procedure of promotion and qualification or to a disagreement with, or suspicion about, the firm in question. Simultaneously, as we saw, this mobility is the usual means of transferring technological skills from one firm to another. But also, notably as regards the most highly skilled workers, such transfers of personnel are also the means of transferring knowledge and openings around markets. In the manufacture of precision equipment, technicians from the client firm will in effect have ongoing relations, not only with the director of the subcontracting firm, but also with the highly qualified worker responsible for the production and finishing of the equipment. This is also the person who will go to the client's workshop to carry out the installation and final adjustment of the equipment. On such occasions, he forms continuing relations with the clients. When he changes employer, he may thus also bring to his new firm profitable commercial introductions into the markets that he has frequented. It is in this sense that we speak here of equivalence of rules, to the extent that the rule that enunciates the proper behavior with regard to mobility, and the correct attitudes on the part of the different protagonists (the skilled worker, the previous employer, the new employer, or even the client or the supplier of machines) is concerned at one and the same time with working relations, the management of technologies, and also that of the product market.

The different rules corresponding to the forms of management of the different resources are not independent of each other, any more than, as we have seen, the resources themselves can be strictly qualified. At the same time, the rules of a given domain are not strictly determined by the rules of another domain. Thus it is, as we very well know, that the rules of employment are not strictly linked to the type of technology in operation, any more than forms of competition and of cooperation in the product market are closely linked to the forms of management of the work force (Saglio, 1990a). This is the reason why we use the term congruence to define this phenomenon of the determination of the coherence between different types of rules.

The logical consistency of the rules concerning the different domains is not a given: at certain moments, contradictions may appear, corresponding to conflicts between actors, and these may produce an evolution of the regulation, a change in the system. So, within the group of small firms in the plastic-molding industry in the Oyonnax region, which we have studied, the dispute that occurred at the beginning of the 1950s produced a lasting modification in the forms of recruitment of injection press workers. From that time on, the type of career that could be followed by these workers was modified, and thus, also, the mode of creation of new craft firms. In the former system, the *piéçard* (a local term meaning piecework employee) could progressively begin to set up on his own account, while continuing to work in collaboration with his previous employer, or even, sometimes, in the same work space. The modalities of wage bargaining, by the negotiation of piece rates, were congruent with this kind of opening up of workers' careers.

The change that appeared after the strike, with a change in the forms of remuneration and wage scales, led to these possibilities of advancement being blocked. From this date onward, immigrant workers operating the injection presses turned progressively into a particular industrial category, not likely to be integrated in the same way as their predecessors. The prewar workers were principally of Italian origin (in 1938 they represented 83 percent of the foreign population of the town), and a number of currently existing firms were set up by Italians or their children. After the strike, the nationalities of the foreign workers present became much more varied, and their integration into the tissue of Oyonnax became more difficult. Simultaneously, the creation of firms ceased to be the work of moldmakers, and became that of more skilled French workers, most often metallurgists and moldmaking engineers.

At the time of such a labor dispute, it is thus the equilibrium and the consistency of the entire web of rules that is modified. This modification does not necessarily affect each of the rules: certain of them may remain congruent with the new equilibrium. In this case, the priority accorded to local firms in subcontracting remained unchanged, being the same after the strike as before, in the same way that the strike did not call into question the distribution of work by gender.

The construction of collective identities

In the concept of industrial relations systems elaborated by J. T. Dunlop, the production of rules, and thus the regulation of the system, is managed by the play of the actors, and not by the domination of any of them, nor by a logic coming from outside to impose itself on all of them. It is the same thing for an industrial system, in the sense that we are using the term here, in that there effectively exists a collective identity which can be referred to by all the actors who participate in the enunciation of the rules and of regulation.

The elements, or the controlled resources, which are utilizable to set up such a collective identity, are highly varied. Very often, notably in the French case, it is the fact of being active in the same product market that defines the identity of the system: one is then in the presence of professional industrial systems, or branches thereof. These may be more or less closed, to the extent that the group of actors exercises more or less tight control over the key resource. Thus regulated professions, the very access to which is strictly regulated by a system of restrictive practices and peer acceptance, are closed professional systems articulated around collective monopolies (Saglio and Thuderoz 1989). But closure is not necessarily acquired and reinforced by statist regulation: the oligopoly of automobile manufacturers in the USA during the years from 1945 to 1975 (Tolliday 1988) was also a closed professional system, even though it was not guaranteed by the legal protection of the state.

Belonging to the same geographical ensemble can be utilized as a collective reference identity for an industrial system. In this case, we talk about a localized industrial system. The local identity to which the actors refer in such cases must not, however, be simply confused with the fact of living and working locally; here again it is a question of a particular social construct, and not of an evident datum. So it is that in the Oyonnax plastic-molding industry, certain firms which, economically, were a part of that branch – that is to say that they produced objects in plastic and were located in the town itself – did not consider themselves, and were not considered by the other actors, as belonging to the local industrial system. They did not maintain favored subcontracting relations with other local firms; they did not have the same practices in terms of the organization of work and of personnel management; and they did not participate in networks of technological exchange. Conversely, certain actors may participate in localized industrial systems without, however, working or living in the locality. Collective identity is thus more complex than the geographical reference alone: it is rather a question of accepting the web of rules, and of acknowledgment by the other partners that one is a member of the group, and not just a question of reference to the formal definitions advanced by the actors: the fact of living in Oyonnax and of working in the plastics industry was not, in the strict sense, either a necessary or a sufficient condition for belonging to the localized industrial system. On the contrary, what was necessary was to accept the web of rules and to be recognized by the other partners as belonging to the system.

The industrial district, in the sense given to this term by Becattini, corresponds to an industrial system bringing together the double identity, local and professional. The firms that participate in this system contribute to the manufacturing of the same product, and belong to the same geographical entity. This kind of connection is frequent, but not obligatory: looking at firms in the Beauce region of Québec, we find we are dealing with a localized industrial system which groups together firms belonging to several different branches of activity.

LOCAL IDENTITY AND PROFESSIONAL IDENTITY IN THE FRENCH HISTORICAL CONTEXT

The structure of representative organizations

In the French industrial relations system, representative organizations, both on the side of the employers and on that of the employees, give considerable importance to local organizations. Employees' organizations, or at least the principal ones, are organized outside the firm along two lines, professional (by branch) and interprofessional. The union brings together all employees of the firms in the branch, and in the locality. As to the interprofessional body, it brings together all of the unions of the different branches according to geographical criteria. At intermediate levels, the two tracks exist simultaneously: the local unions are organized in terms of national unions and branch federations, the interprofessional organizations in departmental and regional unions. The totality of these organizations is unified within each large confederation. Such a structure goes back a long time, having its origins in the dualism of the organizations: on the one hand the federation of labor exchanges, and on the other the trade unions, which later became industry-wide unions, these two structures being brought together at the time of the consitution of the CGT (Confédération Générale du Travail) in 1895.

On the employers' side, the structure of organizations is more difficult to understand. In those organizations founded on the principle of adhesion, and participating in collective bargaining, one finds a triple structure: geographical interprofessional groupings, with the function of representing all the firms in the zone concerned; industry-wide groups representing firms in the same branch, in the sense of collective conventions; and groups representing firms making the same product or using the same technology. These latter do not normally have the function of participating in wage regulation, and are involved only with the defense of the economic interests of the firms concerned. Besides this structure there are other modes of representation of the employers' position, notably chambers of commerce and industry, which normally have no role to play in collective bargaining in the French industrial relations system (Bunel and Saglio 1979).

At first sight, such structures could turn out to be favorable to the development of a local bargaining system. On the employers' side, as on the side of the employees' unions, local bodies exist, organized according to industry and on the interprofessional level. On the other hand, recognition of unions in the firm was, in France, a very late development, and it was only in December 1968 that legislation was adopted to lay down the rules of recognition of union groupings in firms. The local branch level appeared then, historically, as the basic level of representation of the employees, and also as the basic level of the employers' organizations. It was at this level that, starting from before the setting up of the legal structures of the French industrial relations system, the actors worked out compromises to put an end to labor disputes in the second half of the nineteenth century. Such agreements on wages and piece-rates, though they were relatively common, did not last very long, as the legislation in force permitted the employers to go back on them once peace had been restored.

The apparent paradox in the French system is thus that this de facto preference given to solutions favoring compromise at the local branch level has not, however, led to the emergence of numerous industrial districts or localized industrial systems. Whatever the name that is used, observers (Ganne 1989; Houssel 1989) note in effect that those that exist in France have mostly taken root in marginal situations, which are somewhat exceptional.

The explanation of this apparent paradox resides in the difficulty experienced by actors in French industrial relations when it comes to a stabilization of local regulations. As has previously been noted, French employees' unions have had the greatest difficulty in getting their legitimacy recognized by their partners on the employers' side. For a very long time, heads of companies firmly denied the legitimacy of the union in the firm, and this attitude indirectly reinforced the revolutionary orientation of the employees' unionism.

Stabilization through state intervention

From this point on, stabilization of regulation could only be obtained, when it could be obtained at all, by the intervention of a third partner, the state. Historically, it has been in situations where the state turned out to be an important economic partner that the actors have succeeded in setting up a lasting form of regulation of conditions of work and employment. Such situations are encountered when the state is a purchaser of the firm's product. In such configurations, its behavior as a rational purchaser, seeking to lower the prices of products, generates pressure for a lowering of wages, and thus becomes contradictory to the second role of the state, which is that of the guarantor of a certain form of harmony in labor relations.

To attempt to break out of this vicious circle, different solutions were tried out in the last decades of the nineteenth century, and these resulted in the Millerand decrees of 10 August 1899, which are still, today, the basis for regulating the attribution of public market contracts. By these decrees, the state required that companies tendering in the public sector should respect the customary wages and working conditions obtaining in the profession and the region concerned. In the case of an agreement having been reached locally between the legitimate representatives of the workers and the employees, it was thus imposed on firms serving in the public-sector market.

Such procedures, although not imposed by law outside of the public-sector market situation, nevertheless had a regulatory effect on employment and working conditions, as was indeed remarked, some years later, by a labor ministry official: although the decrees of 10 August 1899 concern only contracts negotiated for public administrative bodies, the publicity they are given can go well beyond the field of these contracts. Publication of pay slips by these bodies reveals salary levels and the normal duration of work, as founded on general practice and consent, and in this way everyone, employer and worker alike, knows that these are the normal duration and rate for the region. Employers and workers quite naturally take them, even when they deviate from them, as a basis for their contracts, since they are subject to the regulatory influence (Labour Ministry circular, 5 February 1915, quoted in Oualid and Picquemard 1928: 164, note).

The progressive setting up of such procedures of regulation assisted the reinforcement of branch structures, on the unions' and the employers' side equally. The branch became progressively substituted for the trade as the center around which interests were grouped, and as the foundation of workers' solidarity. But simultaneously, and notably during World War I, the state's involvement in the economic and industrial management of branches led the actors gradually to favor the national level of regulation. Statist policy meant that it was at this level that the actors could influence the articulation between wage regulation and economic regulation, by intervening in industrial policies. In employees' unionism, and notably for the CGT between the two World Wars, nationalization of industries was the element of its political program that corresponded to this economic orientation (Kuisel 1981: 77–80). For the employers, the priority thus accorded to national branch regulation satisfied two classic objectives: on the one hand, to distance the level of negotiation from the firm, and leave the field open at this level for the exercise of the employer's authority with regard to the management of the work force (Bunel and Saglio 1979); and, on the other hand, to promote the organization of economic and industrial regulation in order to avoid dangerous games of market-based competition (Duchemin 1940).

By such means, the two actors can avoid face-to-face situations and confrontational bargaining. To obtain improvements in their salary situation, employees appeal more to state pressure and to legislation than to the construction of a real bargaining system, articulated or not, which would require that each of the partners recognize the legitimacy of the other. To fix the conditions of competition, the employers rely on a statist industrial policy, with the prospect of criticizing it in practice on account of its excessive *dirigisme*. As to the state, which in France has never been able to act simply as a referee, it watches with interest as its roles and prerogatives increase.

The renewal of the local, or the decline of the economic?

The equilibrium of the roles thus established at the beginning of the century, and which were vigorously reinforced by the economic and industrial mobilization of World War I, gives the general characteristics of a large section of French industrial systems, concerning industrial production activities. Of course it is not a question of a single model, and other configurations exist, notably in occupations where collective regulation is particularly marked, such as high status professions, or banks, or in professions whose regulation was introduced after World War II, as was the case notably for the health professions and the social services.

In this model, the essential role has devolved on the central state, which ensures the consistency of regulations and arbitrates the principal disputes, in questions of the management of wage relations as much as in those of economic policy. The internal cohesion of the two other actors is rather weak. The employees' unions are divided, and, even taking them all together, they represent only a minority of workers. It is only very rarely that they can manage to construct an effective closure of labor markets, or to impose themselves as negotiating partners who must be given a voice on the level of firms. Employers' organizations are scarcely more powerful, and their capacity for mobilizing their members remains essentially defensive.

Such a configuration supplies the key for explaining the scarcity of localized industrial systems in France, even though the internal structure of the organizations representing the different protagonists corresponds to the local level. The partners are not sufficiently strong locally to recognize each other and to agree to confine their strategic games to the local framework. In the French system the social and economic partners are mostly weak, and look for legitimacy through the central state rather than through mobilization of their grass roots members on behalf of local negotiations.

In this kind of configuration, the new interest that has developed since the middle of the 1970s concerning forms of regulation that give more importance to the local level, like the reinforcement of the company level, is, in large part, due to a tendency toward disengagement shown by the state. Having given a new impetus, in the middle of the 1970s, to industrial relations bargaining but within the existing framework, the state has progressively abandoned, or at least severely diminished, its role of direct intervention, while at the same time continuing to impose the opening up of the French economy to international competition (Cohen 1989). All the same, this progressive disengagement has not produced a real reinforcement of the other two protagonists at the branch level, where they have traditionally showed themselves to be the strongest and best structured.

It is in this conjuncture that new modes of regulation and recomposition of industrial systems have appeared, sometimes around older models or local identities, reinvigorated by decentralization policies. But in such systems, the weakness of identities and of mutual recognition leads to questions about the stability of the solutions which have been sketched out, since the new equilibrium cannot be considered purely and simply as a transfer of skills which had previously been devolved upon the central state and the national representative bodies. The emergence of this principal level of regulation at the beginning of the century corresponded, for the state itself, to a mode of resolution of a double problem: that of the competitivity of French industry, of course, but also that of the integration of the working class into the French political system (Donzelot 1984). In a certain way, the setting up of this model of regulation can be considered as a Saint-Simonian solution to the social question.

In current French society, the highest stakes in societal matters have shifted. The problem of the political integration of the working class no longer threatens overall social cohesion; and so one can also understand that the supervision of production and of the industrial economy, from being a major political problem, has been reduced to nothing more than a question of good management (and this is notably what has come out of the analysis of the form taken in the 1980s by debates on the nationalization and privatization programs: the least that can be said is that the politically inspired resurgence of traditional arguments in this domain no longer attracts much attention). The emergence of new types of regulation can thus take place on the basis of this depoliticizization of problems of production. In this respect, it can be hypothesized that modes of regulation of a less restrictive nature than those of previous models, and resting on weaker collective identities, may nevertheless turn out to be stable in the future.

Part V

THE INCORPORATION OF LABOR

13

ALTERNATIVE ROUTES TO LABOR FLEXIBILITY

Guy Standing

INTRODUCTION

Regular wage labor is neither the present nor the future for a growing proportion of the population of industrialized countries. Almost perversely, this realization has coincided with a loss of a radical, progressive vision of the Good Society. Instead, the fear is that we are witnessing the creation of a disjointed society made up, in Marx's words (describing preproletarian workers), "by simple addition of homologous magnitudes, such as potatoes in a sack form a sack of potatoes." In the past decade or so, we have seen the limit of Gramsci's brilliant insight, Fordism, which no longer seems the predominant form of the labor process (if it ever was). However, most of us are still searching for an alternative paradigm of the present and an alternative avenue to utopia. Without some vision, however modest, we might as well stay in our gardens or backyards. And perhaps that is a clue, for access to a garden is not just a residue of Thomas More's *Utopia*, but a part of many modern visions, encapsulating the possibility of combining different forms of paid and own account work. In the reality of Europe today, we live in disturbing times, when inequalities are worsening and when the Galbraithian strictures on American society of the 1960s are being writ large in European cities such as Rome, London, and Amsterdam, with extraordinary private affluence (of a few) coexisting with public squalor (of many). That is one context in which to approach the issue of alternative paths for the 1990s.

Another contextual point can be derived from the prevailing vocabulary. The "key words" of an era both define the underlying trends and help in identifying the emerging opportunities for better alternatives. Among the key words of the 1980s were individualism, privatization, flexibility, deregulation, structural adjustment, liberalization, and supply-sideism.[1] As we know, those key words are all part of the ideology that has played all the best tunes of the past decade. It is an era in which the main voices have been profoundly anticorporatist, antisocialist, and antisocial. Margaret Thatcher's famous quip was no slip of the tongue:

> There is no such thing as society. There are individual men and women, and there are families.

It is this rejection of collective responsibility and sense of community and social solidarity, as well as the very active promotion of self-interest, which epitomizes the mainstream of the past decade, even though "Thatcherism" represents its extreme variant.

This chapter is an attempt to reflect on two routes that different societies might follow to varying degrees in the 1990s. These are built on stylized facts about labor process developments, the analysis of which has been pioneered by many of those who attended the "Pathways to Industrial and Regional Development in the 1990s" conference in Los Angeles in 1990, and on our ongoing country studies of labor flexibility in the ILO. Space precludes empirical and documentary support for the stylized facts, although they are fairly well known. Moreover, of course, juxtaposing opposing scenarios does injustice to the nuances of any particular country or labor process. It will be taken as axiomatic that the changes in the labor process (the structure of employment, the forms of labor status, types of contract, forms of payment, etc.) reflect changes in the structure of production, including technological and organizational dimensions.

The stylized facts that we need to keep in mind include the extensive erosion of secure full-time employment, the persistence of mass unemployment and "underemployment" of one form or another, the crumbling of unionized mass production manufacturing, the growth of both "functional" and external labor flexibility, the widening of wage and earnings differentials, and the failure of traditional social security systems (both Beveridge and Bismarck varieties) either to compensate for the labor process fragmentation or to provide the income security anticipated in the social consensus societies of the post-1945 era. Although all these have been mainstream trends that each avoided to some extent somewhere or other, even the most corporatist societies, of the Nordic region principally, have failed to arrest most of them.

Ultimately, the greatest failure of the 1980s was the disturbingly underplayed tune about distribution. One has difficulty thinking of some haven where income, wealth, and other socioeconomic distribution improved; one has no difficulty whatsoever in naming a long list of countries where they worsened, despite the disgraceful shortage of data needed to make comprehensive pictures.

It might be useful to contemplate two possible scenarios that start from recent trends. The first will be called *subordinated flexibility*, representing a continuation of the dominant trends of the 1980s; the second will be called *cooperative flexibility*, although one might call it integrative or social flexibility, in that it represents the vital positive value of labor security and collective regulations. The hope is that we can move on from those key words cited at the outset, and thereby point to a more attractive set of options than implied by the dominant trends in most of Western Europe and North America.

SOME CONCEPTUAL PRELIMINARIES

In a series of papers over the past few years, it has been argued that since the 1970s the postwar social consensus has been disrupted by the erosion of seven forms of labor security or "labor rights," each of which was strengthened in the preceding thirty years or so (Standing 1983, 1989). Those are:

1 income security;
2 labor market security;
3 employment security;
4 work security;
5 job, or occupational, security;
6 labor process security; and
7 labor reproduction security.

These are defined elsewhere, but it is strategically important to distinguish between what have been called *meta* rights and *instrumental* rights. The former are those that are long-term goals reflecting fundamental values, the attainment of which is not realistic in the short term but which should be pursued steadily and consistently; instrumental rights are those that are necessary or helpful in the pursuit of real human values and aspirations. All social and economic policies should be judged by whether or not they enhance the prospect of attaining meta rights. Many critics of the supply-side orthodoxy of the 1980s have erred by mixing up the two types of labor rights, often giving precedence to the defense of labor market security (i.e. the pursuit of "full employment") and to anguished concern over employment security, both of which should be treated as instrumental rather than meta rights. An alternative strategy to the current orthodoxy should be based on the recognition that the labor meta rights are income security and labor process security. One way of interpreting the trends of the dominant orthodoxy of the 1970s and 1980s is that what would be meta rights in social democratic visions of a Good Society were treated as instrumental or secondary rights, or even rejected as rights altogether.

In this chapter it will be taken as axiomatic that the meta rights of a Good Society would include the following:

1 generalized income security for all, consistent with the economic level of development;
2 declining inequality, not the reverse;
3 labor process security, involving the strengthening of a sense of community, solidarity and active participation, through economic democracy; and
4 the right to occupation, and the right to work.

These are stated baldly, in the sense that Gunnar Myrdal used to urge social scientists to state their values at the outset, and also because they may help give a structure to the consideration of key elements of the alternative scenarios this chapter attempts to highlight. In that context, it might be useful to reiterate some crucial conceptual distinctions that are fundamental to the flexibility debate.

Labor process security is taken to mean that workers and particularly their representative organizations have the capacity to determine, or the possibility of influencing, the development of the labor process, defined in terms of working conditions, work structures, skill acquisition and reformulation, and so on. If, for instance, management has almost full control over labor relations, and the workers little or none, then there is definitionally labor process insecurity. One could never attain "full" labor process security, in part because one would find it very hard to agree on any ideal structure of institutional mechanisms. Nevertheless, labor process security is a meta right, a question of degree and of trends, in the sense that reforms and changes in labor practices should be judged by whether they increase or diminish the existing degree of labor process insecurity.

Another distinction that must be preserved is between job and employment security. Too many authors refer to job security when what they mean is employment security. The former exists if workers have secure niches within an enterprise or within the labor process more generally, niches consisting of a protected bundle of tasks that cannot be abrogated arbitrarily. By contrast, employment security exists if workers have protection against arbitrary dismissal from employment. In the 1980s many groups lost both forms of security, some lost only one, and some conceded on one to gain on the other.

One must also distinguish between an *occupation* and a *job*. This is critically important for the subject matter of our deliberations. It is significant that in the English language, the etymologically earliest definition of occupation was taking possession of a piece of territory. An occupation involves a career of learning and the mastery, or possession, of the mysteries of a craft or profession. There is a sense of continuity, a progression, and above all an acquisition of status, control, and autonomy. The status aspect of occupations is pervasive. One still occasionally thinks of an occupation as a "calling." The term stems from a social structure in which the detailed or technical division of labor was relatively undeveloped, when apprentices were introduced to the "mystery" of a craft. In intention, the concept of occupation refers to a positive idea of work, as creative activity, the combination of intellectual and manual activities – conception and execution – in the context of "skill" refinement. One is tempted to believe that for many workers in the latter part of the twentieth century the development of the technical and social divisions of labor, under the guise of "numerical" and "functional" labor flexibility, has debased the notion of occupation.

By contrast, a "job" is a much humbler word, conveying an activity, a limited and limiting piece of work, a narrow set of tasks. Often it has a pejorative meaning attached to it, implying a lack of permanency, a lack of accumulated wisdom or "skill" and an aura of insecurity. Usually it conveys a task or period of employment of short and limited duration. Thus "job work" is another term for "piece work." A job is what one does now, an occupation is what one is. One would not say of a job what John Ruskin said of an occupation:

> The character of men depends more on their occupation than on any teaching we can give them.

If someone has a job, without occupation, the social psychologists might be able to tell us the likely behavioral consequences. Those surely cannot be favorable in influencing the development of community and social solidarity. But by putting labor market and employment security on a pedestal, the tendency to idealize "jobs" becomes overwhelming.

It is scarcely less relevant to our task to reflect on another much-used key word – *skill.* A striking theme of the era is that everywhere politicians, economists, employers, unions, the media, and sundry consultants are agreed on one point; that training is vital and that raising skill levels is the principal means of solving "Eurosclerosis," cutting unemployment, combatting labor market fragmentation, and solving diverse other economic and social ills. To dissent is to risk excommunication from rational society. Training is the answer. If that does not work, then more training is the answer. If the unemployed do not perceive their need, they must be obliged to fulfill their duty to be trained. Already, in some countries being in labor market training schemes is equivalent to being employed, and the payment for the one is sometimes linked to payment for the other.[2]

It is in that context that we need to review our ideas about "skill," which has been a much-used term in the literature on labor flexibility. This, like the notion of a job, is also mixed up with ideas about occupation. In the abstract, a skill implies some combination of creative and manual abilities, a learning process and relevant work experience. To develop a skill *usually* requires training, but not always. More than ever, we need to distinguish very clearly between "vocational (occupational) training" and "job training," and between those and "orientation training." Frederic Taylor knew the differences very well, and the distinctions are even more relevant now. When training schemes are devised in terms of "minimal modules of employable skill," then not only has occupational training been reduced to job training but that has been transformed into a form of labor market regulation and labor control, in which trainees have little chance of autonomy or status.

Skill has three underlying meanings. To many, it reflects *objective* characteristics of work activity. This is skill as technique. Following Georges

Friedman, one might start by dividing "skilled workers" in this sense into "specialists" and the "specialized," whereby the latter merely receive general training, do routinized work and are excluded from the design and comprehension of the production process. By contrast, specialists have far more discretion over when and how to carry out their tasks. But depicting skill as technique is actually rather unsatisfactory. One can be skilled in having certain capacities, but in practical terms be in a job that is not skilled in the specialist/occupation sense of the term. As far as the current era of labor flexibility is concerned, the *way* the erosion of occupational security has proceeded has probably accelerated the displacement of specialist craftsmen by specialized workers over more and more traditional occupational spheres. We will return to this theme in subsequent sections.

A second notion of skill is *autonomy* in the production process; this links the relevant work activity to aspects of labor control, and various authors have actually contended that the skilled are differentiated primarily by the degree to which they have an autonomous status. If workers have no freedom over the use and development of manual capacities then surely those capacities will be restricted and distorted, thus removing a vital ingredient for the pursuit of occupation. It is not clear that many of those calling for more training to achieve labor flexibility are calling for a more autonomous work force.

A third and related meaning of skill relates to *social status*. Some social scientists argue that some occupations are called skilled simply because of custom or because barriers to entry have preserved an artificial or arbitrary skill hierarchy, perhaps complemented by a ritual of training that bears little relation to the task complexity, the general knowledge, or the specific knowledge required. This aspect of skill is closely related to mechanisms by which labor market inequalities are intensified.

These three aspects of skill should be kept in mind in assessments of the flexibility debate and literature.[3] Indeed, if one took an integrated approach to both the notions of occupation and skill one might conclude that, despite the rhetoric to the contrary, trends in connection with skill and training in the 1980s were conducive to occupational deskilling, just as state involvement in the labor process was spreading in many countries, not contracting, while there was growing labor market regulation, not deregulation.

We have dealt at rather tedious length with these conceptual issues, though they do seem crucial. One of the difficulties is that the vocabulary of the era is so deceptively benign, so liberating in tone. Who could be against policies that "provide workers with jobs," "give training," and "raise skills"? One hopes it is permissible to be uncomfortable. Those slogans seem to be answers that beg some rather awkward questions.

SUBORDINATED FLEXIBILITY: LEGACY OF THE 1980s

An essential part of the supply-side, libertarian agenda of the 1980s was that individualistic labor regulations should displace collective protective regulations, the twin justification being the desire to promote individual property rights and the belief that decentralized, individualized labor relations would ensure that markets would "clear" and unemployment fall (or rise) to its natural rate, determined by frictional and structural characteristics of the labor market. Ideally, legalization of employment contracts would impose behavioral norms based on individualistic common law principles. The courts could be relied upon to rule against collective interventions by reference to freedom in the labor market. Unions could indulge in fanfare flagwaving negotiations once every two or three years, when collective contracts were up for renegotiation, but ideally, by the tenets of those who advocate this libertarian path, this would become atavistic to the point of being unnecessary or ritualistic. To the extent that industrial or craft unions persisted they would be constrained to the point of being agents of management as much as agents of redistribution in favor of labor, let alone anything more radical.

Much of this agenda has been realized, to some extent. The individualization of employment contracts has extended way down the pay and occupational status scales and such contracts seem to have become more comprehensive as their practice has spread. While this trend is well advanced internationally, there has also been a steady erosion of collective and protective regulations. *Implicit deregulation* has been occurring by:

a) nonimplementation of protective regulations;
b) inadequate resources and personnel devoted to the task of policing existing regulations;
c) erosion of the capacity to resist among those denied their rights to protection;
d) a growing and cultivated sense of ambiguity among potential beneficiaries about the validity of such rights; and
e) an increasing loss of entitlements to protection by virtue of their labor status.

One major cause of the last form of implicit deregulation has been the widespread drift away from regular full-time employment in large enterprises. Major causes of the third and fourth forms have been deunionization (or more likely nonunionization), the persistence of chronically high unemployment, and the fears engendered by pervasive marginalization and widening inequality. Yet the biggest single factor in the implicit deregulation of the 1980s, compared with the implicit regulation of the 1960s and 1970s, was less tangible: fear changed sides. Whereas in the 1960s employers led in introducing practices ahead of regulations to avoid the threat of disruption, in the 1980s they led "deregulation" in the confident belief that the door was open, that others were going through it and that resistance was enfeebled.

Explicit deregulation has come about from the repeal of protective laws, whether wholly or, more commonly, by the creation of selective or "targeted" rather than general "rights," including the creation of loopholes in the legislative framework. Rights per se have been eroded by the tightening of conditionality, the necessary conditions for protection being made more restrictive and the onus of proof of right often being shifted to the workers and made more costly to pursue. Examples of explicit deregulation include the emaciation of the Wages Councils and the abolition of restrictions on night work for teenagers in the United Kingdom and legislation in the Federal Republic of Germany permitting unions and employers to derogate from working time legislation. There are numerous other cases all over Europe testifying to the underlying trend.

Although there has been both implicit and explicit erosion of protective mechanisms and collective institutions, it would be misleading to characterize this as pure deregulation. Proindividualistic regulations have been displacing procollective regulations. For proponents of what we should call a path to *subordinated flexibility*, the 1980s might be seen as a transitional era, between one based on collective regulation to one of "contractualization." In the process, it is scarcely an exaggeration to say that those who could help themselves have been helped, whereas those who need help have been left to help themselves. It is also somewhat ironic that the current orthodoxy espouses "deregulation" by reference to the desire to create "full" employment. Whereas the old orthodoxy was that regulations were to protect workers in employment, the current one is that regulations should be judged by whether or not they enable more people to obtain jobs. Inter alia, this is one indication of the dubious primacy given to the notion of full employment. If critics of the prevailing orthodoxy continue to put full employment on the highest pedestal of social policy, they should not bellyache if regulations are adapted to promote that goal.[4]

The causes of the growth of various forms of labor and employment flexibility will not be discussed here. But whether or not "the big firms are coming out of the corner," as Maryellen Kelly and Ben Harrison (1989) suggest (with good reason, in my view), the way labor and employment relations have been moving is unlikely to be reversed.[5] The shift away from protected, full-time wage employment to more decentralized, less secure forms of labor relation, including outsourcing, subcontracting, and casualization, owes more to managerial, cost, and technological considerations than to a passing period of recession and structural adjustment. This means that conventional forms of protective regulation will be increasingly avoided, rather than evaded.

Four outcomes of this flexibilization deserve to be stressed – labor force fragmentation, the erosion and restructuring of economic entitlements (including welfare), the increasing need to combat the social and economic exclusion of a minority, and the continuing search for new means of raising productivity, given the partial demise or deficiencies of Taylorist methods in the more flexible labor process in the 1980s and anticipated for the 1990s.

Labor force fragmentation

The underresearched theme of labor process fragmentation is closely associated with growing inequality.[6] We know that wage flexibility has grown and that this has been a euphemism for the ability of firms to lower real wages and widen or change wage differentials more easily. For instance, a survey of employers by the United Kingdom's Confederation of British Industry in May 1989 found that in the previous three years those firms reporting that it was easy to increase wage differentials had risen from a third to over a half, while those reporting that it was hard to do so had fallen from a third to a little over a quarter (CBI 1989: 14).

Besides evidence on wage differentials, there are good reasons for believing that inequality has grown much more than conventional income and earnings data suggest. One reason is the "postcapitalist" nature of many employment relationships, and the ongoing process of labor-force fragmentation. Any conceptualization of that will be questionable, but it is surely essential to have an approximate image of the diverse strata to which new regulatory frameworks would have to apply. After all, one of the fundamental tenets of the postwar social consensus welfare state was the presumption that the labor force was unfragmented, with a norm of a male full-time wage worker in regular employment. In the 1990s, labor regulation and distribution policy will have to take account of something like the following ideal type pattern of fragmentation that has been growing in the "post-Fordist" era.[7]

In the context of more flexible labor processes, one can identify seven strata that deviate from the presumed norm of regular, full-time wage and salaried employment. At the top of the heap is an *elite* who have become "capitalist employees," through profit sharing, the acquisition of subsidized shares, and also by virtue of a growing range of fringe benefits, some of which have a full value way in excess of any taxable monetized value. Thus executives increasingly rely on "performance related" bonuses; in 1989 those accounted for 20 percent of management salaries in Britain, beyond which there were numerous other perks which have been growing in significance relative to other income sources (CBI 1989: 5). In the USA, senior executives commonly receive over half their total income in non-salary forms. This elite have income security, work security, labor market security, employment security, job security, and labor process security, to the extent that they want it. The image one has of this elite is one of internationalization and increasing detachment from the labor process beneath them. In terms of inequality, much of their power lies in the concealed way their income and wealth accumulate, essentially legally. It is hardly comforting that perhaps the main drawback for those involved is that their lifestyle is generally intense and fraught with stress, which ultimately threatens any individual's hold on positions of status and control. At the very peak are a handful of rapacious individuals who seem to live on the margins of sanity, crazed by the pursuit of their n^{th} billion dollars, but below such Brechtian monsters are layers of very wealthy, sober groups affluently detached from national regulatory frameworks.

Below this elite category one can detect what might be called frenetic *proficians*. A notable phenomenon of the past decade has been the spectacular growth of nominally independent "consultants" and self-employed specialists. Key characteristics of this group are their relative youth, their frenetic work schedule and their self-satisfaction. Their expansion has been a feature of "flexible specialization." They have little labor security, but remind us of Tawney's tadpoles – most dying, but some becoming smug croaking frogs – except that a far larger proportion seem to thrive, because they have taken advantage of enterprise flexibility and the advantage to firms of using flexible specialists for short-term purposes. Whether the growth of this stratum will continue into the 1990s is hard to predict, since there may be a tendency for their "market price" to decline and for enterprises to find that as the cost of such workers come down it will be more advantageous to reintegrate those functions "in-house" to take account of economies of specialist knowledge. Yet it seems more likely that the external flexibility granted to the firm and the perceived autonomy granted to the proficians will combine to preserve semiautonomy. After all, these two aspects should create a monetized wedge that could be shared (each valuing the flexibility and autonomy that the more distanced working relationship entails).

A key aspect of the growth of these first two strata is that they tend to be beyond the welfare state and other regulatory institutions, increasingly having access to privatized benefits and neither contributing to nor gaining entitlement to social security benefits. This elitist detachment leads them to give political support to the associated transformation of the welfare state, from what Richard Titmus called the institutional redistributive model to the selective, residual model.

Those two strata represent the upper echelons of "popular capitalism." Below them are those who seem increasingly oriented to serving the interests of the upper strata, *national bureaucrats*, who retain the form of labor security obtained in the expansive times of the 1960s. In some countries this group may have lost some labor security, but it is hard to gain a clear picture of the seriousness of any erosion.

The fourth stratum down is hard to label, but might be called the *capitalist worker* stratum, consisting of those randomly fortunate wage-earning workers who, through share payments or access to successful profit-sharing schemes, accumulate savings that allow them to set up a full-time or part-time business or to live off dividends. The group may be only a tiny fraction numerically, but is ideologically rather important. They have gained in terms of income security and many have less need for employment or labor market security and so be less inclined to oppose implicit or explicit erosion of those forms of labor security.

Fifth down the labor process ladder is the old *proletarian* stratum, made up of unionized, mostly male workers in regular wage employment. This was the Beveridge and Bismarkian norm for the national insurance social security system. For well-known reasons, they have been declining numerically and have had their labor process security eroded in various ways. They have lost visibly and noisily, ironically often blamed and reviled by those who stood to benefit from their strength as much as by those who in short and narrow terms stood to gain from their defeat. Many of them have become a principal source of the growth of the bottom stratum outlined below. In general – despite the almost belated entrenchment of "insider–outsider" theory – there has been a tendency for this fifth stratum to lose in terms of income security, job security, work security, employment security, labor market security, and, most crucially, labor reproduction and labor process security. Part of the decline of their status, bargaining strength, and social cohesion comes from the erosion of job security (as distinct from employment security), stemming largely from a series of concessions to management on job demarcations. If wages are attached to job or task assignments and if workers are forced to concede the whole right to job retitling and rebundling to the will of management then effective wage flexibility can be boosted. Another striking feature of the regular CBI employer surveys in the United Kingdom is that in the past decade between a quarter and a third of all pay settlements have included concessions on working arrangements (CBI 1989: 6, 13). This implies a reduction not only in job security but in income security.

This stratum has also lost employment security for similar reasons, and has been threatened by two-tier or multiple-tier employment contract structures and by two-tier wage structures, which may be transitional phenomena en route to flexible (*not* deregulated) employment.[8] The fifth stratum has also been affected by the growth of working time flexibility. This might be seen as beneficial for both employers and workers, but unless flexible working is regulated by collective negotiation procedures, as is still the rule in Italy and the Federal Republic of Germany, it will be subordinated flexibility, essentially on the employer's terms, as in the case of increasing shift working and the abolition of restrictions on night work and weekend working. In short, one can hypothesize with some confidence that a growing proportion of such workers possess little more than a temporary niche, far more likely than ever before to move or be moved across "occupational" (*sic*) categories and industries. By that route they swell the sixth stratum.

That is what might best be called the *flexiworker* stratum, which has grown enormously in many countries in recent years, encompassing many "nonregular" forms of wage and other relatively low-income quasi-wage labor. The category is not so much deproletarianized as unproletarianized, having few if any of the laborist values of the traditional working class. For them the notion of the Rights of Labor would have few resonances beyond reminding them of fathers and mothers or their history books. They would not so much say farewell to the working class as refuse to acknowledge that they belong to it. The traditional vulnerability of such workers has been accentuated by the explicit and implicit "deregulation" of the time.

The category is particularly heterogeneous – so much so that one is tempted to subdivide it – but what the various groups have in common is an absence of any form of labor security, most of all labor process security.[9] Whereas increasingly the fifth stratum has been expected to be functionally flexible (i.e. losing their crucial job security), flexiworkers are also labor-status flexible, likely to shift between wage and nonwage forms of employment, sometimes combining activities, sometimes not. Official labor force classifications are often hard to apply. Euphemisms are almost amusing. We have seen the emergence of "permanent temporaries," "self-employed employees," and "in-house outworkers." Some firms maintain workers on temporary contracts for many years; some employment agencies put workers on permanent contracts as temporaries, guaranteeing them a retainer and employment status but not any particular job. The fragmentation can have grim effects that merely highlight the underlying insecurity. Recently there was a wretched legal case in the United Kingdom where fishermen who had been working for a firm for twenty years or more were deemed by the courts not to have satisfied the condition that to be entitled to redundancy benefits workers must have been in regular employment for at least two years. As their work in the period immediately preceding their redundancy had become irregular, the obligatory period of regular employment continuity had been broken. The fishermen had become flexiworkers without knowing it.

Flexiworkers need not be unskilled in the technical sense. Thus the spread of teleworking (telecomputing, or remote work) often involves workers with computational skills. But they are likely to have extreme employment insecurity and have little or no labor process security, in that they are isolated from the work process, liable to be ignored in promotions, easily dismissed and unlikely to identify with unions in companies. Apparently, telecomputing has been a mechanism for geographical decentralization of work forces and even for extending the international division of labor. Some firms have set up regional centers in which teleworkers can work for part of the time. The New York Life Insurance Company, having had difficulty in retaining trained staff in its US offices, set up an office in Ireland, where it sends insurance claim forms to be processed and telephoned back to New York. Such remote working has even been used as a form of regional policy to shift jobs to high unemployment areas without the need to shift the plant or offices of firms wanting to remain in strategically placed central locations. The United Kingdom's National Economic Development Council has recently launched information technology work centers in five inner cities for precisely that purpose. In short, by such methods capital remains large and centralized; labor becomes increasingly small and decentralized.

Flexiworkers are likely to be in and out of jobs, whether full-time or part-time, but rarely in them long enough to earn entitlement to occupational welfare, social security, or even privatized insurance benefits, let alone long enough to develop the confidence to join or form unions.

Some of the characteristics of the growth of flexiworkers are too well rehearsed to bear much repetition. But one trend worth stressing is that large parts of the employment function itself may be in the process of being contracted out to intermediary, labor-only agencies – their "output" being labor.[10] This will take many forms, including "turnkey" provision of work teams of complementary groups of workers. One may confidently anticipate a lucrative profession mushrooming in the 1990s – employment contract lawyers – bargaining just like agents for football players or actors over not only salaries and benefit packages for their client workers but over such matters as employee loan arrangements between companies (a growing practice in Japanese firms), zero hours contracts, maximum/minimum working time contracts, on-call contracts, annual hours contracts, and the like.

A symbolic move in the direction of employee subcontracting was the recent abolition of Britain's national dock labor scheme, whereby dockers had obtained some semblance of income and employment security through the register of dockers, which had helped maintain minimum wages and had regulated training and skill standards. For over fifty years this symbol of decasualization had been preserved; now the prospect of labor auctions, wage undercutting, and market clearing is back. But that is only one of many such trends. Temporary employment agencies have grown for secretaries, contract cleaners, building workers, hotel kitchen staff, security services, maintenance workers, electronics company workers, and numerous others. Some of the "flexible specialist," labor-only subcontracting companies have become multinational corporations, and in some cases, the agencies have become monopolist-monopsonists, which gives them strong rent-acquisition possibilities. Already there is an International Confederation of Temporary Work Firms, which recently reported that the number of workers in the European Community on temporary contracts had been growing by between 15 and 20 percent a year in the 1980s, a rate expected to rise with the EC's single market after 1992. Apparently, growth had been greatest in the Netherlands, France, and the Federal Republic of Germany.

One can paint these ongoing developments as either benign or malign. The libertarian would probably contend that competition will force the agencies to offer workers incentives to stay with their agency, and there may indeed be some stabilization of employment relationships between such workers and their agencies, who will decide on their allocation to firm *x* or *y*, with pensions, sick pay, paid leave, and the like. But there is likely to be a worrying absence of job security, in that workers will be shifted around at the will of intermediaries *and* the firms to whom they are contracted. This is subordinated flexibility, which can only be a threat to occupation by virtue of the incessant insecurity and casualization at the point of production. There should also be concern about the skills that such a trend emphasizes and deemphasizes. Skills that will be fostered include mobility, those underdeveloped will include understanding of production processes. This is not flexible specialization but specialized flexibility.

There has already been a proliferation of contractual forms of employment relation, and it is the associated employment insecurity that most easily identifies flexiworkers. There are contracts that cover specific categories of worker and there are contracts that are shaped by the production or management process. Broadly speaking, one might envisage the former as having rather more to do with labor force stratification and the latter with job segmentation. In recent ILO enterprise-level labor flexibility surveys we have been probing how to disaggregate labor status categories. The experience has once again highlighted the inadequacy of dualistic vocabulary. The term, for instance, covers a wide spectrum of usefully distinctive contractual relations. Here is not the place to go into the distinctions, or their implications, but just listing the main forms of employment contract indicates the fragmention process:

1 casual, without oral or written contract, typically day-work;
2 temporary, fixed-term, oral agreement, non-retention – either for a) stop-gap work, or b) job work;
3 temporary, fixed-term, oral, on a continuing work basis – either for a) or b);
4 temporary, fixed-term, written contract, as for 2a or 2b;
5 temporary, age-limited, e.g. for youths aged 16–19 or for preretirement;
6 apprenticeship contracts, with or without assurance of subsequent employment;
7 trainee, probationary, specified or unspecified duration;
8 adaptation contracts, for post-training practical purposes;
9 temporary contracts for first-time job seekers, as in Germany;
10 job-sharing contracts;
11 employment-orientation contracts, particularly for youths;

12 solidarity contracts, involving shared cuts in wages and worktime in downturns;
13 variable time contracts, where the length of the working day or week is adjusted to meet the firm's requirements;
14 part-time contracts, also involving features of other forms of contract;
15 regular, full-time contracts.

This list is scarcely exhaustive, and excludes forms of dependent "self-employment," many of which reflect a contractual relation with a specific company. Flexiworkers fit into many of the first fourteen categories, and while most of those contractual forms have always existed, it is their relative growth that is striking. The scope for employment insecurity and income insecurity in this form of "contractual fragmentation" is enormous, even though the diversity provides potential flexibility for employers and workers that might be advantageous for either or both. The threat to labor process security, or solidarity, must be considerable, for such workers typically cannot relate to their fellow workers, having no common social space into which to bond cohesively, while the regular, full-time workers are encouraged to feel apart because the employment insecurity is borne by this relatively peripheral category. But perhaps the principal characteristic of flexiworkers is that they commonly lack entitlement to both enterprise/occupational welfare and national insurance-based state welfare. Some groups may have access to some benefits, some to others. But the dominant picture is one of exclusion and inadequacy of entitlement. Ironically, of course, they have a greater need for such benefits because they lack all six forms of labor security. This means that they are always threatened by the fear of floating into the seventh stratum.

This fragment, which also mushroomed in the 1980s, has attracted various epithets, the most controversial being "underclass," a term treated as an ideological football by the likes of Charles Murray and Lawrence Mead. But liberals have also sentimentalized its growth, often failing to criticize the productive, technological, and labor structure that generates it.[11] Anyhow, one might best describe it as the *detached stratum*, since the defining characteristic is a detachment from regular economic activity, often involving long-term or chronic recurrent unemployment, and an equally chronic need for state transfers from outside national insurance schemes. It is from this group that the state can draw lowly paid labor to help undermine workers in other strata, and its existence must act as the biggest source of fear for those unwilling to become subordinately flexible.

The "Fordist system" may have broken down, but if so it has done so not just because mass production based on regular wage employment has shrunk but because the regulatory framework has become dysfunctional, given that more and more of the population of advanced capitalist countries have become detached from productive society, that is, detached as workers.

Consumer capitalism depends on workers consuming mass-produced products, but increasingly people's identities are tied only to consumption, not to production. This applies to most of the labor fragments, most of which are not easily organizable because productively they have no collective identity. It has almost reached the stage when one cannot envisage collective *class* action any more, only sectional action, on for instance ethnic, age, or gender issues.

The labor regulatory framework built up in the era of welfare capitalism is also becoming dysfunctional, because more of the population are detached from productive society. They may be completely detached, as in the case of the long-term unemployed and many of those in "labor market schemes." Or they may be behaviorally detached, as with most flexiworkers, who have little continuity of employment and thus little access to nonwage components of working-class income, such as earnings-related benefits that were built up as incentives to continuity of employment, as in the Federal Republic of Germany. Historically, the social insurance welfare system has had a regulatory function, with earnings-related mechanisms expected to increase productivity, through incentive and worker commitment effects, and the firms' return to investment in training. But if the labor process is not generating the sort of employment for which such regulations can function to raise efficiency, then one can anticipate attempts to revise the system to secure a more effective regulatory framework. This is one reason for the drift towards workfare, which we might define as the payment of benefits conditional on some predetermined work-related activity, which includes "training" and work habituation courses. Does subordinated flexibility lead anywhere other than to workfare?

Labor process fragmentation erodes the integrative functionality of both Beveridge and Bismarckian social security systems. They were conceived primarily to foster productivity and ensure income security; they have become means of weakening labor process security and of undermining labor market security, simply because the former relies on common interests and the latter depends crucially on workers having effective freedom in the labor market. The more complex the system, the more the regulatory objectives will be uppermost. Moreover, as argued elsewhere, with labor flexibility and fragmentation, the contributory base tends to shrink and benefit entitlement tends to narrow and grow weaker. Many of those advocating more flexibility also contend that welfare encourages and strengthens behavioral dependency, and in that spirit governments have shifted towards "targeting," means tests, tighter conditionality, tighter behavioral monitoring, and the promotion of "welfare pluralism." In many countries one sees a strong underlying trend towards some variant of workfare, which is scripted to play a major role in the libertarian cinema of the 1990s.

A final labor process trend closely associated with the growth of subordinated flexibility, and to the trends to targeted or pluralist welfare and workfare, is the changing role of "training." This stems from the perceived need to raise productivity. The basic supply-side view is that wage differentials should widen and be more individualized, the orthodox ideology being that the rich need more to give them the incentive to work, while the pool need less to make them work. But along with more flexible payment systems, ostensibly designed to promote productivity, has been a trend that deserves more critical attention than it has received, the growth of what might be called "trainingitis," which derives from the erosion of skill in the traditional senses of that term. The more occupations are split into jobs, the more labor statuses are flexible, and the larger the sixth and seventh labor force strata grow, the greater the need for *job* training and labor market training. In many European countries the role of the state in this has grown enormously, and the script is that a larger proportion of the population will have work "careers" consisting of flexible combinations of short-term jobs preceded and succeeded by training and retraining, a pattern potentially leading to a whirlwind of jobs interspersed with training. Some observers have talked about an "active society" for the 1990s, based on benign presentations of this twilight zone of intermittent training and productive activity. But individualized workers obliged to shift from pillar to post in a fashion determined for them by the state, or by any other regulatory body, are scarcely likely to lead the way to the acta vita idealized so memorably by Hannah Arendt.

In sum, if the libertarian, supply-side path to flexibility persists, the following scenario seems the most likely:

1 a contractualization of the labor process, with proindividualistic regulations constraining collective action;
2 welfare pluralism, with the state as fall-back "safety net" provider, with privatized benefits for the upper strata and with voluntary private services left to fill the gaps left by an incomplete insurance system (a model presented as "the caring society" in the Netherlands) (de Neubourg 1990);
3 privatization of social policy as well as of economic spheres;
4 workfare replacing means-tested and universal transfer payments for those deemed to be "employable";
5 more policing of welfare "scroungers";
6 a steadily growing police presence in civil society to control the losers in an aggressively competitive economic environment;
7 a neo-corporatist state based on an overt employer–government alliance to replace tripartism, with trade unions shrinking and shackled by legislation and their own fragility in the context of flexible labor markets and fragmented productive systems.

Whether or not these are exaggerated as stated, they are sufficiently present as trends to suggest that a search for alternative paths would be reasonable.

TOWARDS COOPERATIVE FLEXIBILITY

There is a nucleus of another route to labor process flexibility, which builds on the corporatist traditions kept alive in the Nordic region, but which combats a critical limitation of the Nordic models. An essential element of any viable alternative must be the avoidance of labor fragmentation and income insecurity. Yet the danger is that critics of subordinated flexibility and supply side trends of the 1980s will continue to give primacy to labor market and employment security. Neither of these is really a meta right, though both should be seen as instrumental to the promotion of other labor rights. Treating them otherwise ultimately undermines their political legitimacy.

The error of the old social democratic "laborist" route to the Good Society lies in a faulty syllogism: there is a right to work; all rights imply duties; therefore, there is a duty to labor. One could give countless examples of unfortunate deductions from this line of reasoning. Thus, to give just one minor instance, it was characteristic that a recent international meeting of trade unionists concluded that the answer to the inadequacy of social security provision associated with the growth of more flexible forms of employment was more full-time jobs (OECD 1989: 7). But one may guess that full-time wage employment is neither desired by the majority nor desirable on efficiency or equity grounds. As noted earlier, if full-time wage employment is treated as a meta right, then one must favor market clearing wages, whatever they may be, and a compromise on various other forms of labor security, notably labor process and work security.

Full employment is always possible. They managed it fairly well in slave societies, which may not be the best recommendation. However, it is not a means of removing labor fragmentation, nor is it any more a reliable means of reversing the inequalities and erosion of labor process security. Recognition of that seems essential if policy reform is to take advantage of current realities without drifting into some atavistic cul de sac, such as the mass creation of "public-sector jobs." In that context, in some places, often in a rather disjointed manner, mechanisms for promoting active rather than subordinated flexibility have been taking shape. We may be in the midst of an era of social experimentation, in the course of which partial reforms may be pieced together to create the basis for a much more flexible lifestyle by the end of the twentieth century. The vision must surely be a social structure in which labor process security and income security are guarded and enhanced as meta rights.

Such reforms are taking a number of complementary directions, and one feels that they would fuse much more successfully and sooner if we could articulate the type of social and labor structure that we would like to create. Here it might be useful merely to outline the types of reform that may be shaping an alternative path to labor flexibility.

First and foremost, unless organizations can be revived to represent the collective voice of the vulnerable segments of society, notably those in the sixth and seventh strata, the necessary impetus to sustainable nonsubordinated flexibility will not emerge or will be dissipated. That is why we need to be concerned with alternative regroupings of unions, and in particular the possibility of communal unions evolving in place of industrial or craft unions. Recently, a variant of this has also been called "associational unionism," since it associates individualized workers who could benefit from collective representation. If workers are "postcapitalist labor" in the sense of not being in stable proletarianized relationships, they will be uncommitted to industrial unionism, just as flexiworkers can scarcely be expected to be committed to craft unionism, besides being hard to organize or to retain. But communal unionism will also not flourish if the organization merely represents an agency for job placements, advisory services, personal loans, and a source of social security for their members, even though all those functions are desirable. If that is all unions become, the state or private commercial firms will always try to turn them into individualistic entities. Only if they are constantly concerned with the primary problem of the era, redistribution, will they develop a pivotal role. There are signs that moves to redefine communal solidarity are growing in significance, and that the promotion of economic democracy, as well as industrial and political democracy, will be high on the agenda in the 1990s. Without economic democracy, one can see no alternative to the type of subordinated flexibility sketched earlier.

More advanced in practical terms is discussion and application of experimental policies to promote flexible lifestyles that build on the fragmenting tendencies in the labor process. Many of those are double-edged, in that they could be either converted into instruments of intensifying subordinated flexibility or integrated with other policies that together promote a more active, egalitarian flexibility.

If there are more part-time employment slots, and if there is a perceived need to have quicker responses to economic restructuring – perhaps as a result of more rapid and pervasive technological innovation – then it makes little sense to hope to buck the trend. It makes much more sense to facilitate flexible work patterns on terms desired by workers. That is one reason for foreseeing an era of social experimentation. Haltingly, one sees the nucleus of what could be called a *social dividend* route to flexibility, called that because ultimately it is based on redistributing the economic surplus through ways other than wage income and welfare (Standing 1989). This route would also give precedence to the right to work over the right to employment, bearing in mind that a right to do something can only exist if there is a matching right not to do it.

One could argue that a social dividend approach is crystalizing in the various experimental policies and institutional developments taking place in various parts of Europe. One thinks of sabbatical year and "time bank" debates in Sweden and Finland, solidarity contracts in Belgium, partial retirement schemes, career break and parental leave arrangements, the *revenu minimum d'insertion* in France, wage earner funds, renewed interest in profit sharing, the renewed growth of cooperatives in Italy and elsewhere (including the former Soviet Union, in a big way), industrial districts in the Federal Republic of Germany and in other countries, and so on. Experimentation is the general principle of the moment. Thus old ideas about time banks have been repackaged as less radical reforms. And, for instance, in Finland a government committee was set up to consider how sabbatical years could be phased into existence, with the state arranging for all workers to have the right to periodic breaks from their main labor force activity (Lilja, Santamäki-Vuori, Standing 1990). Partial retirement schemes and the removal of arbitrary retirement age notions are widely regarded as cautious steps in the general direction of lifetime flexibility, especially in the context of the ageing of European societies, though such schemes can be and have been easily turned into sources of inequity and discrimination unless developed in the appropriate institutional context.

One also notices more constructive discussion of unconditional income transfers, or citizenship income grants. That would decouple labor market status and behavior from income security and facilitate flexible combinations of productive and reproductive activities, helping in the process to legalize the shadow economy and encourage the growth of the "informal" economy. As long as governments lack the courage to promote the genuine right to income security, that is, unconditional, universal, and individual, the "flexible specialization" potential will be restricted, and the potential pursuit of occupation stultified. Basic security from deprivation as a citizenship right will be a necessary condition. It is not the only one, for to create an environment of cooperative flexibility will require labour process security, to prevent the vulnerable from being systematically marginalized, and thus being a threat to the working community, and to combat the coercive potential of contract law replacing collective regulation. There must be regulation to provide the basic safeguards against the structural inequalities that market mechanisms are bound to produce and intensify. That is why work security and payment system security (including minimum wage protection) will remain essential components of any path to an active flexibility society. But, ultimately, the institutional structure that promotes labor process security will be far more significant for that than any number of regulations. New forms of union, new forms of collective agency and new meanings of solidarity will need to emerge.

CONCLUDING POINTS

Since work was begun on this chapter, political events in Eastern Europe and the Soviet Union have given these deliberations a new poignancy. A bogeyman has been removed, exorcised, and after a year or two debates about the direction of social policy may become decidedly more progressive as a consequence. The market-oriented orthodoxy is promoting subordinated flexibility everywhere, but in the wings is an emerging paradigm based on cooperative flexibility. For that to succeed it is essential that we focus on forms of recovering and extending labor process security. Trade unions are almost anachronistic in societies where trades or crafts are made ephemeral. Industrial unions are enfeebled by the multinational, multi-industrial nature of capitalist conglomerates. There must be moves downstream so that unions represent the aspirations and needs of all the groups in our fifth, sixth and seventh labor strata. The only way for such organizations to emerge as class-based organizations is to diminish the fear factor, and the only way to do that is to shift the social security system in the direction of decoupling the labor market from the provision of income security. If one sees the existing welfare system as a regulatory device, then critics of the trend to subordinated flexibility should want to revise it radically. That surely is one of the two keys to creative, cooperative flexibility, the other being economic democracy in some form.

At present, the fragmentation of society is such that political democracy and the prevailing welfare institutions are far more likely to cement the divides than to reduce the marginalization of flexiworkers and detached groups. A majority could almost always be mobilized politically to reduce the rights and the security of the vulnerable minorities. Enabling large numbers to make the transition from the bottom three strata into the others, let alone disintegrating the artificial fragmentation of the labor process, is scarcely feasible unless the institutional basis of income provision and work organization is revised.

In sum, an alternative path to flexibility will have to be based on the promotion of the meta rights of labor process and income security, and will have to evolve through institutional mechanisms geared to create three forms of democracy – political democracy, industrial democracy (i.e. through codetermination to ensure work security, protective regulations, etc.), and economic democracy (i.e. through institutional mechanisms to redistribute economic surplus equitably, involving the collectivization of profits and citizenship income dividends). This would effectively reverse the traditional social democratic and socialist agenda, since rather than nationalize the means of production, it would privatize the management and ownership functions while socializing the surplus. The form of this alternative framework is still very far from clear, but the contours are beginning to take shape.

ACKNOWLEDGMENTS

Views expressed in this chapter are not necessarily those of the ILO. Thanks are due to Kate McFate for useful discussions.

14

LABOR CONVENTIONS, ECONOMIC FLUCTUATIONS, AND FLEXIBILITY

Robert Salais

FLEXIBILITY AND CONVENTIONS: THE PROBLEM

The term "flexibility" is nowadays used to refer to an ensemble of properties which must be possessed as much by the forms of economic regulation as by the enterprises themselves. These properties correspond to several types of empirical phenomena: the development of uncertain markets where a variable demand manifests itself in an unpredictable manner over time; the growing differentiation of products following a closer analysis of the needs of demanders; the tendency toward quality competition (and no longer only price or cost competition); the growing dynamism of small organizations confronted by the bureaucratic rigidity of large ones; and the call for the polyvalence, autonomy, and initiative of workers in production associated with the deployment of flexible, computer-assisted technologies, etc. In sum, while economic thought, orthodox or heterodox, is largely based on the predictability and homogeneity of behavior, flexibility confronts us with an economic universe which increasingly takes on, for its participants, a character of heterogeneity and uncertainty.

The aim of this chapter is to sketch out research into labor flexibility from the perspective of the "economics of conventions" (*économie des conventions*), an approach elaborated in a series of articles recently presented in a special edition of *La Revue Economique* (1989). Three major questions raised by the flexibility of labor will be examined in the attempt to utilize the economics of conventions. But first I shall review briefly the basics of the approach that will be used in the discussion that will follow.

1 The nature of the compromise underpinning the labor relations between employers and wage-earners

For several decades, the relative security of employment in big enterprises, the existence of social rights and unemployment insurance guaranteed by the state, together with the standardization and routinization of work, diverted attention from the uncertainty intrinsic to labor relations. This uncertainty arises out of the conflicting interests of employers and wage-earners, and carries over into established agreements and commitments with respect to work quality and effort. The need for flexibility, which arose out of the crisis of the 1970s, called into question these earlier social agreements and made them appear for what they were: on the one hand, regulations and, on the other, particular regulations belonging to a certain regulatory regime. In other words, such conventions permitted only one efficient mode of regulation among many pathways which could have, in principle, been taken as a response to uncertainty. This led to the necessity to ask oneself which social form assures the resolution of uncertainties specific to labor relations (Salais and Thévenot 1986). The need for flexibility provokes a series of questions: can labor relations function without a certain overall set of legitimating principles in spite of the fact that flexibility makes impossible a uniform model of contract? Under these circumstances, what is the origin of its taken-for-granted character? Does the unilateral authority of the employer provide sufficient basis for this legitimacy and one which assures the respect for commitments? How is one to conceptualize both agreement and conflict within the same system of regulation of labor? What is the relationship between rules of work and norms of productivity?

2 The necessary link between the analysis of labor and that of the product

Labor flexibility calls into question the relations between the management of labor, the product, and its market, in the mass production system. If, for example, sales revenue becomes unpredictable, the wage cannot be fixed in the short term and must depend, in one way or another, on what happens on the market for the product. Long production runs for the same homogeneous product limits labor to certain repetitive and standardized tasks, to a narrowly defined quality of work. Adapting oneself to the variety of quality of demand on a fluctuating market necessitates a redefinition of work qualities. It is clear also that the connection between the quality of work and the quality of the product established by flexible methods is not limited to the enterprise. Rather it brings into play the formation and reproduction of the qualities of the work force in society as a whole. If wage-labor is characterized by the divorce between work and the product of work, then it rests, at the same time, upon their articulation. In what way is this articulation modified by flexibility?

3 The complexity of rules at work and the variety of forms of flexibility

Flexibility calls upon us to think about diversity and complexity. In so far as the variability, differentiation, and heterogeneity of product qualities become the determinant components of competition, our hypothesis, inspired by earlier research into conventions[1] (Thévenot 1989; Eymard-Duvernay 1989a) is that it is no longer possible to reduce all forms of firm organization, work rules, and wage determination to a single model. Rather, we can hypothesize that several models of labor conventions exist and that each one presents a specific form of flexibility at work and is adapted to a particular economic universe. Our ambition is to go beyond the vision of flexibility as an empirical critique of the dominant model of production (a critique which we shall return to at the end of this chapter) toward a more positive and diversified conception. We shall propose a plurality of labor conventions founded upon the qualities of the products of work.

WHAT IS A CONVENTION? ITS APPLICATION TO LABOR[2]

Convention, contract, and norm: points of reference

The word "convention" signifies a taken-for-granted rule commanding spontaneous obedience, the result of a founding agreement (the Geneva Convention, for example) and the founding moment (the *Convention Constituante*). The choice of this term is aimed at underlining the fact that agreement between individuals, even when limited to a short-term contractual transaction, is not possible without a shared framework, an external point of reference going beyond the simple encounter between separate wills. The economics of conventions, then, rejects the neoclassical postulate according to which the agreement of the parties – the labor contract – can only take place via prior written specification of every detail; reflecting identical, optimizing rationalities and intentions. Nonetheless, we reject the opposing sociological postulate according to which agreement results from a submission of individuals to exogeneous norms which either impose themselves from the exterior (being decreed, for example, by social institutions or by labor law, etc.) or are completely internalized in the consciousness of actors without their awareness. (See the extract "Durkheim and the contract", below, in which we suggest that Durkheim develops a more subtle view of contract than usually understood.)

In other words, this middle-ground solution consists in the notion that the convention is a system of reciprocal expectations regarding the competence and behavior of individuals, conceived and accepted as taken-for-granted (Favereau 1986). For efficient coordination to take place, there is no need for these expectations to be written or objectified in any way. Indeed, it would make little difference since their degree of generality or efficacy would not be affected by being written down, for the degree and the modalities of exteriorization of the rules vis-à-vis individuals depends on the system of conventions in force.

Durkheim and the contract

The concept of convention has a larger theoretical relevance than contract or norm in analysis of labour flexibility. It seems to us close to the methodological rules which were outlined by Durkheim in his "The Division of Labour in Society," 1893. Criticizing Spencer and liberal economists, he questioned the contractual relations within the division of labor (see, also, Didry 1991):

> Contractual relations necessarily develop with the division of labor because the latter is not possible without exchange, of which the contract is the juridical form. Doubtless it is an error to believe that all social relations can be reduced to the contract: all the more so given that the contract presupposes something other than itself; nevertheless there are special connections which have their origins in the wills of individuals. There is a *consensus* of a certain type which is expressed in contracts and which, in its higher forms, represents an important factor in the general *consensus* [Durkheim's emphasis] . . . .
>
> But for the general result to be obtained, it is not sufficient that the public authority ensures that commitments are respected. Rather, it is necessary – at least in most cases – that commitments be spontaneously respected. If contracts were only observed because of force or the threat of force, contractual solidarity would be singularly precarious . . . . On the other hand, consider the hypothesis that, in order to eliminate this danger, the free consent of individuals suffices. This is true but the difficulty is not so easily conjured away. For, what can be said to constitute "free consent"? Verbal or written acquiescence is not sufficient proof; forced acquiescence is not impossible. All constraint must therefore be absent. But what is the origin of constraint? It does not consist in the direct employment of violence because even indirect violence eliminates liberty . . . .
>
> In a given society, every object of exchange at every moment has a definite value which one can call its social value. One can perceive quite easily the principal conditions according to which social value varies. These are, above all, the sum of efforts necessary for the production of the object, the intensity of the needs it satisfies and, finally, the breadth of the satisfaction it brings . . . .
>
> This definition given, we can assent that the contract is only fully consented to if the services exchanged have an equivalent social value. Under these conditions, eveyone receives the thing he desires and delivers that which he gives in return for what both one and the other are worth. This equilibrium of wills,

> recognised and consecrated in the contract, produces and maintains itself automatically since it is simply a consequence and alternative form of the very equilibrium of things. It is truly spontaneous.
>
> (Durkheim 1967: 375–7)

Durkheim emphasizes that the contract itself, as the basis of the contractual relationship, cannot be limited to the simple meeting of two free wills. Respect for commitments demands the necessary mediation of a third party who guarantees it and supplies a legal recourse in the case of observed violations. Durkheim calls this intervention by the regulations of a public authority, the "law of contracts." But this is not sufficient: the contractual relationship must be respected beyond free consent. The equilibrium of wills must be "spontaneous" and this is only possible because the "equilibrium of wills" is simply a consequence and another form of the "equilibrium of things." This is due to the fact that the equivalence of social values is inscribed in things. And this equilibrium must be held beyond constraint whether it be the effect of authority or of a particular form of violence. Otherwise the relationship would appear to be of a "singular precarity," hardly borne out by observation.

It appears that, thus conceived, a contract between two individual wills could be considered as part of a *convention* which, at the same time, goes beyond it, legitimates it and regulates it in its detailed aspects. Within this condition, the contract is clearly more than an exterior norm, for the convention which founds the contract simultaneously rests on rules (the law of contracts), on a principle of equivalence of exchanged objects, common to contractors and on a spontaneous agreement (agreement beyond consent and constraint). In fact, the contractual solidarity described by Durkheim presents a number of the traits which we shall call, later in this chapter, the market labor conventions. One will note that the equivalence of social values exchanged are considered not only in terms of quantity, but also, fundamentally, in terms of the equivalence of qualities (as it is expressed in the systematic use of terms such as "the sum of efforts," "the intensity of needs," and "the breadth of satisfactions"). At the same time, Durkheim limits its scope of validity to relations where "the values exchanged are really equivalent" (415); opening up, for Durkheim, the possibility of other modalities of solidarity such as "organic solidarity." For us, this means that we must conceptualize several types of conventions.

Two points, however, should be made concerning the current analysis of labor relations. First, conventions do not exclude logical contradiction. On the contrary, they are a form which enables the coordination of contradictory interests based on opposed logics, but which need to be brought together to be satisfied. Clearly, if the entrepreneur is interested by the product of work and the workers by the wage, only their meeting and their activity in the enterprise are capable of satisfying these two interests. The situation is indissolubly conflictual and cooperative.

Second, the convention, as an empirical phenomenon, does not exclude conflict. For a given regime, it is the framework within which the conflicts are conceptualized and resolved, for it formulates the pertinent variables around which conflicts arise and are resolved. For example, conflict over the size and periodicity of hourly wage increases is an everyday occurrence. Nobody, however, questions the idea that the hourly wage rate is central to the conflict. Yet there exist other forms of remuneration such as piecework or monthly salaries (not based on hours of work), wages linked to the performance of the enterprise, etc. which imply other systems of production, alternative organizations of work, and even other forms of consumption. But the conflict takes place within a convention to which all the participants conform without thinking; a convention according to which the pertinent variable and the fair evaluation of work is the hourly wage rate. The existence of a prior convention only becomes apparent to the participants in moments of crisis when the form of earnings itself is contested rather than the level or rate of increase.

Labor conventions and equivalence

As in most economics, we assume that "equivalence" is the foundation of all exchange, i.e. exchanged objects have the same value for all the parties to the transaction. For this situation to be transformed into the basis of a stable and positive coordination between economic agents, equivalence must rest upon a common principle; a scale and method of evaluation common to all the parties involved. This is an important and necessary assumption. In fact, one cannot conceive of economic coordination founded upon the prior and permanent negotiation of rules between agents, which would imply suspicion and lack of trust. Coordination is either spontaneous or nonexistent.

The application of the concept of convention to labor relations is therefore very complex. The objects that establish the equivalence for individuals in this relation are problematic as are the principles according to which the objects are evaluated.

Figure 14.1 presents the three points which must be brought together by a theory of conventions applied to labor:

1 Labor has no material existence independent of and separate from the person of the worker. It cannot be an object over which a commodity transaction can be made (cf. Marx 1867: V. 2). Only the product of work

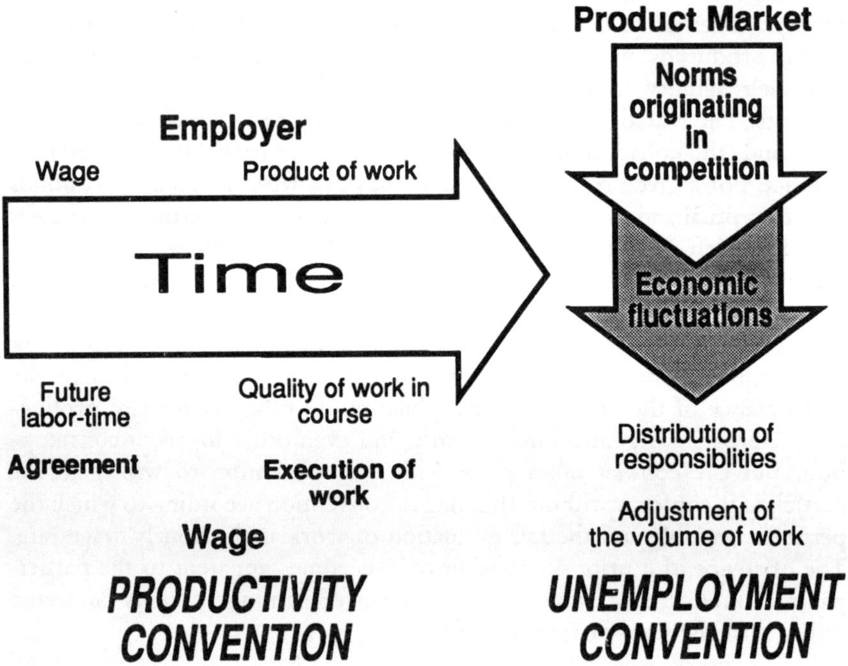

Figure 14.1 The domains of action of unemployment and productivity conventions

can be the support for such a commodity exchange but, except for certain "extreme" labor conventions where the wage relation is disappearing, the product cannot be the pivot of an agreement, because the product is not yet in existence when the agreement has to be concluded. It must be replaced by something else.

2 The labor relation must be concluded between the entrepreneur and the wage-worker prior to the realization of work and the sale of the product. It consists of two successive operations: agreement (hiring) and work or production. Separated in time, these two operations nevertheless cannot fail to be in relation, which influences in return their content.

3 The labor relation unfolds in the overall framework of the market: the value of the product of work anticipated by the entrepreneur is either validated or not by price on the product-market. This difference between the forecast and the result must be capable of being absorbed without disrupting the practice of the current labor relation.

These three points correspond to three moments which are linked in time and which must be theoretically presented as such:

a) hiring
b) work in production; and
c) the realization of the product on the market.

The unity of these moments is realized by the economic cycle of the product of work, considered from the point of view of the labor relationship. The convention of productivity covers the first two moments and the convention of unemployment, the last one.

The convention of productivity

The convention of productivity refers to something not yet in existence, future labor-time. The convention of productivity thus establishes a compromise between two equivalences:

1 During recruitment, the wage is exchanged against future labor-time; this is the first wage/labor-time equivalence in labor relations. Constituted around an immediate transaction, this first equivalence takes on a market form. It is defined around a rule or standard for setting wages.
2 Thereafter the wage-earner puts to work his labor-power in production, transforming labor-time into a product; this is the second labor-time/product of work equivalence. Unfolding in and referring to labor-time, this second equivalence does not, in essence, take on a market form. It can be constituted by various principles of equivalence (to be defined theoretically in the last part of the chapter) which also define the work rules.

A definite social relationship between wage and productivity (defined both in terms of quality and effort) results from this compromise. This compromise is ambivalent because it mixes two equivalences which are different in nature and provokes the uncertainty with respect to productivity. It is thus prominent in the contemporary search for flexibility. At the conclusion of the labor relation, which takes place around a wage/future labor-time exchange according to a definite wage rule, nothing guarantees that the second labor-time/product equivalence will be realized in accordance with the expectations of the entrepreneur.

The unemployment convention

Economic fluctuations in the product market (with respect to quantity, price, or quality) render necessary the existence of stable rules of short-term adjustment on the quantity (and quality) of work engaged via the convention of productivity. The unemployment convention defines these rules of work adjustment, and the respective roles of workers and employers in it.

The assumption can be made that workers' and employers' responsibilities in this adjustment depend on the structure and distribution of property rights. For instance, in industrial labor conventions (see pages 296–8) the entrepreneur owns the resources in capital and organizes the production process under his responsibility; as a consequence, he disposes of the monetary amount of sales. It has thus the right to adjust, under his own responsibility, the number of employees salaried by his firm; but as a counterpart, fired workers can benefit from social indemnities.

Since the convention of productivity equilibrates two conceptions of the fair relationship existing between work and remuneration, short-term adjustment must endanger neither the rule governing the wage, nor the rules governing work effort and quality. It is endogeneous to both sets of rules. For example, in labor conventions founded on rules derived from Taylorism, the short-term wage rate is inflexible, and adjustment to economic fluctuations takes place around the number of work stations occupied. Here adjustment takes an impersonal form such as the "last in, first out" rule or criteria relating to the age of early retirement in the case of overmanning.

Conventions, wage-determination, and work rules

The continuity of the labor relation is threatened by economic fluctuations; demand fluctuations can erode expected profits; the same problem can arise from internal fluctuations of the quality of work and the effort of the workers. Conventions thus must rest upon an "understanding" of fluctuations which creates shared (or "common") knowledge by which participants cope with fluctuations. Both the knowledge of these fluctuations and the coping mechanisms must be invested with a legitimacy such that, if the threat posed by fluctuation is ongoing, disagreement leading to rupture of the convention is rare and usually contained within narrow limits. It is necessary then that they draw simultaneously on both principles of equivalence, in constructing a general set of expectations about legitimate wage and work rules.

Wage and work rules therefore must fulfill two requirements: first, they must mobilize knowledge about the uncertainty of the labor relation and then, they must manage this uncertainty in a legitimate way. They are not, therefore, merely rules of action or of judgment. They are also cognitive rules which provide information and allow participants to define and evaluate (Favereau 1989). They consist of shared values, classifications, variables, criteria, and ratios and serve as instruments for measuring or judging the ongoing functioning of conventions.

Take, for example, the use of the rules based on Taylorism. Organization into discrete tasks renders work measurable and predictable. Jobs in the workplace are based on a classification of concrete tasks – elementary operations – which define the quality of work. A sort of codification associates a variable, abstract amount of time, with each operation in the classification. A job can thus be defined as a sum of elementary operations and an amount of time prescribed for effort. An output norm (effort) is then defined, together with an objective calculation of the wage rate around which wage bargaining takes place (labor-time/wage equivalence). Individual tasks can then be added together to permit the calculation of costs accounting of the firm. The productivity of labor is observable: labor-time can be assigned to the product (labor time/product equivalence) and effort measured. The firm can measure its profit and derive a norm by calculating its difference with the overall sales figure. Thus able to identify appropriate and needed labor-time in the form of concrete jobs in the workplace, the enterprise is able to convert any excess work time in the short term into overmanning or unemployment (convention of unemployment).

MODELS OF LABOR CONVENTIONS

From the product of work and its quality toward work

Labor conventions have their sources in the separation between work and its product;[3] their domain is to define relations and connections between these two areas: work and product. If the evaluation of individual effort rested upon the simple exchange of products, the domain of labor would have no specificity; the analysis of the conventions defining the various forms of equivalence of products would suffice and we would need only study the quality of goods (Eymard-Duvernay 1989a and 1989b).[4] Furthermore, in those cases where work is confused with its product, labor conventions and conventions of the quality of goods would be identical (consider, for instance, an economy solely composed of individual independent producers, without firms and wage-earners).

The approach adopted here begins with the product and advances toward work. It examines the quality of the product and the quality of work and analyzes how labor conventions affect their articulation. The first equivalence of the convention of productivity (labor-time/product equivalence) rests on product quality as it is defined by the organization of the production process and its technology. The rules of work and work activity are, therefore, closely associated with the rules defining the quality of the product and which are imbricated in the organization of production. Our task, here, is to specify the rules which define the quality of work in light of product quality, within the organization of production.

The second equivalence of the convention of productivity is that of the market. The wage rule belongs to this second equivalence. It must be defined in relation to the product and to the standard of quality established on the product market. This proposition is unusual in that it subordinates the role of the labor market in the determination of the wage to the product market. Wage rule determination, in other words, has its theoretical roots in the product market. For a given wage-rule, it is only the wage level and its variations which are determined by the labor market. This is clear in the extreme case where the remuneration of the individual comes directly from the sale price of the product; in this case, the form of the wage is the market price of the product. But the current price will depend not only on demand, but also on the number of individuals offering the same product and the same product-quality on the market.

We are left therefore with two axes to our model of product and labor markets. The first axis concerns work activity and the labor-time/product equivalence. At one end is the specialization of individuals and at the other the standardization of jobs. The second axis concerns the product market and the wage/labour-time equivalence. At one end may be found wage variability and at the other wage short-term fixity; each pole corresponds to a type of market: uncertain or predictable. These two axes define four pure models of labor conventions. These models will be elaborated in the last section of the chapter, but we shall first discuss in greater detail the choice of these two axes.

The first axis

Work rules determination (labor-time/product equivalence) counterposes the *quality of persons* to the *qualification of jobs*. This corresponds to the specialized/standardized product distinction. Making the quality of persons an alternative to the qualification of jobs leads to several oppositions (see, also, the "Idiosyncracy" extract). For the former, the organization of production is based on the specialization of persons and, for the latter on the standardization of tasks (and, beyond that, of work-stations). The quality of work, when it is specialized, is linked to individuals who have a quality based on knowledge and know-how (the mastery of a "trade") which only they possess. Standardization by objectifying the labor-process allows their personal qualities to be built into material and bureaucratic organization of the labor process. The skills needed to carry out tasks are standardized, with individuals no longer possessing specific qualities. We shall see later that, in these cases, the wage can therefore be attached to the job-description and not to the person. The organization of production which, in the former case, is based on a collective of workers, each with their own particular identity, rests, on the latter case when it is standardized, upon the connecting together of identified tasks and work-stations across a particular deployment of equipments. In the first case (specialization), persons represent a valuable, capitalizable resource. In the second (standardization), tasks are costs to be reduced. Standardization, is, in effect, intimately linked to economies of scale as the only way to reduce average unit costs. Skill specificity, on the other hand, is linked to investment in the development of know-how and in the quality of labor. Expressed in the terms of industrial economics, specialization privileges economies of variety, based not on the length of the series, but on the range of products and the mastery of a variety of qualities.

Idiosyncracy and economic analysis

When economists want to study the variety of qualities of labor, they take as their point of departure, the levels of skill (as revealed by the classifications of work-stations in the wage-hierarchy). They consider, therefore, that work is already constituted as a factor of production which can be described objectively and standardized in terms of a classification of quality and of a measure of quantity – in terms of hours of work or number of jobs.

Nevertheless, the recognition of an opposition between quality of persons and qualification of jobs traverses economic analysis in diverse forms. A positive form based on the observation of large, rationalized enterprises, insists on the persistence of the idiosyncratic character of the labor process (Doeringer and Piore 1971) or on the role of effort-conventions which move the X-efficiency away from the production frontier (Liebenstein 1966) or, finally, on the strategic character of individual knowledge in the Hayekian tradition, 1945. A critical form uses the point of view of economic rationality – the neoclassical production function or the model of Taylor – to denounce those elements which prevent the attainment of an optimum and which result directly from the inopportune intervention of persons: the non-observability[5] of performance, opportunist behavior, corporatist alliances leading to monopolies on the labor market and restrictive practices, etc. Remember, for instance, the fundamental analysis of internal labor markets in large enterprises undertaken by Doeringer and Piore:

> Almost every job involves some specific skills. Even the simplest custodial tasks are facilitated by familiarity with the physical environment specific to the workplace in which they are performed. The apparently routine operation of standard machines can be importantly aided by familiarity with the particular piece of operating equipment .... In some cases workers are able to anticipate trouble and diagnose its source by subtle changes in the sound or smell of the equipment. Moreover, performance in some production or managerial jobs involves a team element, and a critical skill is the ability to operate effectively with the given members of the team. This ability is dependent upon the interaction skills of the personalities of the members, and the individual's work "skills" are specific in the sense that skills necessary to work on one team are never quite the same as those required in another.
>
> (Doeringer and Piore 1971: 15–16)

While Doeringer and Piore presumably refer to an alternative model of production where this idiosyncracy positively deploys its economic advantages (this being suggesting by their future work[6]), their followers on this point (Williamson, Wachter and Harris 1975) tend to reduce the influence of such idiosyncratic knowledge to the "local conditions and particular circumstances" conceptualised by F. Hayek (quoted by Williamson *et al.*).

These specialized goods embody the quality of workers who make them and such products compete in markets based on quality and not (or not as much) on price or cost. On the contrary, a standardized good is produced by a known and widely available technology which gives no competitive edge to quality. For such a product competition is based on price. In the first case, information about the quality of the product is directly attached to recognition of the quality of the individual producers' work. In the second case, standardization permits such limits, which are both geographical and social, to be transcended as the standard spreads.

The second axis

Wage rule determination according to the product market opposes the wage variability to the short-term rigidity of wage. It opposes uncertain and predictable markets and separates dedicated and generic products.

The problem of the frequently observed divergence of empirical data from the neoclassical theory of labor market, empirically arises from the possibility that workers will engage in autonomous bargaining over the volume of work and the wage which blocks the instantaneous adjustment mechanism of the market. The short-term rigidity of the wage-level and the level of employment are phenomena foreign to the neoclassical market. It is well known that neoclassical hypotheses treat "work" as any other commodity; it is not separated in theory from its product.

The problem is to give a status to the short-term rigidity of the wage agreed to by the producer, in the theoretical framework of labor conventions. It is not a question of the stability of rules themselves – stability being a characteristic of all conventions – but rather that of the opposition between rules of rigid wages and rules of variable wages (Reynaud, 1986a and 1986b). Following the line of reasoning advanced above (from the product of work towards work itself), we shall look not so much to labor-market imperfections as possible explanation of these rigidities (the standard approach), but rather to types of product and their markets. In this way, we hope to show that for some types of product and market short-term rigidity of the wage can be efficient for the producer and the latter is then obliged to admit that market fluctuations for his product generate no information pertinent to the quality of work invested in it. Otherwise, he would, for example, reduce wages when the price of the product falls and choose more selectively its work force and cut working hours. This condition is due to the fact that since market fluctuations give no information about the product, variable wage-rates have no adjustment role and wage-flexibility, therefore not disposing of any market legitimacy, cannot serve as the basis for a convention between workers and entrepreneurs. It becomes, then, efficient for the producer to postulate the short-term independence of output level for his product vis-à-vis market fluctuations, and wage rigidity is then the only logical and legitimate convention.

Two product and market configurations emerge in these conditions. The first is the opposition between dedicated and general-purpose products and the second opposes uncertain and predictable markets. The bringing together of these two oppositions with the wage rigidity/flexibility opposition sheds light on the nature of the independence of output levels for the product vis-à-vis a fluctuating market.

Consider first the opposite situation. When the quality of the product is such that it is aimed at a specific need (defining what we called, in Salais and Storper, 1989, a dedicated product) the wage cannot but reflect the quality of work, which in turn is gauged by the value accorded by the buyer; the wage thus reflects price and quantity sold. For, in such market, the link between the quality of work and the satisfaction of the buyer is known and established. For example, a fall in sale price signifies a reduction in quality of the product and of the work that goes into it. The variability of price of labor, in this case, reflects the range of qualities of the product according to the range of customers' demands.

Wage rigidity as a rule is economically possible only if the variation in the value of sales signifies for the producer not variations in the quality of work, but only variations in consumer tastes or demands. This is the case when two conditions are met:

1 the producer masters the definition of quality (for example, he succeeded in imposing his standard of quality for all the producers of the product on the market, what we could call an "industrial standard"); and
2 the product is aimed at a large, identical, and anonymous demand (defining what we called, in Salais and Storper (1989), a generic product).

In this case, if demand falls, it is because the design of the product is out of step with the changing tastes of consumers, and not because the quality of work *per se* is inadequate. In other words, the quality of work and the satisfaction of the consumer are two separate issues.

The second way to conceive the wage variability/inflexibility distinction stems from the nature of the product-market. A generic product is, *par excellence*, the basis of a *predictable* market. The large number of potential buyers for a generic product allows the producer to calculate his risk in serving the market (for instance, by the use of quantitative and statistical techniques, regression to the mean ensuring a fluctuation around a medium-term trend). This allows planning, i.e. inflexibility, in the volume of work and in the wage. On the other hand, a dedicated product, aimed at a particular demand, defines an inherently *uncertain* market – no evaluation of the risk associated with demand, around which the producer can plan its production strategy, can be undertaken. The producer cannot make commitments in an uncertain market, nor can he arm himself against fluctuations in the future by (say) expecting to switch from one individual demand to another, either immediately or later on. As a consequence the adjustment of labour costs must be very flexible.

FOUR MODELS OF LABOR CONVENTIONS AND FLEXIBILITY

Four "pure" configurations of labor conventions are outlined here, but it should be remembered that any real situation is likely to contain a mixture of conventions. Nonetheless, each model should call to mind familiar empirical situations.

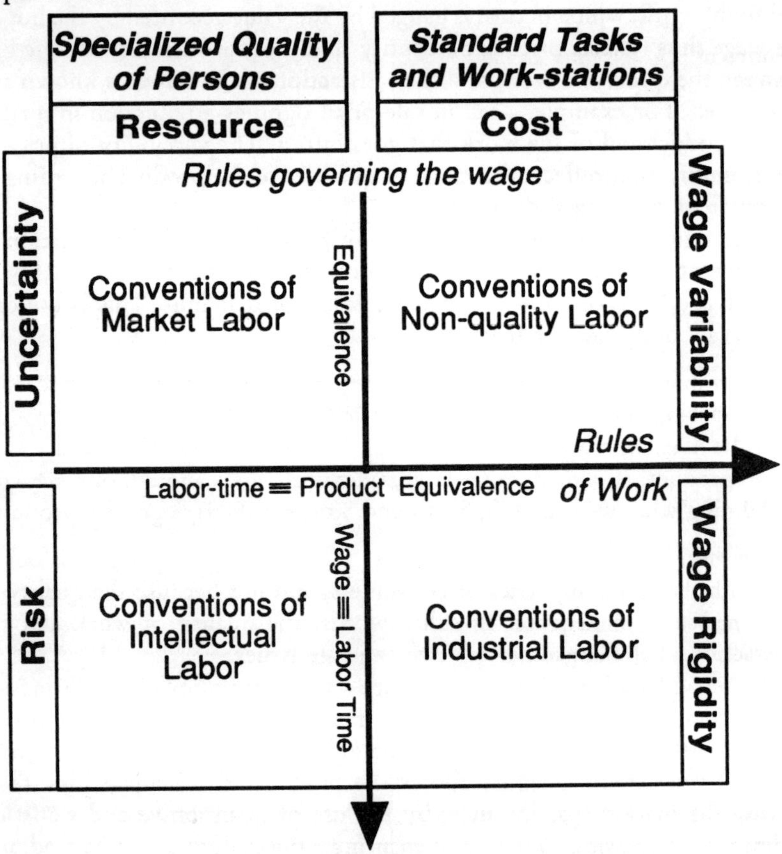

Figure 14.2 Four models of labor conventions

CONVENTIONS OF MARKET LABOR

Economic fluctuation	*market uncertainty with respect to quality*
Evaluation of the quality of work	*price of product on the market*
Forms of wage	*assigned to the person; deduced from the price of the product*
Forms of flexibility	*related to the variety of the quality of the product*

CONVENTIONS OF NONQUALITY LABOR

Economic fluctuation	*local uncertainty with respect to demand*
Evaluation of the quality of work	*no quality*
Forms of wage	*paid by task*
Forms of flexibility	*exteriorization, instability*

CONVENTIONS OF INTELLECTUAL LABOR

Economic fluctuation	*general risk with respect to quality*
Evaluation of the quality of work	*rules of professional or scientific ethics*
Forms of wage	*specialized investments in persons*
Forms of flexibility	*related to a "base" of general knowledge*

CONVENTIONS OF INDUSTRIAL LABOR

Economic fluctuation	*short-term risk of the economic conjuncture*
Evaluation of the quality of work	*system of classification of work posts*
Forms of wage	*function of hourly rate, number of work hours and of work-stations. Existence of an indirect wage.*
Forms of flexibility	*stocks, short-term adjustments in employment (unemployment or short-time)*

Figure 14.3 Economic fluctuations and flexibility of labor

The four models are summarized in Figure 14.2. We shall review each model in turn, describing their respective productivity and unemployment conventions, characterization of fluctuations, and type of flexibility. Each model is equipped with a form of labor flexibility[7] adapted to a given economic environment, as shown in Figure 14.3. Note that "rigidity" is nothing more than a critique aimed at the particular flexibility of one model by the participants in another model.

Conventions of market labor

Pictured in the top-left section of Figures 14.2 and 14.3 are those conventions where labor *quality* manifests itself directly in the product.[8] The product is, both, specialized (workers master the know-how for fabrication) and dedicated (satisfaction of a unique taste). Quality is evaluated by the intensity of demand. The price obtained for the product evaluates the quality of work performed. The contribution of labor tends, in these conventions, to be conflated with the quality of the product and two principles of equivalence are reduced to a single one, that of the equivalence labor-product. The labor market is regulated by the wage, but the latter is in turn a reflection of the price of the product. The "union rate" negotiated by members of a "trade," the price of the product accomplished by a workteam and negotiated with him and the piece rate in certain activities are forms of payment which can be referred to as rules of conventions of market labor (Mottez 1966).

The convention of productivity here is often confused with the nature of the product market, with its numerous, specific, potentially unique qualities of demand. The regulation of the labor market takes place, therefore, not by unemployment (that is, by inactivity of qualities of work), but by geographical mobility from one local labor market to another or by retraining into another quality. In effect, individuals take with them the quality of the product (a quality which cannot be objectified independently of its creator). The convention of unemployment is, thus, a convention of worker mobility and learning of specialized skills.

The community of "the trade" is a historical example of conventions of market labor. In the French rolling mills before the installation of mechanical trains after World War II, the operator was "somebody" (Verry 1955). His authority and prestige were founded on the practical, technical knowledge needed for the manufacture of iron sheets. This enhanced the value of the "tricks of the trade". The training of rolling mill operators took place on the job and followed the hierarchical organization of the workteam. The wage of the group leader was often calculated in terms of its overall output whereas the other members of the team were paid a percentage of this basic remuneration. This scarce, skilled labor, shifting as it did from one mill to another, was snatched up by the Lorrain ironmasters (Noiriel 1984). It was hardly surprising then that the late-nineteenth-century promoters of harmony between capital and labor exalted the liberal values of setting up one's own business, since this resembled the professional ethics of "the trade" (the spirit of independence and autonomy). These values were, in turn, denounced by the advocates of industrial rationalization:

> The worker is no longer that small businessman who, in the determination of the price by negotiation, undertakes – at the same time that he elaborates a definition of the fair price – an apprenticeship in the management of his future small business.
>
> (Mottez 1966)

The relevant economic fluctuation here has to do with uncertainty over the product qualities that will be in demand. This uncertainty is "local" in the sense that the market is defined by a close acquaintance between producers and consumers, this alone permitting the circulation and elaboration of the information necessary for exchanges when specialized goods based on an inherited know-how and specific technologies are in demand. For the producers, the problem is not fluctuating in the volume of demand, but variation in the qualities which compose it.

Flexibility thus consists of having available workers whose range of skills and equipment, whose polyvalence enables them to deal with a variety of quality of products. The price of the product is the only possible standard of quality of work in these conventions, and such flexibility allows sufficient adjustment to price variations. The correct strategy for the producer is orientation toward qualities whose price is rising.

Market uncertainty over quality is mastered to the extent that the definition of qualities by the producer corresponds to that of consumers. In this case, even if certain clients abandon him, he can find others, especially if his product scope serves numerous industries and geographical markets. Thus potential failure in local markets can be dealt with by output shifts, external renewal of the workforce, internal training or by a reorientation towards other market territories where consumers tastes are still consistant with output.

Conventions of intellectual labor

Seen on the lower left section of Figures 14.2 and 14.3, these give priority in both the configuration of the product and the organization of the production process, to investment in specific skills. But, unlike in the preceding cases, where the product imbues work with particular qualities, here the quality of the individual's skills is what imbues the product with its qualities. This quality, moreover, consists in the creation of new knowledge, in the mastery of a unique know-how and in the creation of new markets for products based on innovations. The difference also arises from the fact that these are specialized products which eventually take a general and universal form. The specific knowledge which goes into fabrication of a product with recognized qualities, in effect, creates the market. In the final analysis, such innovation-based products are not subject to competition and thus do not have prices fixed by the market. The economic value of the product is essentially that of the knowledge-based work that goes into it; as a result, there is a single wage/work equivalence.

The productivity convention is based upon internal – scientific – rules governing the quality of work. Individuals, who are the objects of intangible investment, are a fixed resource. Producers are sheltered from the classic conjunctural conditions of the predictable market. The only possible market sanction they face is "structural" – the absence of any market opening for the commodity.

Unemployment is attenuated by the variable or flexible quality of the product of intellectual labor, in other words, by changing the product. When there is no market for the product, this signifies that the quality is not recognized by potential buyers and that the product must be changed by harnessing the knowledge which is specific to the work force in question. This change is possible in the sense that, unlike market labor conventions, these skills are not simply "tricks of the trade", but rather generalizable empirical methods, i.e. scientific methods. Here we encounter the differences identified by Piore (1972), which define the quality of workers according to two convention-systems: higher and lower primary sectors distinguished, respectively, by the ability to adapt general rules to variable concrete situations and by routine familiarity with a given technology, product or firm. High-technology enterprises, whose priority is research and development, and other innovative or creative firms participate in these conventions. In effect, the producer must succeed in giving his creative work a universally acceptable form, unlike the work of the artisan.

Producers do face risk over the qualities of market demand, in which the real issue is whether investments in human capital are concordant with the evolution of demand. In effect, risk concerns the very existence of a market. Flexibility has to do with the quality or configuration of the product. Its broad scope is linked to the extent to which concepts and the methods can be universalized and to the possibilities for adaptation of existing products and industrialization of new products.

Conventions of non-quality labor

Seen at upper-right hand of Figures 14.2 and 14.3, these conventions are based on standardization of tasks and permit the utilization of interchangeable individuals, whose identity has no importance in the quality of work. In effect, standardization suppresses individual qualities. More precisely, worker identity is external to work itself (e.g. family membership or membership of a rural or ethnic community), having no place in the world of work. Personal qualities generated outside the workplace are nonetheless important in determining the appropriateness of an individual to a given job. Thus, the employment of individuals whose skills have, strictly speaking, no market-price permits them to be evaluated by low wages. Wages are low in part because they do not reflect cost of investment in reproduction of skills. This is a basic difference from the two preceding models of conventions, where quality of work directly reflects that of individuals and wages must cover reproduction of these qualities. Such a wage level is a precondition of the stability of those conventions. Here, by contrast, temporal stability comes from the durable existence of a labor surplus (considered in these conventions as a kind of free and natural resource).

A good example of such conventions is the replacement, thanks to standardization, of men by women in many jobs, as well as the massive employment of rural or immigrant labor. In both cases, the nonwork qualities of these groups (i.e. peasant ties or household roles) are more important than skills. Sabel (1982) distinguishes three categories of workers, who illustrate non-quality labor conventions: "peasant-workers," "would-be craftsmen," and "ghetto-workers." The universe of reference of these workers is external to the jobs that they occupy at any given moment. Peasant-workers accept unstable employment and can tolerate unemployment since they can either return to agriculture, earn supplementary income, or be aided by their community. Members of the same community can often assemble themselves in a single enterprise and continue to apply the hierarchical or even clan-based organization of their community under the codified appearance of the relations within an enterprise (Bernoux, Motte, and Saglio 1973). Lacking initial skills values in the market, would-be craftsmen are ready to accept anything in order to penetrate the world of the factory. The absence of any long-term project condemns ghetto-workers to dead-end jobs and to an acceptance of the moral judgment which upholds their incapacity to undertake anything better.

Such conventions typify situations where employment is unstable or precarious. They can also be encountered when products of standardized labor are dedicated to one or more particular demands. A classical illustration of this in industrial economics is the case of subcontractors lacking knowledge specific enough to give them independence on the market. One can also associate these conventions with the "sweating-system," where work (at home or in small family enterprises) has lost the independence of "the trade" without gaining autonomy by access to the market. Production is placed under the direction of the market (price, volume or quantity) by the middleman who places the order. In this way the sweating system plunges workers into uncertainty.

Wage variability is the basis of these nonquality labor conventions. The wage cannot be both a remuneration and a true resource for these individuals, and often assumes the character of a supplementary income. Although the skills of the worker are unimportant, the worker's specific character is important in the external labor market, where it is responsible for the mobilization of labor supplies. The convention depends on these nonwork, non-skilled-based identities. The frequent presence of a compromise between the workplace identity and the other identities of employees is the result. Paternalism, for example, could be interpreted as a compromise between the role of the family and the standardization of work.

The productivity convention does not need to be closely regulated since it is underpinned by an unemployment convention which tailors income from work and hence, labor costs, to fluctuations in demand. This convention transmits the charges of unemployment to the community members to which the workers employed belong. As in neoclassical theory, labor is a variable cost, both in terms of price and quantity.

The economic fluctuation of nonquality labor conventions creates uncertainty with respect to both the price and the volume of demand. Producers involved in these conventions fabricate standardized products dedicated to given, often very large, clients. The volume of work is adjusted in terms of hours of work. It adjusts itself immediately to demand variations and is externalized onto the surrounding community (ethnic or rural community or the family). Flexibility is thus connectd to precarity of employment. It is for these conventions that the compromise with the employment rules imposed by the state, which are rooted in industrial conventions, prove to be delicate, and often lead to violation or subterfuge, i.e. to use of legal loopholes and other forms of employment (temporary work, short-term contracts, subcontracting etc.)

Conventions of industrial labor

These last conventions, seen in the lower right section of Figures 14.2 and 14.3, combine rationalized work management and predictable markets. The industrial convention of productivity rests on a paradox which our familiarity with mass production and the large rationalized enterprise tends to mask. Work, having lost both its specialization and its dedication to an individualized demand, severs the link between labor and its product. The demand to which the firm addresses itself – especially the large firm – becomes independent of the market, which no longer evaluates the quality of the product, nor the quality of work. Quality is assured by incorporation of industrial standards into physical equipment and becomes more or less independent from the workers. Direct productivity incentives for workers disappear and have little to do with direct wages. By what means can they be reestablished? The response is complex and could have an unavoidably historical dimension (see the extract on "Polanyi"). In part, the response has to do with the emergence of social policies directed toward labor.

K. Polanyi and the industrial labor conventions

In industrial conventions, the individual has, in effect, lost all the traditional determinations upon which other conventions for the definition and insurance of the quality of work are based: "the trade," the family, the community, etc. Whence the problem raised by K. Polanyi (1944) when he analyses the crisis of economic liberalism and the "great transformation" needed to overcome this crisis: the deployment of economic and social policies by the state. How, given the progressive organization of the market for goods, does work, which – like money and land – depended upon "non-contractual organizations such as the family, neighbourhood, trade or religion," organize according to "the fiction" of the market?

As Polanyi stresses:

> ". . . labour, land, and money are essential elements of industry; they must be organized in markets; in fact, these markets form an absolutely vital part of the economic system. But labour, land, and money are obviously *not* commodities . . . Labour is only another name for a human activity which goes with life itself, which is in its turn not produced for sale but for entirely different reasons, nor can that activity be detached from the rest of life, be stored or mobilized . . . None of them is produced for sale. The commodity description of labour, land and money is entirely fictitious . . . Nevertheless, it is with the help of this fiction that the actual markets for labour, land and money are organized."
>
> (Polanyi 1944: 72)

This impossibility is nonetheless realized. Polanyi demonstrates that a purely social solution such as the localized paternalism of the Speenhamland system cannot only fail: the attempt to ensure that the wage does not descend below a certain minimum income, notwithstanding the market situation, leads to the lowering of the wage and to the collapse of the incentive to work: "Under Speenhamland society was rent by two opposing influences, the one emanating from paternalism . . . ; the other organizing the elements of production . . . under a market system." (1944: 80). Social institutions are linked, but in a complicated manner, to the quality of work and productivity. Far from being external, social institutions influence these latter two factors from the inside. In the case analysed by Polanyi, the maintenance of a locally restricted social identity was in contradiction with the universal identity of work emerging from the development of large-scale industry and from the extension of markets.

In industrial conventions, the worker becomes the bearer of social rights underwritten by the state and/or by agreements negotiated by coöperative elites. We shall deal with this point by limiting ourselves to the study of unemployment in France (Salais, Baverez, Reynaud 1986). In this case, access to social rights is subordinate to access to salaried work. Individuals must conform explicit norms of productivity for the production of goods. Conventions of industrial labor make the firm play an essential role in the definition of norms, and their observance according to procedures which are considered fair and legitimate by the participants. These procedures are structured around the concept of "employment," consisting of a skill and wage classification attached to the jobs (and not to the individuals), hiring and firing impersonal rules and a system of unemployment compensations. Thus, an overall convention of productivity establishing the link between the quality of work, quality of product, and remuneration is defined by means which are at once general (the state), decentralized (the firm), and intermediate (representative groups, agreements made between different levels of the state, branches, and firms).[9] The products of these industrial conventions in the mass production industries facilitate the calculation of economic risk and offer the organizational capacity to manage collective risk, so that minimum revenue can be maintained even in the face of market fluctuations (job loss, the loss or insufficiency of individual productivity). This protection is effected via social policies embodied in labor law and negotiated by the groups referred to above. In the French industrial convention of unemployment, the responsibility vis-à-vis fluctuations of demand rests on the employers; but this responsibility has two faces in that, at the same moment, entrepreneur is committed to his personnel, but ultimately not committed because the state assumes ultimate financial responsibilities.

Thus, industrial labor conventions *are* congruent with macroeconomic fluctuations, for mass products markets are sensitive to the macroeconomic conjuncture through fluctuations in both consumption and investment. Adjustments in response to economic fluctuations are therefore carried out through short-term adjustments in the quantities produced and sold. Adjustment in response to a prolonged downturn which cannot be absorbed by building up stocks requires cuts in hours of work (short-term unemployment) or in the number of jobs (total unemployment). We must note the relationship between the *mass product* in theories of regulation (Boyer, 1986) and industrial conventions. But these conventions are considered by them only in an indirect way, as the support of an aggregate demand in macroeconomic policy. When they are analyzed, as we did, as the basis of risk-calculation by firms, the congruence of mass product with *demand forecasting*, the efficiency of wage inflexibility and the management of fluctuations as *deviations from a trend* may be seen. By the same token, it appears that these characteristics define markets from which other products (dedicated products), other markets (uncertain markets), and other forms of adjustment of wages and employment are excluded, and ignored.

There is, thus, a flexibility proper to industrial labor conventions, which rests upon the variability of the number of occupied jobs (or, via these jobs, of the hours of work). But it remains adequate only if the industrial definition of the quality of work is not threatened by the evolution of demand. In effect, if the standards of quality are thoroughly modified by the product markets, even the quality of the jobs offered will have to be changed; and what matters will, then, be not flexibility, but investment.

CONCLUSION

As always, the search for alternative answers to old theoretical questions raises an enormous number of issues. We shall conclude, tentatively, with two methodological observations. First, the empirical analysis of concrete labor relations can be considerably enriched by the approach used here. Though we have here only examined pure models, more complex situations could be analyzed in terms of compromises between conventions. For example, the recent increase in use of market conventions might not be the simple flexibilization of mass production, but a complex compromise between at least two systems of labor conventions – industrial and market. Second, the analysis of evolutionary dynamics and crises[10] could be enriched by the analysis of distortions between conventions (lags, incoherences in the choices, and so on). For instance, a firm can move towards markets utilizing some conventions for the quality of products while not adapting its labor conventions for that of labor, given the autonomy of rules and institutions proper to each model of conventions. More generally, we can conceive of historical situations – such as that of much contemporary advanced capitalism – in which social institutions and the rules of employment policy can remain based upon industrial labor conventions while enterprises progress toward more market-based compromises, and examine the stability or unstability of such situations.

ACKNOWLEDGMENTS

I would like to thank Malcolm Mansfield for the translation and Michael Storper for his revision. This version has also benefited from a discussion with Suzanne de Brunhoff.

Part VI

COLLECTIVE ORDER AND INDUSTRIAL POLICY IN POST-FORDISM

15

LEVELS OF POLICY AND THE NATURE OF POST-FORDIST COMPETITION

Patrizio Bianchi

This chapter analyzes the relation between organization of production and competitive behavior and outlines the policymaking aspects of regional development that are implicit in the post-Fordist era. The chapter is essentially based on a European perspective and focuses on the new European Community approach to industrial and regional policies.

FORDIST PRODUCTION AND MARKET STRATEGY

The model of organization of production defined as Fordist is a model of production geared to great volumes of homogenous goods, and therefore realized with an organization of production devoted to generating advantages by specialization. The search for efficiency leads to an organization of productive processes based on the maximum specificity of assets employed and at the same time the maximum rigidity of relations among these assets.

These rigid relations are internally organized in a hierarchical structure, where all production and market decisions are centralized. Within this production organization, components and parts of the end products could be produced separately by small-sized companies which specialize in those productions, but timing and technical standards of this production are entirely determined by the large company, which controls the entire production cycle.

In mass production the search for efficiency implies that fixed investments be "sunk" because they are specified to the production of a given good. This sunkenness is crucial not only to determine the control of production organization but also to control the market organization.[1] According to the Bain–Sylos–Modigliani tradition, revamped by recent literature on strategic deterrence, we can stress that the Fordist model of production plays the role of market-ruler simply because of this sunkenness of fixed investments, that is to say because of the specificity of the assets needed for homogeneous production.[2]

As the entrant is supposed to use the same technology as the incumbent, the incumbent can deter potential entry (by preempting the market demand) through an expansion of its productive size to leave no market space for a rival having at least a minimum efficient scale (Modigliani 1958). This means that the incumbent identifies his size of production not by technological decision but by strategic decision (Gold 1982).

Why do small companies survive in this model?

Modigliani (1958) demonstrates that, in the presence of technological discontinuities, a pre-emptive strategy can be developed by the incumbent with the help of small companies, making explicit the possibility of coexistence within the same industry of different-sized companies having different levels of efficiency. Assuming the existence of one large and several small companies, the large firm may develop a strategy of accommodation (that is to say of accepting entry) toward small companies in order to fill any demand gap and therefore to deter the entry of efficient, that is, large competitors. In case of slow and uncertain growth of demand, the incumbent can accommodate the entry of small companies to buffer the fluctuation, by accepting them during expansion and expelling them during decrease of demand.

On the other hand, given that joint production is excluded, the possibility for a single firm to sell differentiated goods implies the management of parallel processes, each one of which is individually specified for the production of one homogeneous good. In this context the large company could favor the accommodation of small companies producing small quantities of differentiated products, in order to avoid both the change of its production management and the potential entry of rivals supplying the differentiated goods.

Small firms can survive when they are useful to the deterring strategy of the large company playing the role of incumbent. In any case the life of small companies depends on the production and market decisions of the large company. In the late 1950s these assumptions were considered general. Today, we can affirm that these assumptions are consistent with oligopolistic forms of mass production, as a historically determinate production system, not as parts of any universal cycle of development.

According to this perspective, we can use the product life cycle scheme not as a universal description of the evolution of industrial production systems, but simply to sketch different oligopolistic games emerging in the various historic phases of the development of mass production.[3]

In the expansion phase the rate of demand growth is constant and therefore predictable, thus the incumbent can plan all the investments necessary not only for efficient production, but also to deter the potential entry of efficient mass-producing competitors. Because the quality of demand is assumed to be stable, the deterring strategy is credible because of the sunkenness of these investments.

Fudenberg and Tirole (1983) demonstrate that a deterrent action is connected to the irreversibility of capital invested by the same entity that acquired the right of first entry into a market. Eaton and Lipsey (1979 and 1980) stress that such deterring strategy is successful also in a period of predictable increase of demand, inasmuch as the incumbent is always able, by expanding the supply of its existing plant, to anticipate the attack of an external operator that must construct a new plant to enter the expanding market.

When the market approaches saturation, demand tends to decrease and become uncertain so that firms, in order to maintain their levels of activity, should apply policies of revamping, by an accelerated renewal of the products and a segmentation of the market. A homogeneous product cannot satisfy the emerging differences in consumer tastes and any preemptive strategy based on standardized production results in defeat. The sunkenness of investments is a barrier to entry for entrants, but it is also a barrier to exit for the incumbent and it becomes a relevant constraint for a rapid organizational adjustment to the changing demand.[4]

Faced with a rigid incumbent, which is supposed to be organized in order to control a market situation that it has exhausted, small firms can be more competitive, because they are specialized in supplying specific demands. In this light, as Schoenberger (1988: 115) clearly states, "what Piore and Sabel call 'flexible specialization' is a response to market instability, leading firms to target smaller, more rapidly changing, and more specialized market segments (Sabel 1982; Piore and Sabel 1984)."

The role of small business with respect to the large corporation changes: in the stage of demand expansion, the role of small firms was clearly assumed as being dependent on the big company's strategy; in the stage of maturity and decline, small companies could be an alternative to big business, because of their higher capacity to respond to an unpredictable and fragmented demand, whereas the big companies cannot react because of the specificity and sunkenness of their investments (needed in the past to deter entries).

It obviously has to be noted that industrial districts did not arise simply in response to the difficulties of mass production firms. They occupy different market niches, often in completely different industries. The Italian and German cases suggest certainly that local agglomeration of small and medium-sized companies are not the result of large firm decline, but are deeply rooted in the social and economic history.

However, what we wish to stress is that territorial aggregations of complementary companies emerged as a peculiar pathway to industrialization and regional development in a period of market disorder, when the previous model of market and production regulation failed. The independent role played by the territorial aggregation of companies enlightens opportunities offered by a nonhierarchical network of companies, linked by cultural and social roots; Italian and German industrial districts became a lively development pattern, which could be an alternative to the well-established way based on the Fordist model, and therefore these examples become relevant for the American and European debate on reindustrialization (Piore and Sabel 1984).

This attention was magnified by the evidence that the flourishing areas of Europe, from the so-called Third Italy, to Catalonia and Pais Valenciano, from the southern regions of France to the central areas of Germany, were developing according to a model of diffuse entrepreneurship.

In the second half of the 1980s large firms came out of the corner and a new wave of industrial concentration occurred, emphasizing a clear tendency toward reimposing a strong market regulation. At the same time many industrial districts around Europe showed serious competitive problems and started on the difficult process of internal reorganization. However, the recent experience of dynamism of local systems has developed a strong interest in a policymaking approach based on promotion of industrial networks and local partnership.

The European Economic Community has reoriented its regional and industrial policy approach toward structural policies aimed at creating and reinforcing industrial networks, especially in the less-developed regions. Today, these policies are a crucial aspect of the entire program of completion of the European Single Market.

These structural policies are aimed at re-creating local networks, through an active promotion of local partnership; they are based on integrated programs of public actions, which are activated by different authorities. In order to understand the nature of this policymaking approach, it is necessary to analyze deeper the different aspects of industrial districts and their position in the new competitive environment, characterized by the revenge of the big businesses.

INDUSTRIAL DISTRICTS AND PRODUCTION SYSTEMS

The extensive literature treating the industrial districts has thus analyzed territorial aggregations of small units of production in two ways:

a) in terms of the geographic aspect of aggregation, emphasizing above all the social and environmental factors of agglomeration; and
b) in terms of the production aspect of aggregation, emphasizing the organization of production as the crucial factor in the aggregation between firms.

These two aspects of the industrial district come together in analyses of specific situations in which skill and experience have formed aggregations on a local level constituting the common link between specific production activities. In the case of the Marshallian Industrial District (MID) the system of production, acting as a single agent on the market, consists of a set of cooperative relations between specialized agents for phases belonging to the same production cycle.[5]

As Becattini (1989) stresses, recalling the classical Marshallian view, the crucial characteristic of the MID is the set of cooperative relations among local agents who see themselves as part of the same organism, defined in historical and social terms; the diffuse entrepreneurship is the main aspect of this territorial agglomeration.

The specific advantage of these production organizations with respect to general agglomerations of small and medium-sized firms is given by a division of labor, based on territorial, historical, and cultural traditions, represented in what has been defined, following Marshall, as an "industrial atmosphere" (Marshall 1879). This is the crucial element that binds together aspects of territory and production. We can redefine this industrial atmosphere as the set of intangible assets belonging to the system of production as a whole.

The cost advantage for the MID of industrial atmosphere compared with systems of production which are organized differently (for example, the traditional large, centralized firm or generic agglomerations of individual small firms) is that it reduces the transaction costs between the agents which interact within it, lowering the costs of information on reliability of subcontractors without raising in the same proportion the costs of coordination of them; the division of labor based on diffuse entrepreneurship and on stable social partnership increases static, and possibly, dynamic flexibility, because continuous adjustment of the production organization is made possible.

These advantages cannot not be individually appropriated, precisely because of this historic embodiment in the social structure of the production system. It is an externality for individual companies established in the district, but also constitutes for them a barrier to exit from the district system if they need a drastic change in their strategic behavior in order to react to dramatic changes in the rules of competition.

As sketched above, during a period of demand instability and market uncertainty, industrial districts had the possibility to undertake hit-and-run entries taking advantage of the relative rigidity of the large firm as regards its ability to extend its capacities to fit the changes in demand.

This success was based on the opposition between an incumbent (the large firm) which has the characteristics of mass production and an internal structure which is hierarchical and rigid, and a multiproduct entrant, the district, capable of carrying out continuous productive adjustments more quickly than the incumbent and thus able to enter and leave a specific market or a specific segment very rapidly, responding to the variation in the demand curve of the specific product, selling the product at highly competitive prices.

As soon as large firms undertake a policy aimed at stabilizing the demand curve, by means of trademark policies and preemptive emission of new goods, ability to carry out hit-and-run entries by districts is substantially reduced. In this new competitive context, it is crucial to regain the role of incumbent, in order to carry out a segmentation of the market with regard to which the firm can play a dominant role. This implies for the incumbent an organization of manufacturing, which could be also decentralized to small companies, but a clear centralization of the strategic activities of market control.

Recent literature on the strategic deterrence and empirical researches on market trends have clearly shown that the deterring economies of scale are today on distribution, advertising, marketing services, and product innovation rather than on manufacturing; dominant companies can again use capital as a deterrent against potential entrants, but "capital" is not just productive capacity for manufacturing homogeneous goods, but a variety of strategic investments (R & D, product differentiation and advertising, flexible manufacturing systems, marketing and branding, distribution and assistance). All these are new barriers to potential competition, and they are very effective because are oriented to filling any market segments and at the same time to addressing and leading consumer needs.

Thus, in the second part of the 1980s, large corporations reorganized their activities not simply to be more flexible – through a deverticalization of their production lines, and by applying technologies able to increase economies of scope without reducing economies of scale – but they moved to take control of the market through the acquisition of distribution and assistance networks, well-known brands, and a substantial increase of advertising and marketing activities.

In Europe, this stage has also been characterized by a dramatic increase in mergers and acquisitions, that is not completely justified by efficiency but seems to be a response to the need for establishing dominant positions with a view to the completion of the Single Market. This dominant position can be shared by various groups, which use strategic alliances and joint ventures to capture the control of a market which is segmented by the same action of the companies. Many examples, from food products to domestic appliances to armament production, offer clear proof of the trend toward concentration occurring in Europe.

In this competitive phase, the obstacle to the adjustment of already well-established districts is constituted precisely by the difficulty of drastically changing the organization of production set up on the basis of a social system. Thus, in order to avoid declines of the productive system that could induce a generalized disease for the territory, policy actions aimed at reestablishing competitive conditions, by retraining labor forces, information about technologies and market trends, promotion of export consortia were experienced by several regional governments in Europe. Centers providing structural services have been activated in several Italian districts through the cooperation among local public authorities, entrepreneurial associations, private companies (Table 15.1).

This approach to industrial policy was largely opposed by the national governments, which were still oriented to provide direct subsidies to individual companies, rather than to intervene to recreate the local environment. However, this policymaking approach based on the stimulation of endogenous forces was recognized as the crux for the reform of the structural policy approach of the European Community.

INDUSTRIAL DISTRICTS AND PUBLIC POLICY

This new stage of European integration is definitely based on the promotion of market forces; this approach requires the implementation of four sets of policies:

1 the removal of the residual non-tariff barriers and the harmonization of market standards throughout the Community, in order to create a single market for goods, service, labor, and capital;
2 an effective competition policy to avoid restrictive practices, monopolization, excessive concentration, and unfair public aids to national companies;
3 the public support for innovation, especially for very innovative sectors and small companies; and
4 structural policies to create a positive environment in which new industrial and service activities, particularly in the less-favoured areas of the Community might develop.[6]

The completion of the Single Market offers a strong opportunity for industrial concentration. The imposing study on the Cost of Non-Europe provided a clear basis for mergers looking for a better exploitation of economies of scale (Commission 1988c; Cecchini 1988).

The White Book (Commission, 1985) and the Padoa Schioppa Report (1987) also stated that a process of deregulation to favor the creation of a larger market space is by and large not inducing more efficiency, whereas a better utilization of economies of scale is not transferred to the consumers, but is acquired by companies, which are able to monopolize segmented markets. Thus, the completion of the Single Market requires not only an effective competition policy at the European and national level, in order to guarantee against the abuse of dominant positions, but also the active promotion of the development of new competitors.[7]

This policy requires actions either to create or to re-create a local environment able to provide conditions for industrial and regional development. This approach to industrial policy, no longer based on granting funds to individual companies but on the promotion of integrated programs to stimulate local endogenous forces and consolidate a diffuse

Table 15.1 Centers promoted by public authorities providing services to small and medium-sized firms in the regions of Italy

	Name	*Location*	*Sectors*	*Services offered*	*Year founded*	*Employees*
1.	TECNOTES	Piemonte	All	Training, Information Research and Planning	1985	32
2.	CESTEC	Lombardia	All	Information, Projects and Research, Training	1979	11
3.	ARST	Friuli V. G.	All	Documentation and Secretariat	1978	18
4.	CERIT	Friuli V. G.	All	Training, Information, Appl. Research, Certification	1979	30
5.	CATAS	Friuli V. G.		Seating, Tests and Trials, Research, Furniture	1969	8
6.	Centro Reg. Serv. PMI	Friuli V. G.	All	Company diagnoses	1984	6
7.	CRAD	Friuli V. G.	All	Pollution control, Research & Studies, Dosimetry, Information-Technology consultancy	1971	5
8.	BIC	Liguria	All	Assistance in setting up new Firms	1975	3
9.	DITEL	Liguria	All	Studies and Planning	1980	10
10.	ASE	Emilia Romagna	All	Information desk	1986	2
11.	ASTER	Emilia Romagna	All	Documentation, Data-base link-up, Projects	1985	6
12.	CEMOTER	Emilia Romagna	Ground Mov.Tech.	Certification and Research	1982	13
13.	Centro Ceramico	Emilia Romagna	Ceramic	Tests and Trials, Research, Training	1976	27
14.	CERCAL	Emilia Romagna	Shoes	Fashion trends, Training	1983	3
15.	CERMET	Emilia Romagna	Metals	Documentation, Analysis, Technology consultancy	1985	4
16.	CESMA	Emilia Romagna	Agricult. Machinery	Studies and Research	1983	3
17.	CITER	Emilia Romagna	Textile, Clothing	Fashion trends, Information, Research	1980	8

18.	QUASCO	Emilia Romagna	Building Constr.	Information, Research, Training	1985	7
19.	RESFOR	Emilia Romagna	Mech, Sub-contract	Information, Promotion	1986	1
20.	CESECA	Toscana	Footwear	Information, Training	1984	4
21.	CESVIT	Toscana	All	Promotion, Research	1985	1
22.	CSM	Toscana	Furniture	Training, Research, Promotion	1982	3
23.	LUCENSE	Toscana	All	Studies	1984	3
24.	CEDIM	Marche	All	Documentation, Data-base Link-up	1985	6
25.	CESCAM	Marche	Footwear	Training	1985	2
26.	COSMOB	Marche	Furniture	Information and Consultancy	1984	4
27.	ISELQUI	Marche	Musical	Tests and Trials, Research instr., Textile, Shoes, Furniture	1980	50
28.	Centro Laser	Puglia	All	Research and Training	1979	6
29.	CSATA	Puglia	All	Research and Planning, Training	1969	100
30.	CRAI	Calabria	All	Research and Planning, Training	1979	64
31.	CRES	Sicilia	All	Applied Research, Information, Training	1974	10
32.	SIRAP	Sicilia	All	Studies and Projects	1983	7
33.	TECNOSERVIZI	Sicilia	All	Assistance, Training, Information desk	1984	8

Source: Laboratorio di Politica Industriale – Nomisma, 1987.
Note: Research carried out in 13 Italian regions (Piemonte, Lombardia, Friuli Venezia Giulia, Veneto, Liguria, Emilia Romagna, Toscana, Marche, Umbria, Lazio, Puglia, Calabria and Sicilia).

entrepreneurship, is now accepted by the European Community as a general approach to industrial and regional policymaking.[8] It is considered a necessary complement to the antitrust policy for providing an efficiency response to the completion of the Single Market, but also as a necessary measure to guarantee equity and stability in the process of industrial development of the Community (Padoa Schioppa 1987).

A crucial aspect for the implementation of this new policy approach of the European Community has been the reform of the structural funds of the Community (Commission 1989b). The structural funds of the Community are the European Social Fund, the European Regional Funds, and the European Agricultural Fund-Orientation and Guarantee; also to be considered are the European Investment Bank interventions (with the New Community Instruments, which is especially aimed at small and medium-sized firms) and the funds for the Integrated Mediterranean Projects, the Integrated Development Actions, and the Special Action such as the industrialization of Portugal and the interventions in disaster areas (Commission 1989b).[9] This approach has been reinforced after the Maastricht Agreement on the European Economic Union.

Local partnership is the key issue of the reform of the structural funds (Commission 1989b; 14–15). This principle is crucial not simply for the European Regional Development Fund intervention in less favored areas of the Community, but for any action of the Commission concerning small and medium-sized companies (Commission 1988b), and extensively for any policy having a territorial impact (Commission 1987).

The Commission gained substantial advantages from the experience of the Integrated Mediterranean Projects (IMPs). IMPs experimented first with the new approach based on a trilateral relationship among Community, national, and local authorities, on the basis of a scheme of integration of the different interventions provided by the three levels of policymaking (Leonardi 1990).

These policies are termed "structural" because they are aimed at influencing the relations of production between companies, that is division of labor itself; they are alternative to "individual" interventions, which are simply directed at supporting single companies.

There are two types of intervention which can be considered. On the one hand, there is the possibility of intervening in areas already having a local network of specialized producers, but facing a restructuring process, and areas not having any specific agglomeration of local entrepreneurs.

In the case of interventions in a local system not particularly characterized by a preexisting district, the intervention will consist of several integrated actions:

a) aimed at making the system itself explicit, encouraging agents to see themselves as part of a single system;
b) giving direction to the specialization of the local agents in order to permit greater articulation of the division of labor based on mutual reliability;

c) providing an easy access to market and technology information; this action is usually based on the creation of a center, jointly managed to provide services, which cannot usually be offered at the local level, but which are necessary to compete at the international level.

These actions are intended to create an industrial atmosphere, to reinforce common interests, to strengthen local partnership, and therefore to establish an industrial district. These interventions have been carried out by local government, supported by national and Community authorities, and have the task of taking over and managing collectively local agency or service centers, devoted to reduce the cost of coordination and acquisition of the specific entrepreneurial capacities which are necessary for the formation of a diffuse entrepreneurship. They cannot be activited individually and are considered as necessary to the development of the activity of the individual firms. When these interventions concern less favored areas of the Community, these interventions can be designed and implemented by an integrated program, involving municipal, regional, national, and Community projects; this program can be financed mainly on the European Regional Development Fund, and the other structural funds of the Community.

As regards interventions in already existing districts experiencing periods of crisis, the intervention of industrial policy must be aimed at industrial atmosphere itself and (working backward with respect to the procedure described above), must act on the costs of information and coordination. Intervention in industrial atmosphere means pinpointing, within the wide range of shared know-how, on the one hand those particular aspects most closely linked to the product, on which the division of labor is based, and on the other those functional aspects connected with the overall organization of production.

In this perspective, the possibilities of intervention on a district, whether it can be defined in strict terms as a MID or else more generally as a system of quasi-markets, are much greater, and require a precise analysis both of the relations of production that characterize the local system and the relations of the market with which the particular district is in conflict. The latter means in practice identifying the precise nature of the changes on the market to see whether characteristics of conflict emerge which are based on a break with the hitherto existing rules of competition.[10]

Policy interventions, then, have specific characteristics in relation to the particular situation of the relevant market. We are dealing with an action aimed either at the definition or at the redefinition of the collective intangible assets connected to the specific organization on which the relations of the production system have been built in the past and which form the basis for possible lines of common organization for the future.

In relation to an already consolidated district, however, interventions tend to be directed at adjusting that specific technical, market and social knowledge which constituted the relative advantage of that particular district in conditions of stability and which can become the obstacle to change.

In the case of a new district, interventions will constitute the basis for the development of a new ordered system of relations aimed at giving direction to the aggregation of firms. This aggregation can then become a quasi-market system of production characterized by a wealth of common knowledge which can reduce the costs of information without increasing the costs of coordination.

In the case of a preexisting district which is facing a crisis, interventions such as the setting up of production service centers, can become a means of carrying out a selection and promotion of certain agents operating in the area and can serve to modify the existing production relations, also transforming the existing system into a system of quasi-hierarchy under the leadership of a limited number of agents chosen by means of this selection process.

The technicalities of public intervention must all be customized in relation to the district as a specific production system, whose traditional geographical or cultural area, however, can become larger or smaller over a period of time. A local industrial policy intervention can, in fact, lead to systemic adjustments which necessitate the integration of preexisting agents with agents which are traditionally outside the cultural and territorial area covered by the district.

In order to formulate and implement a structural policy program, it is useful to make a distinction between the geographical aggregation of firms and the system of production relations already in existence and operating in a particular competitive situation. It is necessary to define the division of labor existing among manufacturing and service agents with respect to which integrated interventions can be planned.

The network analysis must be extended to include all the agents who in effect interact within the system of production relations even if they are agents (contractors, subcontractors, universities, research and training centers, financial institutions, entrepreneurial associations, public administrations) which remain outside the area of the district itself. On the other side, it is necessary to make a thorough analysis of the territory, to understand the historic, cultural, and environmental conditions, and the social and economic trends, structuring the industrial activity in the area. Thus, design and implementation of an integrated policy have to be rooted in a multidisciplinary research frame.

FINAL REMARKS

This policy approach is coherent with the general attitude of the Community toward market forces, and it is introducing a competition among

European regions, based on the local supply of environmental and institutional conditions for industrial development.

This is a crucial aspect of a more general reorientation of the Community. The European Single Act reversed the previous model based on the creation of a supranational power, which was supposed to overcome national authorities, but framed by the effective veto powers of the national governments. The European Single Act sets up a competition between countries based on their ability to formulate legislation which can favour their own production structure (Padoa Schioppa 1987).

This competition between the member countries, then, becomes competition based on the efficiency of the respective public administrations with regard to the prior setting up of conditions favourable to local development and to the subsequent carrying out of interventions aimed at encouraging the growth of integrated firm systems. In this sense, the ability to plan and carry out integrated interventions based on the territorial aggregation of companies becomes an essential requisite for industrial policy.

This approach implies not simply the transfer of powers to the local authorities, but the promotion of a multilateral policy game, bringing into play the regions as primary actors of the policymaking process.

The new industrial policy approach, enacted by the European Single Act, is explicitly oriented to supporting the local aggregation of new businesses and to promoting industrial partnership and diffuse entrepreneurship, as basic forces for the development of the European economy.

16

DIVERGENT PATTERNS OF BUSINESS ORGANIZATION IN SILICON VALLEY

AnnaLee Saxenian

This chapter examines the politics of business organization in Silicon Valley. It contrasts two patterns of institution-building among the region's technology firms: one which is externally oriented and seeks to insure the competitive position of the industry's largest producers by influencing federal policy, the other which is integrative and enhances the flexibility of small and medium-sized firms by providing collective services and fostering the innovative recombination of resources.

Externally oriented organizations, such as the Semiconductor Industry Association (SIA) and the American Electronics Association (AEA), are nationally recognized as representatives of Silicon Valley because of their active lobbying in Washington, D.C. Integrative organizations such as the Semiconductor Equipment Manufacturers International (SEMI) and the Software Entrepreneurs Forum (SEF), by contrast, operate at the sectoral and regional level and remain largely outside of the public eye.

These contrasting patterns of business organization reflect the hybrid character of the Silicon Valley economy, an industrial region in which mass production corporations coexist with a dense agglomeration of small and medium-sized specialist producers.[1] While mass production and flexible specialization – and a variety of intermediate forms – have coexisted in the region for decades, each faces distinct problems of reproduction and has generated distinct forms of collective action among local producers.

Associations of business in the US are traditionally studied in their role as pressure groups, yet some of the most active business associations in Silicon Valley today operate at the regional level and engage in little, if any, lobbying. Rather than attempting to shape policy or influence elections, these associations play an essential coordinating role for the specialist producers in a fragmented industry by providing collective services and fostering the exchange of information and technology.

The experience of Silicon Valley highlights some of the political obstacles to the elaboration of such integrative institutions in the American context, obstacles which grow out of the historic hegemony of mass production (and its institutional practices) and the strength of individualistic traditions in this country. The failure of the region's specialist producers to develop a political agenda suggests the vulnerability of the regional economy, and has relevance to other regions with similar flexible, decentralized productive systems (Scott and Storper 1988; Maillat and Lecoq 1990; Piore and Sabel 1984).

The chapter begins with a brief overview of the Silicon Valley economy and the problems of coordination in a decentralized, region-based productive system. It then analyzes the history and activities of four of Silicon Valley's leading business associations, locating them along a spectrum from externally oriented to integrative. The third section examines the conflicts that have emerged in the region over issues such as trade policy and the Sematech consortia, arguing that externally oriented groups dominate US technology debates. The concluding section suggests that Silicon Valley remains vulnerable because the individualistic world views of its specialist producers limit their ability to develop an alternative political vision and elaborate systematic regional and sectoral coordinating institutions.

SILICON VALLEY: AN INDUSTRIAL HYBRID

While Silicon Valley is famous for large electronics firms such as Hewlett-Packard Co. and Apple Computers, the regional economy is also populated by thousands of small and medium-sized specialist enterprises.[2] Each of these firms focuses on a narrow phase of the production process and relies on the external economies of the region and its dense concentration of skill and technology to introduce new products for diverse and changing markets. This decentralized industrial structure is reproduced through an ongoing process of new firm formation: entrepreneurs continually leave existing firms and recombine local resources in order to pursue new technical and market opportunities.[3]

As they grow, these enterprises develop linkages with customers and suppliers outside of the region, yet most remain embedded in the technical and social infrastructure of the regional economy as well.[4] Silicon Valley's most flexible and specialized semiconductor firms, for example, collaborate closely with local computer systems producers to develop innovative, high value-added products for semicustom and niche markets. The region's flexible mass producers of computer systems producers rely on face-to-face relations with local contract manufacturers in order continually to introduce new products for fast-changing markets. Even the mass producers of semiconductors, which seek to minimize the unit costs of general-purpose products by achieving high-volume production, depend on the dense concentration of skilled labor in the region.

These different productive strategies – and a range of intermediate ones – have coexisted in the region for decades. They are complementary in many ways: producers of specialty computer systems, for example, depend on access to large quantities of low-cost memory chips, while the mass producers of these commodity semiconductors require customized manufacturing equipment. Silicon Valley is thus an industrial hybrid, with a mix of productive strategies ranging from relatively autonomous high-volume producers to highly interdependent flexible, specialized firms.

The patterns of business association which have emerged in Silicon Valley over the past several decades reflect the hybrid nature of the regional economy. Silicon Valley's mass producers of semiconductors, for example, have formed externally oriented trade associations whose primary purpose is to lobby the federal government for favorable trade policy and related legislation. Their suppliers, the specialist producers of semiconductor equipment, by contrast, have created integrative organizations which provide collective services, set standards, sponsor trade shows, and foster networking and information exchange.

Externally oriented organizations like the SIA behave as classic pluralist pressure groups. Their structure and activities are easily accounted for in the literature of American politics, which focuses on the lobbying activities of corporations and their trade associations. Yet some of the most active business organizations in Silicon Valley today engage in little, if any, lobbying. Associations like SEMI, which play an integrative or coordinating role for the firms in a region or sector, have received far less scholarly attention in the US.[5] Historians are only now uncovering the records of America's nineteenth-century industrial districts, and we still know little about their institutional foundations (Scranton 1989).

This focus on business associations as pressure groups reflects the historic dominance of mass production in the US. The central regulatory problem in a system of mass production is to maintain a consistently high level of demand to amortize the costs of heavy fixed investments and to insure stability in the supply and price of inputs (materials, components, and labor) so that production lines remain fully occupied. Such coordination is typically accomplished by the governance structure of the large, vertically integrated corporation and the macroeconomic activities of the federal government.

The key coordinating institutions in a mass production system are thus the large corporation and the nation-state, and associations of business are oriented primarily toward shaping national policy. In the US, where the vertically integrated mass production corporation is widespread, business organizations remain relatively weak (particularly in comparison with their European counterparts) and individual corporations are increasingly important lobbyists and political players in their own right (Yoffie 1984).

The industrial decentralization which characterizes a flexible, region-based system of production, by contrast, requires stabilization at the local level (Scott and Paul 1990). When production is fragmented among a multiplicity of specialist firms, there are many potential sources of market failure. The price system does not convey sufficient information to producers, particularly in an environment of immense complexity and rapid technological change. Moreover, no individual firm can bear the costs of gathering this information. Institutions which provide market information and foster the diffusion of technology are thus essential to preserving the dynamism of the system.

Specialist firms must also rely on external mechanisms to insure the continued availability of resources. The small and medium-sized firms in a regional economy like Silicon Valley have little incentive to invest in long-term training, education, or research and development programs – activities which are traditionally achieved by the hierarchy of the large corporation – because the fluidity of local labor markets and openness of information flows limit their ability to capture the benefits of their investments. Yet the region's dynamism derives precisely from the ability of local entrepreneurs to specialize and to recombine such resources rapidly.

Extra-firm institutions are thus essential to insure not only the availability of market and technical information, but also the reproduction of a skilled labor supply, investment in long-term research, and the continued supply of capital. Some of these services are provided by the private sector in Silicon Valley: by market research firms, technical consultants, venture capitalists, and universities. However other services, such as training and long-term research, require collective action and even the provision of public resources.

A variety of public and private institutions play such a coordinating role in Europe's industrial districts. In Germany's Baden-Württemberg, for example, the trade associations coordinate joint research projects, support industrial research in local universities, and provide a forum for negotiation and conflict resolution among specialist firms (Herrigel 1988). In the Third Italy, local governments and producers' associations support industrial parks, vocational training, regional research centers, and market research. They even finance and run "service centers," which supply information on technology and markets to specialist firms in a particular sector (Brusco and Righi 1989).

This is not to suggest a direct correspondence between technology, institutions, and politics: a variety of different institutional arrangements can meet the needs of mass production and flexible specialization. Nor is it to imply a one-to-one connection between a firm's organization, strategy, or technology and the political stances of its managers and executives.[6] However, the sectors of Silicon Valley's technology industry have organized different associations to address their distinct problems of reproduction – associations which have often defined conflicting political agendas.

PATTERNS OF BUSINESS ORGANIZATION IN SILICON VALLEY

Four Silicon Valley business organizations exemplify the differences – and potential tensions – between the region's externally oriented and integrative associations. The SIA is a classic American trade association. It seeks to define its members' interests as those of the region and the nation in order to gain supportive trade policy and funding. The AEA is an intermediate case. It played an active integrative role in Silicon Valley during the 1960s and 1970s, but followed the lead of the SIA during the 1980s and redirected its resources almost entirely towards lobbying for national industrial policies.

SEMI and the SEF, by contrast, coordinate the activities of the region's small and medium-sized specialist firms. They foster innovation and collaboration by developing a common technical language and standards to allow engineers and firms to work together, by providing collective services such as trade shows, market forecasts, management training, and technical seminars, and through forums which strengthen the social and professional networks in Silicon Valley.

The Semiconductor Industry Association

The Semiconductor Industry Association is the most prominent business organization in Silicon Valley. To many outsiders it embodies the interests of Silicon Valley, if not all of American high technology – even though only 13 of the region's technology firms are members.

The SIA was founded in 1977 by the chief executives of Silicon Valley's four largest semiconductor firms, all friends who had worked together at Fairchild Semiconductor. While the initial aim of the SIA was to provide a voice for the region's new merchant semiconductor firms against established and politically active firms from outside the region (particularly Texas Instruments), it quickly shifted its attention toward the threat of foreign competition.[7] Frustrated with the perceived ineffectiveness of existing electronics associations such as the broad-based AEA, SIA founders sought a more focused approach.

Executives from SIA companies traveled to Washington regularly during the late 1970s to lobby for favorable government policy and to testify concerning Japanese trade practices and policies which they saw as unfair.[8] These efforts met with little response initially, but the association quickly gained political sophistication. They began positioning the interests of the SIA as those of the nation by linking the health of the semiconductor industry with national competitiveness and security concerns. By the mid-1980s, the SIA had built a powerful political presence in Washington.[9]

With this single-minded focus on lobbying, the SIA behaves more like a traditional American trade association than any other business organization in Silicon Valley. It has clearly had the greatest influence on policy. Within less than a decade from its formation, the SIA had achieved the passage of the landmark bilateral trade agreement with Japan (which reflected virtually all of the SIA demands)[10] and gained a commitment of $500 million in support from the Department of Defense for the collaborative manufacturing consortium, Sematech.[11]

The SIA demands – particularly the trade accord – reflect the agenda of a mass production sector in crisis. Faced with massive losses in commodity semiconductors during the early 1980s due to Japanese competition, the established chip makers sought a political response. The SIA mounted an extensive lobbying campaign and six member firms filed legal actions for trade relief. The resulting trade agreement was designed to shore up demand for the commodity memory products of the SIA's large chip makers.[12] Price floors were designed to eliminate alleged Japanese practices of illegal price cutting, while guarantees of market share in Japan were intended to expand demand for the output of US producers. In short, the agreement sought to stabilize the mass market for US commodity chip producers.

Since that time the SIA policy agenda has included continuous efforts to insure that the conditions of the trade agreement are enforced and to insure extension of (and even increase in) support for Sematech. The SIA experienced its only major failure in 1989 with the folding of the proposed memory chip consortium, U.S. Memories, which it actively supported. The SIA also lobbies on related issues such as antitrust reform, tax policy, and intellectual property rights; and the organization compiles and disseminates semiconductor trade statistics and forecasts.

The SIA's successes are typically explained by its ability to speak efficiently and quickly, without the dissensus of a large organization (Yoffie 1988). The association currently has only 37 members, and is dominated by a smaller group of activist firms. Membership is limited to US-based producers, not surprising given the nature of their concerns – yet there are more than 250 specialist semiconductor producers in the US which do not belong to the SIA. In short, the SIA represents the interests of the established US producers of commodity semiconductors (although it includes a few producers of semicustom chips as well). Nonetheless, the association has succeeded in positioning itself as the voice of Silicon Valley to the outside world.

The American Electronics Association

The AEA is the nation's most broadly based electronics industry association because it has continually expanded to include new sectors as they emerge. It thus represents all companies that design, manufacture, or conduct research in electronics, components, and related information technology products. AEA membership exceeds 3,500 firms nationwide, more than one-third of which are located in California. The AEA is headquartered in Silicon Valley, also the home of the largest of its 21 regional councils.

When the AEA (then the West Coast Electronics Manufacturers Association, WEMA) moved its headquarters to Silicon Valley from Southern California in the early 1960s, it identified itself with the region's fast-growing new technology firms.[13] Unlike the older Electronic Industries of America, which represented established East Coast electronics firms and was based in Washington, the AEA made an explicit commitment to "be where the new companies are" and to build a solid base there before expanding elsewhere. The AEA thus consciously forged an independent identity among Silicon Valley technology companies distinct from the "old-line" electronics businesses of the East. (By 1980, the AEA had greatly surpassed the EIA in membership, with close to 3,000 member firms compared to the EIA's 300.)

The AEA's focus during the 1960s and early 1970s was almost exclusively on fostering the development of the local electronics community by assisting the management of emerging firms, rather than on lobbying on behalf of established corporations.[14] It sponsored seminars to promote the exchange of ideas and information, and management training sessions on subjects ranging from finance and high-tech marketing to production and export assistance. These activities were welcomed by the managers of Silicon Valley firms, who typically had technical rather than managerial backgrounds or training.

The AEA also encouraged the growth of the social and professional networks which bind Silicon Valley firms together by hosting frequent meetings for local CEOs and managers. In the words of a local newspaper:

> Perhaps the AEA's most significant contribution to the electronics industry is what it did to foster networking. Most top executives of young, fast-growing electronics companies are relatively inexperienced in some important management areas. The AEA, with its frequent seminars and monthly meetings of company presidents, provided an excellent opportunity for those executives to meet and learn from their peers.[15]

The article goes on to point out the crucial role the AEA played in integrating the specialized firms in a highly fragmented industrial structure:

> Electronics companies are uniquely systems-oriented. Almost no firm manufactures from the ground up a stand-alone product. A company either draws on other people's components or makes products that fit with other people's products into a system. Friendships made through the AEA help the companies develop products that work together.[16]

This integrative role is confirmed repeatedly in interviews with Silicon Valley managers: one CEO reported that she had found most of her customers through AEA functions, others value the opportunity to stay in touch with old friends and colleagues from prior employers, still others see it as a source of market information, customers, or partners.[17] In fact, when the AEA expanded beyond Silicon Valley, its leaders recognized the importance of these networks. They developed regional councils which replicated the original Silicon Valley structure.

The AEA abstained from national politics throughout most of the 1970s. It did not even open an office in Washington, D.C. until 1980. Moreover, when the association finally hired a Washington lobbyist in 1975, he was relocated to Silicon Valley for five years to gain first-hand understanding of the industry.

The organization's first major policy victory came in 1978, when CEOs of AEA companies traveled to Washington to lobby for a reduction in the capital gains tax. But it was only after 1984, when the US electronics industry recorded its first trade deficit ever, that the AEA increasingly directed its attention toward the federal government. Following the lead of the SIA, and championing the cause of international competitiveness in electronics, the AEA began lobbying for federal policies to reduce the cost of capital, increase investment tax credits, increase funding for R&D, and improve engineering education.

This decision to engage actively in lobbying reflected a decisive shift in the AEA. During the 1980s, funding for education and management training programs was dramatically cut, while the public affairs budget tripled. In 1987, the AEA Public Affairs Office was moved from Silicon Valley to Washington in order to increase the association's presence there. The association now supports three full-time staff members in Washington and four more in its Tokyo office. While its 21 regional councils continue to organize activities for local communities, the central focus of AEA efforts has clearly shifted to lobbying for national policies to improve the competitiveness of the electronics industry.

Despite the potential advantages of its broad membership base, the AEA is often criticized for being ineffectual.[18] It has repeatedly failed to take decisive stands because of the conflicting political agendas of its member firms, particularly the commodity semiconductor firms and the computer makers. The cross-cutting membership, which makes the AEA the sole organization which could represent the shared interests of the technology firms in Silicon Valley, has thus undermined its political strength.

The Semiconductor Equipment and Materials International

The Semiconductor Equipment and Materials International (SEMI) serves the specialist firms in the semiconductor equipment and materials industry, an industry which, even by the standards of Silicon Valley, is highly fragmented and technologically sophisticated.[19] Many of its members produce equipment for only one specific step in the exacting process of semiconductor manufacturing, and many have only one or a handful of products. This specialization means that the top 14 US semiconductor equipment companies account for only 55 percent of total industry sales (Congressional Budget Office 1987).

SEMI thus has a very broad membership base of 1,750 corporate members and it is dominated by small entrepreneurial firms. Two-thirds of its members have fewer than 100 employees and under $10 million in annual sales, and only 5 percent have more than 500 employees or over $50 million in sales (SEMI 1990). And while its membership is concentrated in the US and it is headquartered in Silicon Valley, SEMI is open to foreign members.

SEMI was founded as the Semiconductor Equipment and Materials Institute in 1970 by three Silicon Valley-based semiconductor equipment vendors who were dissatisfied with the minimal attention they were receiving at WESCON, the regional electronics trade show.[20] The newly formed organization conducted its first trade show, called SEMICON, in 1971, to promote the products of semiconductor processing equipment, materials, and service firms. Today SEMICON shows are held not only in California but also in several other US regions as well as Europe and Japan. Many of the industry's small equipment ventures depend on the trade shows for their survival.[21]

In addition to conducting trade shows, SEMI sets standards, promotes technical exchange, and provides education programs for firms in the industry. Establishing standards and solving shared technical problems is especially important in a highly fragmented, technically complex, and rapidly evolving industry. In 1973, for example, there were close to 2,000 specifications for silicon wafers in use by scores of US silicon vendors, and the wafers were manufactured in a variety of different shapes. This lack of uniformity created problems of wastage, inventory, and planning for both vendors and customers. Despite skepticism from semiconductor manufacturers, a standards committee defined and publicized specifications for emerging 3-inch wafer lines. By 1975 over 80 percent of all new wafers met SEMI specifications.

SEMI's Standards Committee now oversees the definition of specifications for the materials, equipment, and processes used in semiconductor manufacturing. Standards setting is a strictly voluntary process of defining consensus specifications, and the use of all SEMI standards is voluntary. More than 3,000 engineers from the equipment industry and their customers work on the standards-setting process, which is coordinated by more than 100 international committees, subcommittees, and task forces. According to many engineers, the process of standard setting is at least as important as the standards themselves, because it helps to forge close understandings and working relationships between suppliers and end users and enhances industry cooperation.

SEMI also organizes education programs for its members. Volunteer committees of SEMI members organize symposia, dinner meetings, information seminars, and conferences. For example, technical symposia are held at the SEMICON shows to keep members informed on research and technological advances in the industry. These forums foster technical interchange between customers, academics, capital providers, and SEMI members, and are widely credited with helping diffuse technology and know-how among firms in the industry.

Finally, SEMI's annual three-day Information Services Seminar (ISS) provides market forecasts for the equipment and materials businesses and the overall semiconductor industry. This provides a vital service to small firms because, as SEMI describes it:

> The vast majority of SEMI member companies are very small, with annual sales of less than $10 million, making them too small to support any kind of ongoing market research judgments on the condition of our rapidly changing industry.
>
> (SEMI 1990)

SEMI is thus a highly participatory organization, whose operating style is one of small groups of engineers helping each other behind the scenes rather than in the public eye.

SEMI abstained from involvement in public policy throughout the 1970s. In 1984, it initiated a government relations program in order to keep its members informed about federal government activities which directly affected them. However the organization explicitly maintains a low profile in Washington, only lobbying about immediate concerns such as export licensing regulations. SEMI did not support the SIA's trade demands, although when SEMI/Sematech was formed as a chapter of the semiconductor manufacturing consortium, 84 companies joined as individual members (rather than through SEMI at large) in order to qualify as potential recipients of Sematech equipment contracts.

Through its trade shows, standards setting, and education programs, SEMI enables its member firms to remain specialized and survive in a technologically sophisticated and fast-changing industry. By setting standards that build a common framework for competition, organizing trade shows to underwrite the costs of marketing, establishing education programs which spread technical information, and promoting networking and interchange with customers, SEMI preserves the specialization and flexibility which allow its members to innovate.

The Software Entrepreneurs' Forum

The Software Entrepreneurs' Forum (SEF) is an association of over 500 Silicon Valley-based software developers, consultants, and software service providers. The SEF plays a central role in supporting the thousands of entrepreneurial software firms in the region. Explicitly regional in focus, it plays only a minimal role in formal politics. The organization is best seen as an attempt to institutionalize the social networks and the culture of information sharing and collaboration which characterized Silicon Valley during the 1960s and 1970s. In fact, it is reminiscent of Silicon Valley's earlier hackers organizations, such as the Homebrew Computer Club.[22]

The SEF was founded in 1983 by a group of 14 software "hackers" who recognized that more than programming talent was needed to publish and sell software. While they knew how to design and develop a product, they still needed to get the software to the market – which required good documentation, packaging, production, advertising, financing, and legal advice. The founders thus defined the formal mission of the SEF as providing the software entrepreneur with the resources needed to sell products and services. According to one member:

> SEF is a group of entrepreneurs who have learned that by working together to network and build their business and technical skill base they can preserve their fierce independence . . . and create software on their own terms.[23]

Today the SEF is the largest regionally based organization of its type in the nation, and has spawned at least five other organizations modeled after it in other states. It sponsors monthly members' dinner meetings, special interest groups, and week-long skills seminars. These forums encourage members to meet with industry leaders, share technical information, develop business skills, establish commercial and social relationships, and learn about new opportunities. The SEF also publishes a monthly newsletter, which keeps members abreast of meeting and seminar highlights and reports on new product releases and other significant news announced by members.

The SEF dinner meetings have become a Silicon Valley institution. They typically feature industry leaders speaking on issues ranging from industry trends and strategies for building a business to up-to-date advice on legal and financial issues. These meetings not only provide information, but also help to create a shared language and culture among the members of a highly fragmented industry. The special interest groups, which are organized around particular functions or operating systems (such as Marketing, Networking, Hypermedia, PC/Windows, and Macintosh), similarly offer an opportunity to obtain technical information and business skills first-hand from prominent professionals in the field.[24]

SEF dinner meetings and seminars are typically well attended and highly participatory. Many local entrepreneurs and start-ups credit their successes to the information and advice gleaned from SEF seminars as well as the relationships developed at these forums. According to the vice president of one software start-up: "I joined SEF because I was impressed by the unusually high level of member participation, which greatly exceeds that of most trade or professional organizations."[25]

The SEF is a non-profit organization run by a board of directors (who meet monthly), a board of advisors (who meet yearly), and a paid executive director. The board of directors supports and develops relationships with other state and national organizations that affect software entrepreneurship. There is a legislative representative and liaison to the Software Services Association (SSA), which maintains a lobbyist to monitor state legislation that affects the software industry. Recently, for example, the SEF undertook a write-in campaign to protest a bill that would penalize software companies which did not display support information at the point of sale.

But such political activity is distinctly secondary to SEF's primary agenda of providing technical, professional and social support for local software developers and entrepreneurs. One successful member says of SEF:

> It's invaluable. For me, its been a place to listen and learn. When you're a small company you can't afford to make mistakes; you have to learn from others. And you can't learn everything from books.[26]

As the SEF continues to grow, it plans to expand its seminar and interest group offerings in order to preserve the sense of community that has been at the core of its success. According to SEF President Stephen Friedman, "I'd like the next 500 members to feel the same level of community and involvement as we share today."[27] The SEF thus plays a crucial coordinating role through its support for networking and information exchange among the enterprises in a highly competitive and fragmented industry.

Other Silicon Valley business associations

While these four groups are not the only business organizations representing Silicon Valley's technology firms, the other associations in the region all follow a similar pattern. The Computer and Business Equipment Manufacturers Association, for example, is a classic Washington-based trade association which represents established computer firms, while the recently formed Computer Systems Policy Project, a pressure group for eleven of the nation's leading computer firms, was explicitly modeled on the SIA.

The Silicon Valley-based International Disk Equipment Manufacturers Association (IDEMA), by contrast, was modeled on SEMI and provides integrative services for the hundreds of specialist firms which supply materials and equipment to the highly fragmented disk drive industry. Finally, the Santa Clara County Manufacturing Group is an explicitly regionally based association dedicated to preserving the physical infrastructure of the region for technology producers.

THE POLITICS OF BUSINESS ORGANIZATION IN SILICON VALLEY

While mass production and flexible specialization can be complementary forms of production, their coexistence generated a series of political tensions in Silicon Valley during the 1980s. Throughout the 1960s and most of the 1970s, the region's high-technology producers shunned political activity altogether. The fierce independence of these entrepreneurs was reflected in *laissez-faire* attitudes. As they began to lose market share to Japan during the late 1970s, however, the region's commodity chip makers turned to the federal government to address their trade grievances.[28]

The SIA quickly became the only organization in the region with a clear policy agenda, and thus shaped the political debates in Silicon Valley during the 1980s. The AEA increasingly followed the lead of the SIA, focusing on lobbying for national policy to support large technology firms; while the region's integrative organizations, like SEMI, repeatedly rejected SIA initiatives, but failed to define an independent vision – and often rejected politics altogether.

The conflicts among the Silicon Valley business community emerged with passage of the Semiconductor Trade Accord. The agreement, which represented the triumph of the mass production sector in crisis, was intended to insure stable and profitable markets for commodity chip makers by eliminating the price collapses they experienced as a result of recurrent cycles of overproduction.[29] However, it had immediate adverse impacts on other sectors of the regional economy. Computer systems firms, in particular, were directly damaged both by the resulting increases in memory chip prices and subsequent shortages of computer memory (as US producers backed out of the memory chip business). Other local firms which rely on commodity semiconductors – from disk drive makers to contract manufacturers – were similarly hurt.

Not surprisingly, none of the other business associations in the region supported the SIA's trade policy demands. Even the region's specialty chip makers shared the systems firms' view of the trade agreement as a protectionist measure aimed at supporting a handful of uncompetitive firms. One executive from a local computer company accused the US government of being prepared to sacrifice one industry for another, claiming that "We have twelve companies that benefit from this. We have about 5,000 losers."[30]

The SIA became increasingly isolated from the wider electronics community in Silicon Valley during the 1980s – just at the moment it was gaining widespread attention and national political support.

Trade protection was not the only possible response for the commodity chip makers. They could have learned from the strategies of the region's specialist chip makers, which remained profitable even during the industry's worst downturn in 1984–6 and introduced innovative, higher value-added products and processes; or they could have learned from Japan to interact closely with their suppliers to improve their manufacturing processes. Either strategy would have allowed them to collaborate with – rather than alienate – their customers and suppliers.

Sematech reinforced the antagonisms in Silicon Valley. While Sematech was hailed nationwide as a pathbreaking form of industrial cooperation, the consortium received mixed reviews at home. None of the region's specialist semiconductor firms (the most technologically innovative producers in the nation) joined Sematech, largely because its prohibitive fees exclude all but the largest firms.[31] In fact, only 14 of the 293 semiconductor producers in the US are members. And while a small group of semiconductor equipment firms were belatedly included in the effort under the auspices of SEMI/Sematech, computer and software firms are still excluded.[32] Even the AEA did not support federal funding for Sematech.

Some critics argue that although the purpose of Sematech, to improve semiconductor manufacturing processes, is laudable, its prohibitive fee structure and narrow membership base are problematic in a highly fragmented industry. They have not, however, mobilized to promote an alternative, more inclusionary model of collaborative research or manufacturing.

Others view Sematech simply as the vehicle for a handful of aging firms with political influence use public resources to shore up their market positions. In the words of the CEO of a local semiconductor firm: "Sematech is a carefully constructed lobby effort" of large producers "to deliberately and systematically exclude smaller companies."[33] These critics articulate the laissez faire view shared by many Silicon Valley entrepreneurs, that government intervention which interferes with the workings of the free market must be avoided.

The debates over trade policy and Sematech clarify the conflicts between the mass producers and the more flexible and specialized firms in Silicon Valley. SIA policies privilege member firms at the expense of customers and suppliers, as well as competitors. They thus mirror the autarchy inherent in their approach to mass production, one in which producers remain at arm's length from both their customers and their suppliers.[34] Moreover, their exclusive emphasis on lobbying for national policy reflects the assumptions of a traditional mass production system, where the key coordinating tasks are performed at the national (or the firm) level.

The integrative policies pursued by organizations like SEMI and the SEF, by contrast, provide coordination at the regional and sectoral level. They support the specialist enterprises in highly fragmented industries by providing technical and market information and by fostering the recombination of skill and know-how. They implicitly recognize the importance of openness and collaboration to their flexibility, and seek to institutionalize those interdependencies.

Yet while the region's specialist producers regularly oppose the programs of the SIA, they have refrained from playing a broader political role in the region or the nation. SEMI and the SEF thus remain entirely reactive in policy debates. In many cases they do not simply reject SIA politics, but political activity altogether. Rather than defining a coherent alternative strategy (one which reflects their integrative approach), they have adopted the traditional *laissez-faire* view of American business, criticizing all forms of public activity. Silicon Valley's specialist producers thus celebrate entrepreneurship and individual achievement at the same time that their integrative institutions provide the collective services and foster the relationships without which such entrepreneurship would be impossible.

CONCLUSION: THE VULNERABILITY OF SILICON VALLEY

The divergent patterns of business organization in Silicon Valley underscore the vulnerability of the regional economy. The region's externally oriented associations have effectively articulated a political program which serves the needs of the region's large technology producers. They are organized to lobby for industrial policy measures, such as funding for HDTV research or antitrust reform, in the name of national competitiveness.

Silicon Valley's integrative organizations, by contrast, have failed to articulate the importance of the region and its dense networks of small and medium-sized firms to the nation's competitiveness. Viewing their successes as the result of individual effort and entrepreneurship alone, the specialist firms resist efforts to define political strategies which might insure the continued dynamism of the region's decentralized productive system. Perhaps it will take a crisis to mobilize the region's specialist producers into political action – but by then it may be too late.

The industrial fragmentation which accounts for the flexibility and adaptive capacity of the Silicon Valley economy thus represents its greatest weakness. More extensive and explicit coordinating institutions are necessary if decentralized regional economies such as Silicon Valley are to survive in the challenges of intensified international competition.

While the region's integrative institutions play an important role by setting standards, running trade shows, and promoting information sharing and networking, they remain limited. Associations such as SEMI or even the AEA could take the lead in enhancing the regional infrastructure by providing training, education, joint R&D, and by insuring the availability of long-term capital. Insuring that representation cuts across industry lines might also foster recognition of the extent of common interests among the firms in the region.

Over time, such institutions could provide a forum for resolving conflicts between local firms (small and large, firms in different sectors) as well as the capacity to plan and respond systematically to challenges such as external

DESCRIPTION

This is an ex-library book and may have the usual library/used-book markings inside.This book has hardback covers. In fair condition, suitable as a study copy. Dust jacket in fair condition.

technological shifts.[35] In short, they would provide a region-wide capacity for systematic thinking, debate, and response to shared challenges. Just as mass production and flexible specialization coexist, so externally oriented and integrative institutions can be complementary. The challenge for Silicon Valley's producers will be to create a political vision which transcends individualism and articulates the importance of the region and its relationships to their technological dynamism.

17

CONCEPTUAL FALLACIES AND OPEN QUESTIONS ON POST-FORDISM

Danièle Leborgne and Alain Lipietz

INTRODUCTION

In the second half of the 1980s, hopes were kindled about the emergence of a new and seemingly progressive model of capitalist development. Post-Fordism appeared as a candidate for the way out of the crisis of dominant postwar economic model, Fordism. These new hopes emerged from old fears about what seemed to be the dominant (and regressive) solution in the first half of the decade, "neo-Fordism." This change in mood among radical economists and geographers relies upon the fact that in the world battle between two ways out of the crisis of Fordism, the countries that seem to be victorious (e.g. Japan, Germany, the Nordic countries) have developed more positive forms of compromise between capital and labor than the losers in competition (e.g. USA, UK, France). Moreover, this change in the mood expresses a more constructive attitude: instead of complaining that things are getting worse and worse, propositions are addressed to unions, communities, political forces, and "enlightened" managers designed to promote a positive-sum compromise.

The new conventional wisdom of the first half of the 1980s may be summarized as follows. New technologies tend to foster the separation between highly skilled designers and engineers, and poorly skilled operatives. Since these operatives should be hired at the lowest price (and not at an "efficiency wage"), it would be useless for capital to respect the wage level of the collective bargaining system, and to finance a costly welfare state. The consequence is that advanced capitalist societies would evolve towards "two-tier," "hour-glass," Brazilian-type societies. Moreover, the shift of routinized production towards the Third World would lead to a deindustrialization of the developed world. In the upper tier of the latter, the yuppies and golden boys of high-tech and finance would flourish, managing principal world-scale corporations, while women, ethnic minorities, and deskilled male workers would survive by working in the service of the upper tier. That would be a form of Fordism without the advantages of Fordism: a neo-Fordism.[1]

After 1983, the gloomy atmosphere of the "monetarist shock" was replaced by the world recovery symbolized by the miracle economies of Japan, Korea, the Third Italy, and Baden-Württemberg. A new discourse, "post-Fordism," emerged to give these successes a theoretical framework. Post-Fordism is based on a shift from "mass production" to "flexible specialization," in which craft work takes place in networks of interconnected, independent, specialized, and flexible firms, which are in a position to react quickly to changes in fashions on the world market and to implement efficiently new technologies. Based on preexisting local resources of capital and skills, new growth regions emerge through the spinoff of innovative firms or through disintegration of big companies.[2]

Neither the objective situation nor the debate is so clear-cut. First, many scholars collapse the two models into one single "craft-flexible post-Fordism"[3] and, in so doing, they eclectically reflect a still confusing reality. Second, claims about the positive aspects of post-Fordism (i.e. reskilling of the labor force, human-scale productive systems) are called into question by evidence of growing neo-Fordist regressive tendencies in many parts of the world economy, including the USA, the UK, and southern Italy. Moreover, the world concentration of capital raises some doubts about the importance of network production.[4]

Strangely enough, both sides in the debate refer to French regulation theory, especially through the reference to the Fordist paradigm. "Fordism" was a term coined by Antonio Gramsci in his analysis of historical blocs as political keystones of development. Yet most participants in the post-Fordist debate reduce it to a debate about new production systems and, contrary to most French regulationists, hold that there is already a solution, whether neo-Fordist or post-Fordist, to replace Fordism.[5]

In this chapter, we will begin by referring back for our definition of Fordism, after which we will discuss some common fallacies in the post-Fordism thesis. We focus on post-Fordism rather than on neo-Fordism because the latter has really been implemented in the USA and UK, and it is now seriously challenged by something else that some have labeled post-Fordism, and which is certainly more interesting, both theoretically and politically. We discuss five open questions with respect to the notion of post-Fordism.

1 A model of development is not only a technological paradigm. Fordism was not just "mass production," nor could post-Fordism be defined merely as "flexible specialization."
2 There is no single way out of the crisis of Fordist labor relations. Neo-Fordism and post-Fordism are conflicting solutions, because neo-Fordist "flexibility" and post-Fordist "craft" cannot be mixed at will.
3 Industrial organization is also an open issue. There are still contests between big corporations and industrial districts, as well as between different forms of industrial networks.

4 Macroeconomics still matters, and ecology matters more and more; the free market will not be the form of regulation of post-Fordism.
5 There is no reason why a single model of development should be world-hegemonic. After-Fordism will very likely not consist of the victory of only one of the conflicting models, but a core–periphery in which the two coexist.

A DEVELOPMENT MODEL IS NOT SIMPLY A TECHNOLOGICAL PARADIGM

Much of the post-Fordist discourse is based upon the idea of a return of craft production. By craft production is usually meant both the supply-side (skilled workers committing their knowledge into the fine-tuning of a flexible technical apparatus) and a demand side (customized commodities serving volatile and fragmented demand). New technologies are said both to provide an answer to the need for flexibility of the demand, and in turn to require a specialized labor force. And this leads, so it goes, to flexible specialization as a model of capitalist development, and it in turn is equated with post-Fordism per se. Fordism, by contrast, was based on mass production, with its rigid assembly lines requiring low skills, and producing standardized commodities.

This presentation of history is based on a dangerous technological determinism. A system such as Fordism is a complete way of living, rooted in a historical bloc, yet these analyses reduce it to the logic of a production system. While it is necessary, for any model of production to achieve stability, that there should be a consistency between the organization of production, labor relations, the pattern of consumption, and even social ideologies of progress, such consistency is a chance discovery of history, the result of a process of social and political learning (Lipietz 1985). Conflicting solutions are possible on the basis of a given set of technologies and they are not determined by the technological paradigm. Let us illustrate this point by taking up the example of Fordism.

As a model of development, Fordism has three main dimensions.

- As a general principle of organization of labor or technological paradigm,[6] Fordism is nothing more than Taylorism plus mechanization. Taylorism implies a strict separation between conception labor, which is the task of the Organization and Methods Office (O & M) on the one hand, and the execution of standardized and formally prescribed tasks on the shopfloor on the other hand. Mechanization embodies the collective knowledge of O & M into the material apparatus. Reduced to these dimensions Fordism induces mass production. Yet the implementation of this technological paradigm in the 1920s raised a serious question. Who would buy the outputs of mass production? Henry Ford is now famous for having proposed an answer: wage earners should be the main customers. Hence the name of Fordism for the *macroeconomic* pattern which was eventually implemented twenty years after the Great Depression, which was the underconsumption crisis stemming from mass production.

- As a macroeconomic pattern (or regime of accumulation), Fordism implied that the gains in productivity arising from its principles of production were matched, on the one hand, by the increase of investments financed by profits, and on the other hand by the increase in purchasing power of the waged workers. As a result, both the share of wages in value-added and the value capital/output ratio were roughly constant, the rate of profit was essentially stable, and the markets for production goods and consumption goods both grew in pace with productivity. But – and here Henry Ford had less to say than Keynes and Beveridge – how could such a steady increase in purchasing power be achieved? Here rules of coordination between economic agents had to be established. Fordism cannot exist in a single firm.
- As a system of rules of coordination (or as a mode of regulation), Fordism implied long-term contracts as the basis of the wage relation, with rigid limitation of layoffs, and a tying of wage increases to prices and overall productivity. Moreover, a large redistribution of revenues through the welfare state and social security insured a basic income for wage earners. The counterpart was the acceptance by the unions of managerial privileges, on the shop floor. As a result, both the principles of labor process organization and the macroeconomic pattern were respected.

The crisis of Fordism has been subject to various explanations. In our analysis, its problems can be traced to supply and demand (Glyn *et al.* 1988; Lipietz 1985, 1989b). The most obvious reason appeared on the demand side. Competitiveness equalized between the United States, Europe, and Japan. The search for economies of scale induced internationalization of production processes and markets. The increase in the price of raw materials imported from the south (notably oil) stimulated competition for exports in the early 1970s. The regulation of the growth of domestic markets through wage increases was thus limited by the necessity to balance external trade. Faced with this demand-side crisis, the reaction of international elites was mainly Keynesian until the end of the 1970s. Then the mood of the international elites of the capitalist world changed. Demand-side management of the crisis had avoided a great depression, but a major and deeper limit became obvious: the fall in profitability.

This was due to a number of causes on the supply-side: a slowdown in productivity with increase in total labor price (including welfare payments), an increase in the capital–output ratio, and an increase in the relative price of raw materials. In those conditions, Keynesian strategies such as an increase in the real wage (however limited) and loose monetary policy could only induce inflation. Hence, the rise of monetarism (i.e. an attack on demand-side growth policies) and the shift to supply-side policies. In our view, the key to the supply side is industrial relations, which in turn encompasses aspects of the industrial production paradigm and the mode of regulation.

Even within the regulation approach, the supply-side problem of Fordism is subject to two interpretations. First, according to a Kaleckian analysis, the rise in the relative price of labor and raw materials was the result of the long boom of the "Golden Age" of the postwar period (Itoh 1980, Armstrong, Glyn, Harrison 1984). This "full-employment profit-squeeze" analysis became the official explanation in OECD reports by the end of the 1970s. Profits were too low because workers were too strong, and this was so because the rules of the game had become too "rigid." Liberal flexibility policies were adopted by the UK, then by American administrations, and were eventually implemented by a number of OECD countries, including even the socialist-communist French government in the 1980s (Lipietz 1990).

Yet a complementary explanation to the Kaleckian full-employment profit-squeeze theory stresses the exhaustion of the efficiency properties of Taylorism. In this view, it is not ultimately rational to eliminate all possibility of initiative on the part of the worker on the shop floor; "responsible autonomy" appears as a better principle, especially as the implementation of new technologies, or the shift to "just-in-time" management require the involvement of operators and their cooperation with managers and designers (Friedman 1977).[7] This is precisely the labor process strategy selected by many successful firms in Japan, Germany, and Scandinavia.

Faced with these events, the notions of neo-Fordism and post-Fordism are on unequal ground. On the one hand, the neo-Fordist analysis acknowledges the main tendencies of the late 1970s and early 1980s, such as a progressive unraveling of the Fordist rules of the game (concession bargaining, development of secondary labor markets, cuts in the welfare state), enhanced by the monetarist shocks of 1979 and 1981, and inducing a dramatic transformation toward a new regime of accumulation, with a polarization of incomes, and a shift of social demand towards export and/or toward the overconsumption of the upper tier of society (Davis 1986). Rather little is said about the changes within the technological paradigm, except to note a shift towards a "computer-assisted Taylorism" (Mathews 1989). This analysis reflects rather well the way the USA dealt with its supply-side crisis.

On the other hand, the first theorizations of "flexible specialization" tried to capture the fact that some firms, regions, or countries were getting the better of the new world configuration brought about by monetarism. In a world of market volatility where customized commodities were favored by the new upper-middle classes, flexibility of *supply* was required, and specialized skills appeared as a better way to meet small batch demand. Since post-Fordism later on appeared as a new solution for the supply side of the crisis, it is indeed strange that these early theorizations insisted mainly on the demand side of the crisis (and wrongly characterized it as a crisis of "underconsumption").[8] The change in industrial relations appeared as an adaptation to a change in the regime of accumulation: post-Fordism in the plant as an adaptation to neo-Fordism in the world economy. This led to a failure to analyze the consistency of specialization and craft on the one hand, and flexibility in labor and product markets on the other hand. Moreover, flexible specialization as a new form of work organization was accorded a kind of neutrality in its macroeconomic effects: there would be no more problem of effective demand.

Yet, labor relations must be *consistent*. First, the different elements of the industrial relations system must be mutually consistent: wage contracts, labor organization, and the social reproduction of appropriately skilled labor forces. Second, they must be consistent with the broader shape of social life in concrete territories: general goals and accepted rules of life in a common society. Third, they should be consistent with the macroeconomics of some regime of accumulation or social structure of accumulation, both at the national and international level. Last, but not least, the global ecological situation is now imposing strong constraints on the generalization of most models.

We will reurn to the problems of macro consistency. But first let us scrutinize the consistency of flexibility and craft within labor relations itself.

TWO PATHWAYS OF CAPITAL/LABOR RELATIONS

When they characterize the *actual* evolutions of labor relations, scholars rightly notice that there are tendencies to flexibilize the wage contract (so as to reduce the direct and indirect cost of labor force), *and* tendencies to mobilize craft skills, know-how, and the intellectual activity of the labor force, in order to increase productivity, to adapt continuously the product to demand, and to increase the efficiency of a costly fixed capital. While a closer adaptation of supply to a volatile demand is a by-product of the latter, it represents a fundamentally different view of the way out of the supply-side crisis of Fordism. Are the two views compatible, as is sometimes held, where neo-Fordism and post-Fordism are combinable (Moulaert and Swingedouw 1989)?

The two views of the way out of the crisis may be thought of as two pathways of development of industrial relations, with a common starting point in Fordism: Taylorism as direct control by the management on the activity of workers, rigidity of the wage contract, as shown in Figure 17.1.[9] The first pathway fosters an evolution from rigidity toward flexibility, the second one from direct control toward responsible autonomy. Stated differently, the first (vertical) axis refers to the external labor market, to the ties between firms and the labor power hoping to be hired. The second (horizontal), axis refers to the internal labor market, the forms of organization of cooperation/hierarchy within the labor process.[10] It should be emphasized that compromises on the regulation of the "internal labor market" are not necessarily negotiated "inside" the firm.

On the first, "external" axis, there are several dimensions of rigidity and flexibility. Many dimensions of the external labor market may be regulated: the direct wage, hiring and firing, the social wage. The external market may be more or less organized. Moreover, such rules may be implemented at the level of individuals, professions, firms, sectors, or society. We shall not consider these details, however, because they are overdetermined by what happens on the other axis.

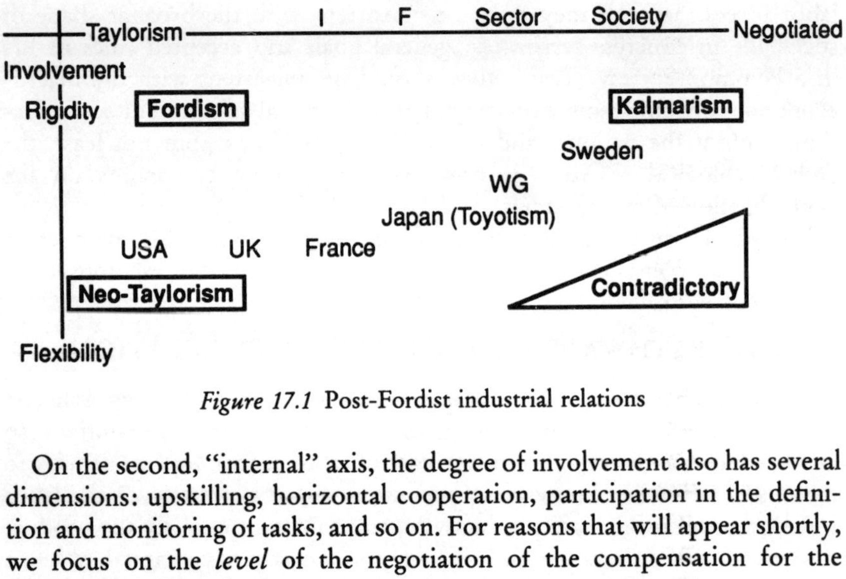

Figure 17.1 Post-Fordist industrial relations

On the second, "internal" axis, the degree of involvement also has several dimensions: upskilling, horizontal cooperation, participation in the definition and monitoring of tasks, and so on. For reasons that will appear shortly, we focus on the *level* of the negotiation of the compensation for the involvement of workers.

1 Involvement may by *individually* negotiated and rewarded by bonuses, career advancement, etc. This is limited by the *collective* character of the required involvement in most processes of cooperative production. "Individually negotiated involvement" (I in Figure 17.1) may be extended to a team or a shop.
2 Involvement may be negotiated firm-by-firm, between management and unions (F in Figure 17.1). Here the firm and the work force share the rewards of the specific skills accumulated in the learning process. This

implies an "(external) rigidity" of the wage contract, whose limit case is the "life-time" contract.[11]

3 Involvement may be negotiated at the sector level (S in Figure 17.1), thus limiting the firm's risk of competition from firms using cheaper labor, and inducing them to share skilling institutions. The external labor market is likely to be more organized, that is broadly speaking more rigid, with sector-wide collective bargaining, and a greater socialization of labor revenues.
4 The involvement may be negotiated at the level of the whole society, with unions and business negotiating at the regional or national level the social content and distribution of the production. In this case, it is understood that unions are responsible for worker performance on the shop floor. Here, the external labor market is likely to be at least as well organized as in the most advanced Fordist-corporatist cases.[12]

Nonetheless, collective involvement of the workers is unlikely to be possible if there is no commonality of destiny of firms and their work forces, i.e. in the context of external flexibility *at any level*. The outer limit of consistency between flexibility and involvement appears as the triangle marked on Figure 17.1. Collective involvement and flexibility are incompatible, and it is here that neo-Fordism and post-Fordism are so often conflated.[13] It follows that there are two privileged lines of evolution i.e. two *real* pathways of development "after Fordism", which are shown in Figure 17.1:

a) External flexibility associated with hierarchical direct control: we are back to some form of Taylorist organization of the labor process, *without* the social counterparts of the "golden age" of Fordism. Let us call this paradigm *neo-Taylorism*.
b) External rigidity (organization) of the labor market associated with negotiated involvement of the workers. Let us call this paradigm *Kalmarian*, in honor of the first car factory reorganized according to the involvement principle in a social-democratic country: Sweden in 1974.

Looking back on the recent experience of the OECD countries, there is a range of experiences, with the USA and the UK privileging external flexibility and ignoring involvement, or introducing individually rewarded involvement (as in France), Japan practicing negotiated involvement at the firm level, Germany at the sector level, and Sweden being closer to the Kalmarian case.[14] The firm-level negotiated involvement (typical of the Japanese case) allows for a *dualistic* coexistence of the two paradigms within a dualistic labor market (e.g. according to gender).[15] This mixed situation could be labelled *Toyotism*.

The experience of USA shows that it is difficult to negotiate involvement at the shop or even plant level in a broader context of liberalism, with its high level of external flexibility. In the USA, individually negotiated involvement may, however, be developed.[16]

INDUSTRIAL ORGANIZATION: STILL AT A CROSSROADS

One of the main empirical tendencies of the 1980s was that "Big is Beautiful." Never in the history of capitalism has there been such a tremendous international centralization of capital. There are no more than five important European car makers and only two major alliances of aircraft-engine makers in the world. This has occurred in the context of a large increase in world industrial output since the early 1970s, at the twilight of Fordism. Nonetheless, advocates of a post-Fordism transition insist that non-vertically integrated forms of industrial organization are becoming dominant, with the rise of industrial districts. In this view, the new pattern of economic ownership should be the clustering of small- and medium-sized, specialized and flexible firms, based on a territorial sector-specific supply of skilled workers. The paradigmatic examples of such a tendency would be the specialized productive areas of Third Italy, Silicon Valley, and so on. How is it possible to reconcile such different diagnoses? A first explanation may be sectoral specificities. Examples of industrial districts are usually found in the clothing, shoe, ceramic, software design, or customized electronic hardware industries, all of which are labor-intensive. On the contrary, concentration would be expected in capital-intensive industries, since even when economies of scope are important, they are increasingly internalized and based on costly flexible equipment. Internal economies of scale thus still matter when *total* output is considered. Still, even in highly centralized sectors, there are strong tendencies towards vertical disintegration, even within a given firm. It is from this latter phenomenon that the idea emerges of an organizational convergence between disintegrating corporations and quasi-intergrating industrial districts (Sabel 1989).

Here, a number of different issues and forms of quasi-integration may be distinguished. First, what do we mean by "quasi"-integration, and what forces are encouraging this form of industrial organization? Most proponents of the notion of post-Fordism simply assume the superiority of small or medium-sized firms as the outcome of "specialization" made possible by the new technologies. In the Coase-Williamson paradigm, there is only one alternative for the firm: make or buy, market or hierarchy. It is advantageous to buy when the price of independent production plus the transaction cost is less than the costs of production and hierarchy within the firm. Since flexible specialization is based on production systems which enjoy important external economies of scope and significant costs of organization, vertical integration would efface the advantages of economies of scope (because of the rigidity and specificity of intermediate production) and would dramatically increase the costs of organization in the form of hierarchy. Hence the victory of market over hierarchy.

There is a grain of truth in this analysis. As Leborgne (1987) has shown, flexible machinery entails a new solution, a compromise between integration and disintegration in terms of production and organization costs, in which the *specialized firm* (or plant) minimizes the costs of organization, maximizes economies of scope through "modular" integration, and allows for the tight control on quality, at the same time that it fosters innovation. Nonetheless, minimizing costs is not the main problem for capitalist production. Maximizing profits, or more precisely the rate of profit, is the great imperative forgotten in the Coase-Williamson paradigm. In this respect, there exist important differences between "make" and "buy." In the first case, the principal firm invests the fixed capital and controls the value added by its own workers. In the second case the principal shares with the contractor the cost of fixed assets, but it has to give up the surplus-value of the contractor's workers.

This points up the specific advantages of *quasi*-integration. As Leborgne (1987) and Laigle (1989) have shown, "quasi" refers to both the production process *and* valorization process. In the valorization process, the validation of fixed capital and labor committed to production by the subcontractor is achieved via routinized interdependence between the two firms (one being a regular supplier/customer of the other). Both of them depend upon the productive complementarity of their assets and know-how. Quasi-integration minimizes both the costs of coordination (because of the autonomy of the specialized firms or plant), and the costs of information/transaction (because of the routinized just-in-time transactions between firms). Moreover the financial risks of R & D and investments are shared within the quasi-integrated network.[17]

Yet there are important differences between types of quasi-integration. In a first (extreme) case, the buyer has at its disposal the know-how of the subcontractor (because the product is standardized, or because the quasi-integration is the result of disintegration of the principal firm). The former is then in a position to induce the latter to invest without guaranteeing its outlets, and also in a position to contract at a price which transfers to it the value added produced by the subcontractor. This case may be properly labeled "vertical quasi-integration." By contrast, when partnership and strategic alliance link a supplier with specific technology to a regular customer of another sector of the division of labor (for instance Mercedes and Bosch, or Aerospatiale and SNECMA), we have a case of "horizontal quasi-integration." The general case, of course, is intermediate: "oblique quasi-integration."[18] For instance, the link between Renault and its seat-maker is oblique because Renault orders for specific goods which are part of the process of production of cars (and were formerly produced in integrated car factories), but the seat-maker is fully responsible for the process of production of the seats (including subcontracting the production of metal frames, fabrics, and so on). As Laigle (1990) has shown in the case of the French car industry, the concrete network of quasi-integration is a mix of horizontal, vertical, and oblique ties (Figure 17.2). The more horizontal the tie, the better the bargaining power of the supplier, but also the more research and development in its product (this being both the cause and the effect of its power), hence the higher quasi-rent earned by the network.[19]

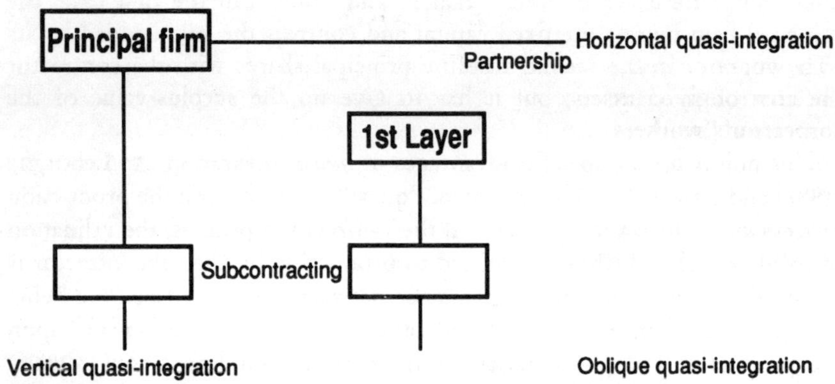

Figure 17.2 Post-Fordist industrial organization

There is a striking resemblance between Figures 17.1 and 17.2. In both cases, the vertical axis represents a form of "flexibility," which increases the share of surplus value accruing to the principal on the basis of direct control over the dominated agent. In both cases, the horizontal axis represents a form of autonomy of this agent, implying a higher efficiency of the principal–agent couple. In other words, quasi-integration is always cost-minimizing *à la* Williamson, but VQI is revenue-maximizing only for the principal, while OQI and partnership are revenue-maximizing for the whole network.

Our thesis (developed in Leborgne and Lipietz 1987, 1989) is twofold. First, this apparent correspondence is real. The tendencies toward negotiated involvement in labor relations and toward partnership in industrial organization seem to be different expressions of the possible formation of an historical bloc which is modernizing certain economies. There are territories where dense, horizontal forms of quasi-integration between firms develop in pace with negotiated involvement within firms, while in other places vertical disintegration (mostly toward other low-wage regions) develops in pace with neo-Taylorism.[20] Second, these two combinations correspond to two different pathways of development, and the first is presently outperforming the second.

Thus, care should be taken when heralding the benefits of industrial districts. In many cases, a hidden subordination through VQI of most firms to a principal firm is likely to emerge (e.g. of the firms of the Third Italy to those of the First Italy). The nature of quasi-integration determines whether the district is a progressive or a regressive evolution from the staff-and-line industrial organization of Fordism. Subcontracting to poorly integrated districts remains consistent with neo-Fordism.

MACROECONOMICS STILL MATTERS AS WELL AS ECOLOGY

The euphoric diagnoses of post-Fordist "possibilities for prosperity" have been strongly criticized on the basis of their lack of concern with social aspects of "really existing flexible specialization" (as Brezhnev used to speak of "really existing socialism"). As Jensen (1989) and Pollert (1988) have pointed out, there is a real danger that "negotiated involvement" in "yeomen democracy" would be restricted to a yeomanry, a male aristocracy of skilled workers.

A clear distinction must be made between an analysis of existing reality and its problems, and the creation of political options for gender and racial equity. Negotiated involvement of workers entails in itself neither social polarization nor equalization. Here, the *scope* of the negotiation will be determinant, and that will be subject to a major political and cultural struggle.[21] In the present situation where the word "flexibility" is used in a confusing way to extol the virtues of neo-Taylorist, Toyotist, and Kalmarian models, conceptual clarification is first needed. The first class of models, which includes individually negotiated involvement, will be favored where an individualistic vision of society is accepted, and will induce social polarization. Kalmarian models on the contrary will be favored when anti-egalitarian social consequences are rejected. Toyotism (that is, firm-level negotiated involvement in "good" firms and neo-Taylorism elsewhere) will be accepted in hierarchical societies acknowleging "natural" differences between individuals and genders. Flexible specialization as a technological paradigm has little influence there.

Nonetheless, the theory of flexible specialization is to be criticized when it professes a benign neglect of macroeconomic issues:

> A shift away from mass production would restore the neoclassical equilibrating mechanisms that were probably most prominent in the

> early-nineteenth century American economy. . . . It would be possible to maintain full employment without resorting to wage determination systems linking purchasing power to the rate of increase of productive capacity.
>
> (Piore and Sabel 1984: 276).

This is an amazing statement based on a completely unrealistic assumption about declining economies of scale in a highly capital-saving paradigm. In fact, flexible specialization, as we have already noticed, is more likely to be based on costly assets, resting on huge and risky investments in research and development. If a massive demand (even a fragmented one) is not secured through some form of regulation of the sharing of gains of productivity, the old instability of pre-Fordist capitalist economies is likely to come back. Macroeconomics still matters.

Macroeconomic constraints are well kown by economists; they can be logically deduced. We may thus be very brief. First, any model of development must be profitable. Second, it must be competitive. Third, demand must match supply.

Assuming that "involvement" is more productive that neo-Taylorism, the first constraint suggests that, when a great part of the surplus has to be reserved to debt servicing, there is little room for "negotiating involvement" because the wage has to be as low as possible. The existing debt load on a given country is thus inducing neo-Taylorism. For the second reason, countries which are already pursuing neo-Taylorism, and are less productive than "involving" ones, are also handicapped in the search of a better capital-labor truce. This does not imply that the roll of the dice is already over: equivalents may be proposed to workers (such as rigid wage contracts). Still such countries as the USA, the UK, Brazil, and Eastern Europe will have difficulties in shifting toward the Kalmarian paradigm. On the contrary, Scandinavia, Germany, Japan, and even Korea (which is developing specific forms of "involvement" in the male sector of its labor force), may be considered to be marching toward the twenty-first century in a good position.

As far as demand is concerned, the Kalmarian paradigm offers greater possibilities to regulate social demand than neo-Taylorism.[22] The latter will be associated with a cycle of "exhilarating" periods of boom (with an increase of profits and of the revenues of the upper tier of society) and periods of depression (due to overinvestment or "cooling off" policies): the return of the business cycle, as opposed to the more regular pattern of Fordism.

This brings us back to the problem of "yeomen democracy," even when workers' involvement is negotiated firm-by-firm. This is what Aoki (1988) has labelled "the dilemma of worker democracy." In this case, the surpluses of productivity are specific to the firm, and the resulting quasi-rent of the firm exists insofar as differential productivity is maintained vis-à-vis other competitors. The negotiated compensations (in terms of higher wages, reduction of labor time, or lifetime career) are limited to this quasi-rent. In this condition, when cyclical downturns on the market occur, insiders and management will be allied against newcomers, subcontractors, and competitors. This tends to consolidate a "workers' aristocracy" at the top of the generalized, pragmatic, meritocratic hierarchy in the whole society.[23] This hierarchy may develop into a completely dualistic structure (negotiated involvement/neo-Taylorism), especially when gender or ethnic differences come into play. In this case, any increase of the aggregate domestic demand would be restricted by the need for competitivity of this dualistic structure, since it is dependent on external outlets.[24] Things are different when a *general* (and global) growth in hourly labor-cost is a rule imposed upon any competitor. Then the issue is not the sharing out of specific quasi-rent (or "extra surplus-value" in Marxist terms), but a general redistribution of gains in social productivity ("relative surplus-value"). So, not only does macroeconomics matter, but it has a strong effect on the social character of post-Fordism.

Up to now, nothing has been said about the nature of the redistribution of the gains in productivity accruing from "negotiated involvement": growth in purchasing power? In free time? Here a major new constraint appears at the dawn of twenty-first century: ecological sustainability. Capitalist development not only degraded work under Taylorism (Braverman 1974); we now understand how much it degraded the earth, according to Marx's prophecy (and this is also true in the state capitalism of the East). In fact, capital/labor conflicts have been allayed for the most part at the expense of nature, hence at the expense of future generations. The hole in the ozone layer and the greenhouse effect are among the consequences of the great industrial boom of the Fordist and Soviet models. The recovery of the 1980s has increased the frequency of industrial disasters, and it is worsening the global ecological crisis. The limits of any productionist model are fully revealed, at the local or global level, and the necessity for future models of development to be sustainable, that is ecologically consistent, is ever clearer.[25]

If we assume the continued development of ecological concern all around the world (and this is the case in the West, in the East, but less in the South), sustainability will become a new factor of legitimation of future capital–labor compromises. As a result the counterparts of gains in productivity, if they are negotiated, will tend to be granted in the form of an increase of free time rather than in the form of an increase in real wage (as in Fordism). That solution may prevail in the most socialized form of Kalmarian compromises.[26] Another solution, consisting of alleviating the damage to the environment through ecologically sensitive industries, would be favored by an overconsuming upper tier attached to neo-Taylorism or Toyotism.

This would raise the relative prices of consumption goods to the detriment of the lower tiers of the society. Ecology would provoke a major distributional conflict.

TOWARD A CORE–PERIPHERY CONFIGURATION IN POST-FORDIST TIMES?

We have been accustomed to thinking that one model of capitalist development should be hegemonic in any given historical period. This was true at the time of British hegemony (competitive capitalism with extensive accumulation), and it was true at the height of Fordism. In these times, countries which failed to adopt the dominant model were locked out of world industrial competition. In its Golden Age from 1950 to 1970, Fordism defeated and marginalized Sovietism and corporatist forms of import substitution in the Third World. As far as manufactured tradable goods were concerned, Adam Smith's law of "absolute advantage" seemed to prevail: Fordist models of production outperformed all others. Thus, non-Fordist countries were restricted to primary goods exports and could only develop industries via protectionism.

The crisis of the 1970s changed all that. New industrial powers emerged from the South, with new models of development[27] combining Taylorism and flexible forms of the wage contract. Nowadays, industrial protectionism has changed direction: high-wage developed countries are protecting themselves against "unfair" competition of these countries with slighly lower productivity but much lower wages.

In the first half of the 1980s, supply-side, neo-Taylorist policies were designed to cut labor costs. Yet, as we have already seen, more advanced social compromises performed very well in the "post-Fordist" countries, either in the form of Toyotism or Kalmarism. Moreover, the trade deficit of the more neo-Taylorist countries (such as the USA and UK) opened huge outlets for the more competitive production of Japan and West-Germany (Lipietz, 1989b). Such a macroeconomic world situation is not stable. Too many countries have a debt to pay: the USA, the UK, the Third World, and Eastern Europe. Should they try to do it through a new offensive in "external flexibility," with more unemployment and lower wages, a new world depression could threaten the prosperity of negotiated involvement.

Protectionism may be a solution, but it is not the main tendency. On the contrary, the desperate attempts of Third World to pay its debt, the necessity for the United States to balance its external accounts, and the integration of the ex-socialist world into the world economy will increase world competition. From the experience of the 1970s and 1980s, it is doubtful that either neo-Fordism or post-Fordism will completely eliminate the other. Thus, the main question is how the two models will function within the same trading system, such as the EC (and its eastern satellites).

We may assume that in labor-intensive routinized sectors, neo-Taylorism may dominate negotiated involvement at a sufficiently low wage. Then, as envisioned by Ricardo's theorem on comparative advantage, countries (or regions) will tend to specialize in sectors where they are comparatively best "endowed," whether in flexibility and low wage – or in involvement. In that case, the total amount of demand will be limited by wage competition due to the coexistence of low-wage, low-involvement regions and higher-wage, higher-involvement regions. The greater the possibility to practice neo-Taylorism, the smaller will be the islands of "yeoman' democracy" *à la* Piore and Sabel, and lower overall global prosperity.

The most likely scenario is the formation of a new world economic hierarchy. It will not oppose an industrial Fordist core to a primary-goods-producing periphery, but will present itself as a de facto new industrial division of labor. The core economies will be those having adopted a Kalmarian compromise in much part of their production, with a possibility of internal dualism (Kalmarism/neo-Taylorism). They will dedicate themselves to high-technology and highly skilled production. The periphery will consist of economies organized according to the neo-Taylorist paradigm, and dedicated to routinized and labor-intensive activities. Intermediate regions (with very wealthy high technology and financial sectors) may develop out of individually negotiated involvement (the "Californian type" defined in Leborgne and Lipietz, 1987).

In this new world industrial hierarchy, some former core economies may become semi-peripheral, including the UK, France, and even many states of the USA. Conversely, Japan and Germany will consolidate their places in the core, with some other former core economies like Sweden, and some formerly peripheral ones catching up, such as Korea following the trajectory of Finland. A part of Eastern Europe would be involved in this process, while the rest of it would become a peripheral to Germany and Japan. As for the Third World, a part of it is becoming marginalized, and another part is becoming integrated into the world economy as a neo-Taylorist (or even primitive Taylorist) subcontracting area around the more advanced growth poles.

CONCLUSIONS

History is alive again. On the ruins of Fordism and Stalinism, humankind is at a crossroads. No technological determinism will light the way. The present industrial divide is first and formost a political divide. The search for social compromise, around ecological constraints, macroeconomic consistency, gender and ethnic equality, all mediated by the nature and degree of political mobilization will decide the outcome.

In the field of industrial relations, external flexibility will compete with negotiated involvement. In the field of industrial organization, oppressive worldwide subcontracting will compete with dense, territorially based partnership. The macroeconomics of the future may be based on a downward spiral of social and ecological competition, leading to recurrent financial, business and environmental crises, or an ecologically sustainable and macroeconomically stable model. As usual, reality will consist of a mix, with a core–periphery structure along geographical, ethnic, and gender lines. Radical economists and geographers may be part of finding the better pathway, both by identifying the possibilities for prosperity and by criticizing unrealistic optimism for flexibility as a panacea.

NOTES AND BIBLIOGRAPHIES

2 FORDIST AND POST-FORDIST INTERNATIONAL DIVISION OF LABOR AND MONETARY REGIMES

Notes

1 It is not possible here to differentiate exactly between both notions; cf. Hübner 1989; Lipietz 1986.
2 I discussed different types of crises and of crisis-theory in Altvater 1981; see also Hübner (1989).
3 Although there were additional inflationary impulses, such as the expenses for waging the Vietnam war.

Bibliography

Aglietta, M. 1979. *A Theory of Capitalist Regulation. The US Experience*, London: New Left Books.

Altvater, E. 1981. "La crisi del 1929 e il dibattito marxista sulla teoria della crisi." in *Storia del Marxismo, Volume terzo Il marxismo nell'etá della Terza Internazionale*, II, edited by Eric Hobsbawm, Georges Haupt, Franz Marek, Ernesto Ragionieri, Vittorio Strada, Corrado Vivanti. Torino: Einaudi.

——. 1983a. "Der Kapitalismus in einer Formkrise." In *Argument Sonderband* AS 100, Berlin (West): Das Argument-Verlag.

——. 1983b. "Der Kapitalismus vor einem Aufschwung?" In *Wirtschaft und Gesellschaft*, 8. Jahrgang, Nr. 2/82, Festschrift für Theodor Prager und Phillip Rieger: Wien.

——. 1987. *Sachzwang Weltmarkt. Verschuldungskrise, blockierte Industrialisierung, ökologische Gefährdung – der Fall Brasilien*, Hamburg: VSA-Verlag.

Altvater, E., J. Hoffmann, W. Semmer. 1979. *Vom Wirtschaftswunder zur Wirtschaftskrise.– Ökonomie und Politik in der Bundesrepublik*, Berlin (West): Olle und Wolter.

Altvater, E., K. Hübner. 1989. "The End of the U.S. American Empire? Monetary and Real Aspects of the United States' Hegemonial Crisis." In *Political Regulation in the "Great Crisis"*, edited by Werner Väth. Berlin: Edition Sigma.

Amin, S. 1975. *Accumulation on a World Scale – A Critique of the Theory of Underdevelopment*, New York and London: Monthly Review Press.

Armstrong, P., A. Glyn, J. Harrison. 1984. *Capitalism since World War II*, London: Fontana Paperbacks.

Atkinson, P., Chouraqui, J.-C. 1985. "The Origins of High Real Interest Rates." *OECD Economic Studies* 5. Autumn: 7–55.

Balassa, B. 1984. "Trade between Developed and Developing Countries: The Decade Ahead." *OECD Economic Studies* 3, Autumn: 7–25.

Beaud, M. 1987. *Le système national mondial hiérarchisé*, Paris: Agalma La Découverte.

Boyer, R., J. Mistral. 1983. "Le Temps Present. La crise (II), Pesanteur et potentialité des années quatre-vingt." *Annales (Economies, Société, Civilisation)* Juillet-Août: 773–89.

Braverman, H. 1974. *Labor and Monopoly Capital. The Degradation of Work in the Twentieth Century*, New York and London: Monthly Review Press.

Chan-Lee, J. H. 1986. "Pure Profits and Tobin's q in Nine OECD Countries." *OECD Economic Studies* 7. Autumn: 205–32.

Chan-Lee, J. H., H. Sutch. 1985. "Profits and Rates of Return." *OECD Economic Studies* 5. Autumn: 127–67.

Chenery, H. B., A. M. Strout. 1966. "Foreign Assistance and Economic Development." *American Economic Review* 56.

Cooper, R. N. 1987. *The International Monetary System. Essays in World Economics*. (second printing), Cambridge, Mass.: MIT Press.

Cuomo Commission Report. 1988. *A New American Formula for a Strong Economy*, New York, London, Toronto, Sydney, Tokyo: Touchstone.

Czempiel, E.-O. 1989. "Die Reprivatisierung der Weltwirtschaft." *Vereinte Nationen* Oktober: 149–55.

Dobb, M. 1966. *Organisierter Kapitalismus*, Frankfurt am Main: Suhrkamp.

Economic Report of the President, transmitted to the Congress. 1988, 1989, 1990. Washington, DC: Government Printing Office.

Fleming. 1962. "Domestic Financial Policies under Fixed and Floating Exchange Rates." International Monetary Fund, Staff Papers, 9, Washington, DC.

Foster, J. B. 1988. "Fordism as a Fetish." *Monthly Review*. March.

Glyn, A., B. Sutcliff. 1971. *British Capitalism, Workers, and the Profits Squeeze*, Harmondsworth: Penguin.

Gordon, D. 1988. "The Global Economy: New Edifice or Crumbling Foundations?" *New Left Review*, 168: 22–64.

Gramsci, A. 1967. *Philosophie der Praxis*. Frankfurt: Fischer.

Herr, H. J. 1989. "Weltgeld und die Instabilität der 70er und 80er Jahre". In *Internationale Geldwirtschaft*, edited by Hajo Riese, Heinz-Peter Spahn. Regensburg: Transfer: 106–54.

Hilferding, R. 1910, repr. 1968. *Das Finanzkapital. Eine Studie über die jüngste Entwicklung des Kapitalismus*. Frankfurt und Wien: Europäische Verlagsanstalt.

Hübner, K. 1989. *Theorie der Regulation. Eine kritische Rekonstruktion eines neuen Ansatzes der Politischen Ökonomie*, Berlin: Edition Sigma.

Jouvenel de. 1933. *La crise du capitalisme americain*, Paris: Gallimard.

Kennedy, P. 1987. *The Rise and Fall of the Great Powers*, New York: Random House.

Keohane, R. O. 1984. *After Hegemony, Cooperation and Discord in the World Political Economy*, Princeton: Princeton University Press.

Kern, H., M. Schumann. 1984. *Das Ende der Arbeitsteilung? Rationalisierung in der industriellen Produktion*, München: Beck.

Keynes, J. M. 1936. *The General Theory of Employment, Interest and Money*, London: Macmillan.

Kindleberger, C. P. 1986. *The World in Depression 1929–1939* (Revised and Enlarged Edition), Berkeley, Los Angeles, London: University of California Press.

Kindleberger, C. P. 1981. "Debt Situation of the Developing Countries in Historical Perspective (1800–1945)", Schweizerisches Institut für Außenwirtschafts-, Struktur- und Marktforschung (Hrsg): Längerfristige Aspekte der internationalen Verschuldung von Entwicklungsländern, Zürich: 372–80.

Krugman, P. 1987. "Prospects for International Debt Reform." *International Monetary and Financial Issues for Developing Countries*, New York: UNCTAD: 259–310.

Lipietz, A. 1986. *Mirages and Miracles*, London: Verso.

Lutz, B. 1989. "Normality, Crisis, or Stagnation. Reflections on the Current State of Capitalist Economies". In *Political Regulation in the "Great Crisis"*, edited by Werner Väth. Berlin: Edition Sigma: 13–24.

MacEwan, A. 1984. "Interdependence and Stability: Do the Levels of Output in the Advanced Capitalist Countries Increasingly Move Up and Down Together." *Review of Radical Political Economies*, 16: 57–79.

Maddison, A. 1987. "Growth and Slowdown in Advanced Capitalist Economies: Techniques of Quantitative Assessment." *Journal of Economic Literature, Vol. XXV* June: 649–98.

——. 1989. *The World Economy in the 20th Century*, Paris: OECD.

Mandel, E. 1968. *Die EWG und die Konkurrenz Europa – Amerika*, Frankfurt am Main: Europäische Verlagsanstalt.

Minsky, H. 1982. "The Financial Instability Hypothesis: Capitalist Processes and the Behavior of the Economy." In *Financial Crisis*, edited by Charles P. Kindleberger, Jean-Pierre Laffarge. Cambridge: Cambridge University Press: 13–39.

——. 1986. *Stabilizing an Unstable Economy*, New Haven and London: Yale University Press.

Modelsky, G. 1981. "Long Cycles, Kondratieffs, and Alternating Innovations: Implications for U.S. Foreign Policy." In *The Political Economy of Foreign Policy Behavior*, edited by C. W. Kegley, Jr., P. McGowan. Beverly Hills: Sage.

Mundell, R. A. 1968. *International Economics*, New York and London: Macmillan.

——. 1971. *The Dollar and the Policy Mix*, Princeton: Princeton University Press.

Parboni, R. 1986. "The Dollar Weapon: From Nixon to Reagan." *New Left Review*, 158: 5–18.

Report on Leveraged Buyouts and the Pot of Gold: Trends, Public Policy, and Case Studies. 1987, 1989. Prepared by the Economic Division of the Congressional Research Service for the use of the Subcommittee on Oversight and Investigations of the Committee on Energy and Commerce, U.S. House of Representatives, Washington D.C.

Sachverständigenrat zur Begutachtung der gesamtwirtschaftlichen Entwicklung, Jahresgutachten 1989/90: Bundestagsdrucksache.

Scharpf, F. W. 1987. *Sozialdemokratische Krisenpolitik in Europa*, Frankfurt, New York: Campus.

Scherrer, C. 1990. "Die Produktivitätsschwäche der US-Industrie. Eine regulationstheoretisch geleitete Ursachenanalyse am Beispiel der Auto- und Stahlindustrie." Mimeo, Freie Universität Berlin Department of Political Science.

Scott, B. R., G. C. Lodge (eds). 1985. *U.S. Competitiveness in the World Economy*, Boston: Harvard Business School Press.

Servan-Schreiber. 1967. *Le défi américain*, Paris: Denoel.

Stanger, M. 1988. *Krisentendenzen der Kapitalakkumulation. Theoretische Kontroversen und empirische Befunde*, Berlin: Edition Cassa Blanca.

Strange, S. 1986. *Casino Capitalism*, Oxford: Basil Blackwell.

——. 1988. *States and Markets. An Introduction to International Political Economy*, London: Pinter Publishers.

Thompson, W. R. 1983. "The World Economy, the Long Cycle, and the Question of World-System Time." In *Foreign Policy and the Modern World System*, edited by P. McGowan, P. Kegley. Beverly Hills: Sage.

Tobin, J. 1969. "A General Equilibrium Approach to Monetary Theory." *Journal of Money, Credit and Banking* Feb.: 15–29.

U.S. Department of Commerce. 1990 *U.S. Industrial Outlook Prospects for over 350 Manufacturing and Services Industries*, Washington,: U.S. Department of Commerce.

Wallerstein, I. 1979. *The Capitalist World Economy*, Cambridge: Cambridge University Press.

Williamson, J. 1988. "The Case for International Monetary Reform." In *The International Monetary System and Economic Development. A Challenge to the International Economic Cooperation*, edited by Christian Dräger and Lothar Späth. Baden-Baden: Nomos Verlagsgesellschaft.

Wood, R. 1986. *From Marshall Plan to Debt Crisis. Foreign Aid and Development Choices in the World Economy*, Berkeley, Los Angeles, London: University of California Press.

World Bank. 1985. *World Development Report*, Washington: World Bank.

3 FORDISM AND POST-FORDISM: A CRITICAL REFORMULATION

Notes

1 This chapter has gained from discussions with Simon Bromley, Simon Clarke, Benjamin Coriat, Richard Hyman, Klaus Nielsen, Ove Kai Pedersen, and Dick Walker. The reading on which it draws was mostly completed whilst I held an ESRC personal research grant; I completed the chapter at the Center for Organisation and Management (COS) in Copenhagen with support from the Danish Social Science Research Council. Thanks to all three bodies for their support.

2 Perez treats Fordist mass production and continuous process industries as the *carrier* industries of the fourth Kondratieff; its *motive* branches were those involved in producing the low cost oil and energy-intensive materials used in Fordist production; it *induced* growth in complementary sectors (Perez 1983: 362–3, 369).

3 It is because these micro- and macroeconomic sources of flexibility have been overwhelmed by its general crisis that Fordism now seems so rigid. Among the micro-sources are layoffs, short-time and overtime working, outsourcing, variations in stock levels, switching between home and foreign markets; the most important of the macro-sources are contracyclical demand management policies and the stabilizing effects of welfare payments.

4 The technical composition of capital refers to the ratio of fixed to circulating capital – more specifically to the relation among means of production, material inputs and work in progress, and labor power.

5 It is symptomatic that Ford's "five dollar day" was introduced to stem massive turnover and absenteeism rates among his assembly-line workers.

6 Galbraith discussed this in *The New Industrial State* (1967); this work anticipated many later studies of the Fordist regime.

7 "Integral economy" is analogous to Gramsci's concept of the "integral state,"

i.e., the state in its inclusive sense. This was defined as "state = political society + civil society." Thus we could define an integral economy as "regime of accumulation + mode of regulation." See also Jessop (1990b).

8 On how discourses and strategies of flexibility are "rewording" the world of work in Sweden, see Elam and Börjeson (1990).

9 Richard Hyman expressed this worry to me and it is addressed below.

10 There are already many other critiques of the concept but few which first draw the necessary distinctions among possible levels of Fordism and/or between its structural and strategic aspects. For some representative criticisms from different perspectives, see Foster (1988); Hirst and Zeitlin (1990); Meegan (1988); Sayer (1989); Williams et al. (1987).

11 Kaplinsky defends this focus on the car industry on the grounds that: "first, it is the largest single industrial employer in the world economy and, second, since the first decades of the twentieth century it has set the pattern for the development of the dominant form of labour process which has subsequently diffused to other manufacturing sectors" (Kaplinsky 1988: 452).

12 Lipietz has suggested an even more general definition of Fordism as involving "essentially Taylorization plus mechanization" (Lipietz 1987).

13 Some difficulties in the Fordist literature stem from failures to meet these criteria; many criticisms leveled against this literature are rooted in an unnecessary conflation of different possible definitions.

14 Such arguments clearly require changes in the definition of Fordism as an accumulation regime – albeit different changes in each case.

15 Much of Boyer's work is inspired by Kaldorian models: for example, Boyer and Petit 1981; or Boyer and Ralle 1987. Other Parisian regulationists also use models that are similar to, based on, or develop Kaldor-Verdoon arguments.

16 Thus Mike Davis criticized Aglietta's pioneering work on US Fordism for ignoring the "1.6 *trillion* dollars devoted to the permanent arms economy since 1946" (cited in Foster 1988: 28).

17 Several parallels can be drawn between the US New Deal and the Nazi New Order as forms of *internal welfare state*: with the *Volksempfaenger* (radio), *Volkswagen*, and *Autobahn* serving equivalent functions to mass household consumer durables, the model T, and the US highway program (Roobeek 1987: 134).

18 Siegel explains the neglect of fascism in terms of "a procedure that reduces to a mere preparatory stage about two-thirds of the sixty years since the inception of the Fordist era in the 1920s" (1988: 4).

19 This argument was put by Haug (1988); he also noted how some post-Fordist theorizing also referred to continuities, reinforced trends, and so on – thus making it ambiguous what was involved in post-Fordism.

20 Once again the Japanese model poses problems here: if the post-Fordist future was pioneered by Japan and if Japan lacked a Fordist past, what does the concept of post-Fordism signify when applied to Japan?

21 This presupposes agreement on the crisis of Fordism – whether as labor process, accumulation regime, mode of regulation, or mode of societalization.

22 Many early comments on Fordism in crisis foresaw only a movement from Fordism to neo-Fordism: according to Kenney and Florida, this involves "the use of new technologies to generate economic renewal within the general context of fordist institutions and social relationships" (Kenney and Florida 1988: 147n). Typical examples are Aglietta's pioneering study of American Fordism (1979) and Palloix's comments on the neo-Fordist labor process (1976).

23 This seems to be Kaplinsky's line when he writes: "there is now widespread

recognition that there are alternatives to the Fordist system, which are also widely recognized to be more productive." One reason that JIT developed as an alternative was the small scale of the early Japanese market for cars and the need to keep investory costs down (1988: 455).

24 Reverting to the labor process in the automobile industry, for example, we may note that one study suggests that the British industry is moving towards the German model whereas the American industry is moving towards the Japanese (Juergens, Malsch, and Dohse 1988: 12).

25 Even flexible manufacturing systems are too costly for most small firms to install and require user firms to push for higher volumes and more product variety than markets might bear (cf. Sayer 1989: 673, citing Schoenberger 1987).

26 The notion of "Spaethkapitalismus" involves a pun on the German concept of "Spaetkapitalismus" (or "late capitalism"). It refers to Lothar Spaeth, the modernizing Christian Democratic Minister-President of Baden-Würtemburg and a leading advocate of a post-Fordist future for Germany.

Bibliography

Aglietta, M. 1979. *A Theory of Capitalist Regulation: the U.S. Experience*, London: New Left Books.

Aglietta, M., and A. Brender. 1984. *Les metamorphoses de la société salariale: La France en projet*, Paris: Calmann-Levy.

Amin, A., and K. Robins. 1990. "Not Marshallian Times." University of Newcastle upon Tyne: Centre for Urban and Regional Studies, unpublished paper.

Badham, R., and J. Mathews. 1989. "The New Production Systems Debate." *Labour and Industry*, 2 (2): 194–246.

Beniger, R. 1986. *The Control Revolution: Technological and Economic Origins of the Information Society*, Cambridge, MA: Harvard University Press.

Boyer, R., and B. Coriat, 1986. "Technical Flexibility and Macro Stabilisation: Preliminary Insights." Paper prepared for the Conference on Innovation Diffusion, Venice, 17–21 March 1986.

Boyer, R., and P. Petit. 1981. "Progrès technique, croissance et emploi: un modele d'inspiration kaldorienne pour six industries européennes." *Révue Economique* 32 (6): 1113–53.

Boyer, R. and P. Ralle, 1987. "Rupture de la croissance et formes d'emploi." In *Aspects de la crise: tome I – Présentation générale et première partie*, edited by Christine Andre, Robert Boyer, Robert Delorme, Daniele Leborgne, and Pascal Petit. Paris: Centre d'Etudes Prospectives d'Economie Mathématique Appliquées à la Planification.

Bromley, S. 1988. "From Fordism to Post-Fordism: Old Wine in New Bottles?" Sheffield: unpublished typescript.

Bromley, S. 1990. *World Oil and United States Hegemony*, Cambridge: Polity.

Clark, J. 1989. *Sustainable Corporate Growth: A Model and Management Planning Tool*, New York: Quorum Books.

Coriat, B. 1990. *L'atelier et le robot*, Paris: Christian Bourgeois.

Elam, M. 1989. "Puzzling Out the Post-Fordist Debate", *Industrial and Economic Democracy*, 11 (i).

Elam, M. and M. Borjeson. 1990. "Strategies of Flexibility or Rewording the world of Work in Sweden?," unpublished paper, Department of Technology and Social Change, Linkoping University, Sweden, February 1990.

Esser, J. and J. Hirsche. 1989. "The Crisis of Fordism and the Dimensions of a

'PostFordist' Regional and Urban Structure." *International Journal of Urban and Regional Research* 13 (3): 417–37.

Florida, R., and M. Feldman. 1987. "Housing in U.S. Fordism: the Class Accord and Postwar Spatial Organization," *Working Paper 87–33*, School of Urban and Public Affairs, Carnegie Mellon University, Pittsburgh.

Foster, J. B. 1988. "The Fetish of Fordism." *Monthly Review* 39 (1): 14–33.

Franko, L. G. 1974. "The Move Toward a Multidivisional Structure in European Organisation." *Administrative Science Quarterly* 19 (4): 493–506.

Galbraith, J. K. 1967. *The New Industrial State*, Harmondsworth: Penguin.

Gramsci, A. 1971. *Selections from the Prison Notebooks.* London: Lawrence & Wishart.

Grou, P. 1987. *The Financial Structure of Multinational Corporations*, Leamington Spa: Berg.

Harvey, D. 1989. *The Condition of Postmodernity: an Enquiry into the origins of Cultural Change*, Oxford: Blackwell.

Haug, W. F. 1986. *Die Faschisierung des buergerlichen Subjekts*, West Berlin: Das Argument Verlag.

Hirsch, J. and R. Roth. 1986. *Das neue Gesicht des Kapitalismus*, Frankfurt: VSA.

Hirst, P. Q., and J. Zeitlin. 1990. "Flexible Specialization vs. Post-Fordism: Theory, Evidence and Policy Implications." See Chapter 4 of this volume.

Jenson, J. 1990. "Learning to be Fordist: Conflict and Consensus in Postwar France," paper prepared for the Seventh Conference of Europeanists, Washington D.C., 23–25 March 1990.

Jessop, B. 1986. "Der Wohlfahrtsstaat im Uebergang vom Fordismus zum Post-Fordismus." *Prokla 65*: 4–30.

——. 1990a. "Regulation Theories in Retrospect and Prospect." *Economy and Society* 19 (2): 153–216.

——. 1990b. *State Theory: Putting Capitalist States in their Place*, Cambridge: Polity Press.

Juergens, U., T. Malsch, and K. Dohse. 1988. "Moderne Zeit beim Autobau: Entwicklungstrends und Gestaltungsoptionen." *WZB-Mitteilungen 42*,Wissenschaftszentrum Berlin, December: 12–14.

Kamppeter, K. 1986. "Fordismus in der Landwirtschaft. Technologische und gesellschaftliche Polarisierung als Kennzeichen gegenwaertiger Agrarentwicklung in der BRD und in Mexiko." *Peripherie* 22/23: 112–36.

Kaplinsky, R. 1988. "Restructuring the capitalist labour process: some lessons from the car industry." *Cambridge Journal of Economics*: 12 (4): 451–70.

Kenney, M., J. Curry, and W. R. Goe. 1988. "Contextualizing Agriculture within Postwar US Society: Fordism as an Integrative Theory," paper for the International Conference on Regulation Theory, Barcelona, 16–18 June 1988.

Kenney, M. and R. Florida. 1988. "Beyond Mass Production: Production and the Labor Process in Japan." *Politics and Society*, 16 (1): 121–58.

Kristensen, P. H. 1990. "Denmark's Concealed Production Culture, its Socio-Historical Construction and Dynamics at Work." In *Technological Innovation and Organizational Change – Danish Patterns of Knowledge, Networks, and Culture*, edited by F. Borum and P. H. Kristensen. Copenhagen: Copenhagen Business School New Social Science Monographs.

Kundig, B. 1984. "Du Taylorisme classique à la 'flexibilisation' du système productif. L'impact macro-économique des differents types d'organisation du travail industrial." *Critique de l'économie politique*: 26/27.

Lipietz, A. 1982. "Towards Global Fordism?" *New Left Review*, 132: 33–47.

——. 1985. "Akkumulation, Krisen und Auswege aus der Krise: Einige methodische Ueblerlegungen zum Begriff 'Regulation'". *Prokla* 58: 109–37.

——. 1987. "La régulation: les mots et les choses." *Revue Economique* 38 (5): 1049–59.

Luescher, R. M. 1986. "Henry und die Kruemelmonster." *Widerspruch* ii.

Maier, C. S. 1970. "Between Taylorism and Technocracy: European Ideologies and the Vision of Productivity in the 1920s." *Journal of Contemporary History* 5 (2): 27–61.

Meegan, R. 1988. "A Crisis of Mass Production?" In *The Economy in Question*, edited by John Allen and Doreen Massey. London: Sage.

Palloix, C. 1976. "Neo-Fordism and the Labour Process." In *Capital and the Labour Process*, edited by the Conference of Socialist Economists, London: CSE Books.

Perez, C. 1983. "Structural Change and Assimilation of new Technologies in the Economic and Social Systems." *Futures* 15 (5): 357–75.

Roobeek, A. J. M. 1987. "The Crisis in Fordism and the Rise of a New Technological Paradigm." *Futures*, 19 (2): 129–54.

Sayer, A. 1989. "Postfordism in Question." *International Journal of Urban and Regional Research* 13 (4): 666–93.

Schonberger, R. J. 1987. "Frugal Manufacturing." *Harvard Business Review* 87, September–October: 95–100.

Sparke, P. 1986. *An Introduction to Design and Culture in the Twentieth Century*, New York: Harper and Row.

Siegel, T. 1988. "Introduction (to Fordism and Fascism)." *International Journal of Political Economy* 18 (1): 2–9.

Storper, M. and A. J. Scott. 1988. "The Geographical Foundations and Social Regulation of Flexible Production Complexes." In *The Power of Geography*, edited by John Wolch and Michael Dear. London: Allen and Unwin. Cited from mimeo.

Streeck, W. 1987. "Neue Formen der Arbeitsorganisation im internationalen Vergleich." *Wirtschaft und Gesellschaft* 13 (4): 317–35.

Tedlow, R. S. 1990. *New and Improved: the story of mass marketing in America*, London: Heinemann.

Tolliday, S. and J. Zeitlin, (eds). 1987. *Between Fordism and Flexibility: The Automobile Industry and Its Workers*, Cambridge: Polity.

van der Pijl, K. 1984. *The Making of the Atlantic Ruling Class*, London: Verso.

Williams, K., T. Cutler, J. Williams, and C. Haslam. 1987. "The End of Mass Production?" *Economy and Society* 16 (4): 404–39.

Williams, K., C. Haslam, A. Wardlow, and J. Williams. 1986. "Accounting for Failure in the Nationalized Enterprises – Coal, Steel, and Cars since 1970." *Economy and Society* 15 (2): 167–219.

Wolf, F. O. 1987. "Illusionaerer Fordismus." *Das Argument* 162: 220–9.

4 FLEXIBLE SPECIALIZATION VERSUS POST-FORDISM: THEORY, EVIDENCE, AND POLICY IMPLICATIONS

Notes

1 This exposition is based primarily on Piore and Sabel (1984), Sabel and Zeitlin (1985), Sabel (1989a), and Sabel (1990).

2 It will be evident that this definition of flexible specialization as a form of craft production entails a revaluation of the conventional stereotype of the latter as the manufacture of luxury goods in tiny volumes using hand tools and obsolete

methods. The historical basis for a more positive interpretation of the technological dynamism of craft production is presented in Sabel and Zeitlin 1985. For a related approach which seeks to distinguish between craft production and "diversified quality production" on the basis of the volumes involved, see Streeck (1987) and Sorge and Streeck (1988).

3 As Piore and Sabel acknowledge (1984: 4–5), their notion of regulatory requirements of the technological paradigm is borrowed from the French regulation school, but used in very different ways.

4 For fuller discussions of the reorgnization of large corporations and the process of "double convergence," see Sabel (1989a), Sabel *et al.* (1989), and Sabel (1990). This process implies, as Sabel (1989b) observes, not only that corporate operating units are coming to resemble the constituent elements of the industrial districts, but also that there are increasing numbers of exchanges and alliances between large-firm subsidiaries and their small-firm counterparts in the districts themselves.

5 For an extended account of these ideas, see Sabel (1989a: 53–9) and Sabel (1989b); and for the potential role of a reorganized labor movement in the transformation of the welfare state, see Kern and Sabel (1990) and Sabel (1990).

6 For a fuller account of flexible specialization as a development strategy, see Sabel (1986); and for a thoughtful critical discussion, see Schmitz (1989). For an extended attempt to apply this approach to the problems of a small semi-developed economy on the European periphery, see Murray (1987).

7 Cf. Williams *et al.* (1987); Pollert (1988); Wood (1989). For an insightful discussion of the critical debates surrounding flexible specialization, see Badham and Mathews (1989).

8 Piore and Sabel (1984: 26–8, 219, 279–80); Sabel and Zeitlin (1985: 137–8); Sabel (1989: 40).

9 These arguments draw on the work of the international working group on "Historical Alternatives to Mass Production" sponsored by the Maison des Sciences de l'Homme in Paris. See Sabel and Zeitlin, forthcoming; and for a discussion of contemporary problems of technological hybridization, Sabel (1990).

10 Cf., for example, Elam (1990). Thus flexible specialization is not, among other things, an optimistic general theory of the labor process which can be counterposed to Braverman's deskilling thesis: cf. Thompson (1989: 218–29); Wood (1989).

11 Piore and Sabel (1984: 261–3); Sabel (1989a: 37–40); Sabel *et al.* (1990); Sabel (1990).

12 Cf. especially Pollert (1988).

13 Even in the automobile industry, for example, national responses to the Fordist model varied considerably: see Tolliday and Zeitlin (1986).

14 For different national responses to "Keynesianism," see Weir and Skocpol (1985) and Hall (1989).

15 See Piore and Sabel (1984: ch. 6).

16 For a review of the evidence, see Hirst and Zeitlin (1989b).

17 A variant of the "post-Fordist" argument which gives great emphasis to the concept of "post-modernity" is David Harvey's *The Condition of Postmodernity* (1989). We have not considered this text in detail since it adds little to the arguments found in *Marxism Today*.

18 See the discussion in Hirst (1989: ch. 1).

19 See in particular Freeman and Perez (1988); Perez (1983, 1985); and Freeman *et al.* (1982).

20 For a valuable overview of the many variants of the regulation approach, see Jessop (1990). In our account of the Parisian regulationists' ideas, we have concentrated on the recent work of Robert Boyer, Alain Lipietz and their collaborators. Key texts we have used include: Boyer (1979, 1986a, 1986b, 1988a, 1988b, 1988c, 1988d, 1989); Lipietz (1987, 1989); Leborgne and Lipietz (1988).
21 See especially Boyer (1986a, 1986b: ch. 1, (1989).
22 See particularly Boyer (1986, 1988a); Lipietz and Leborgne (1988); and Jessop (1990: 42–3). Lipietz also identifies the technological paradigm as a third constituent of the mode of development, defined as "the general principles which govern the evolution of the division of labor" across the economy as a whole. But Lipietz does not specify how this relates to the overall framework of the regulationist analysis, and it remains unclear whether the concept is used in the manner of flexible specialization or of Freeman and Perez: see Leborgne and Lipietz (1988: 264–5).
23 For detailed discussions of the nature of possible post-Fordist modes of development, see Boyer (1988b, 1988c: chs. 11–12, and 1988d); and Leborgne and Lipietz (1988).
24 On the regulationists' attempts to avoid teleology and functionalism, see Lipietz (1987: 16); Boyer (1986a: 59, 95); and Jessop (1990: 67–75).
25 See Jessop (1990: 49–50, 67–9).
26 See particularly the general formulations of their approach in Boyer (1986a); Lipietz (1987: ch. 2); and Lipietz (1989: ch. 2).
27 Compare, for example, the nuanced discussion of the development of Taylorism in France in Boyer (1984) with the balder formulation in Boyer (1988a: 82).
28 Boyer (1986b, 1988c); Lipietz (1987).
29 Boyer (1988c: 230–2, 270–1, 1988d: 404–6); Leborgne and Lipietz (1988: 267–8).
30 Cf., for example, Boyer (1988d); Lipietz (1989).
31 For a fuller discussion of these issues, see Hirst (1985 and 1990).
32 See, for example, the theoretical arguments in Sabel (1982).
33 See Sabel and Zeitlin (1985 and forthcoming).
34 Good examples of such normative–empirical advocacy using flexible specialization concepts are Mathews (1989a and 1989b).
35 Lipietz (1987: 9–10). William of Baskerville is a character in Eco's novel *The Name of the Rose*.
36 See Whitehead (1926).
37 For an argument that tries to show that conceptual realism is central to Marx's epistemology in *Capital* and that this position is theoretically unsustainable, see Cutler *et al.* (1977).
38 For a penetrating account of the theoretical objections to treating classes as social actors, see Hindess (1987).
39 See Boyer (1988b: 623–7; 1988c: 229–32; 1988d: 404–15).
40 Boyer (1978: 107–8, 111).
41 For British studies highlighting the differences between wage rates and earnings for the period before 1939, see McClelland and Reid (1985: 157); and Gillespie and Whiteside (1989).
42 Aglietta (1976); Boyer (1979, 1986a: 65, 1986b: 227); Boyer and Mistral (1982). On the basis of similar calculations, however, Mazier *et al.* (1984) argue that the transition to intensive accumulation in France did not occur until after 1945.
43 See, for example, Fridenson (1987) and Montmollin and Pastr (1984: pt. I). Even for the United States, the impact of Taylorism on work organization and labor management can easily be overstated: see Nelson (1975 and 1980). The same methodological problems apply to the otherwise extremely stimulating study of

historical changes in the nature of French unemployment by Salais *et al.* (1986), which uses principal component analysis rather than more conventional econometric tests. See Salais *et al.* (1986: 85–94) and Salais (1989: 274–6); and compare the much more cautious and historically sensitive discussion by Benedicte Reynaud in Salais *et al.* (1986: 132–7).

44 The classic statement is Friedman (1953).

45 For an ingenious but ultimately unsatisfying attempt to use existing data to test the flexible specialization hypothesis for the US manufacturing, see Luria (1989). Thus, for example, Luria uses share of value-added in manufacturing output (VA/M) as an indicator of product batch-sizes, on the assumption that VA/M rises proportionately as batch sizes fall, and explores its relationship with labor productivity (value-added per employee) for SIC industries over the past two decades. But there may be many reasons for an industry to be characterized by a high VA/M ratio besides small-batch production, while the use of this indicator also depends on the absence of any significant shift in the relative productivity of small and large-batch production, the very question to be examined.

46 See, *inter alia*, Sorge *et al.* (1983); Jones (1982 and 1989); Maurice *et al.* (1986); and Adler (1989).

47 For a selection of case studies written from a flexible specialization perspective, see Best (1989); Brusco and Sabel (1981); Friedman (1988); Herrigel (1989); Hirst and Zeitlin (1989b); Katz and Sabel (1985); Lorenz (1989); Lyberaki (1988); Michelsons (1987, 1989); Piore and Sabel (1983); Regini and Sabel (1989); Sabel (1984); Sabel *et al.* (1989a, 1989b); Sabel and Zeitlin (forthcoming); Storper (1989); Tolliday and Zeitlin (1987); Zeitlin and Totterdill (1989). The major statements of the flexible specialization approach such as Piore and Sabel (1984); Sabel and Zeitlin (1985); and Sabel (1989), are all built up from a comparative analysis of such case studies.

48 See Michelsons (1987); Storper and Christopherson (1987); Christopherson and Storper (1989); Salais and Storper (1990).

49 For a discussion of current debates about the interpretation of texts in the history of political ideas, see Tully (1988).

50 Compare, for example, Piore and Sabel (1984); Tolliday and Zeitlin (1986); Friedman (1988) and Sabel (1989) with Williams *et al.* (1987), Wood (1988), Sayer (1986 and 1989), Kenney and Florida (1988 and 1989).

51 For a major case study, see Cusumano (1985).

52 See Tolliday and Zeitlin (1986); Sabel (1989: 37–9); Regini and Sabel (1989: 33–44); Sabel *et al.* (1989b); and Sabel (1990). But cf. also Sayer (1989: 685–9); Kenney and Florida (1988 and 1989).

53 On Emilia-Romagna and the "Third Italy" more broadly, see, in English, Brusco (1982); Sabel (1982: 220–6); Piore and Sabel (1983); Zeitlin (1989b); Goodman *et al.* (1989); Pyke *et al.* (1990). Contrasting interpretations can be found in Murray (1987) and Amin (1989). On Baden-Württemburg, see Sabel *et al.* (1989) and Herrigel (1989). For a more general debate on the interpretation of regional case studies in relation to the flexible specialization hypothesis, see Amin and Robins (1990) and the responses by Michael Piore, Charles Sabel and Michael Storper in Pyke *et al.* (1990: ch. 12).

54 See Hirst and Zeitlin (1989a and 1989b); Lane (1988). Attempts to use predominantly British evidence to criticize the flexible specialization hypothesis therefore badly miss the mark: see, for example, Pollert (1988).

55 For a survey of debates on industrial policy primarily in this narrow sense, see Thompson (1987).

56 For Boyer's response to criticisms of regulation theory for its lack of a coherent policy program, see Boyer (1986a: 105–9).
57 For criticism of Keynesianism and state planning strategies, including the use of MITI as a model for "picking winners," see Hirst and Zeitlin 1989: Introduction. For an example of advocacy that Britain adopt a MITI-style strategy, see Smith (1984).
58 For a characterization of the British Conservative's approach to macroeconomic policy between 1982 and 1988 as "electoral Keynesianism," see Hirst (1989: ch. 4).
59 On MITI, beyond the classic account of Johnson (1982), see Friedman (1988) and Okimoto (1989).
60 For a useful discussion of the definitional problems of the category "small firms" in different national contexts, see Sengenberger and Loveman (1988).
61 For an account of the failure of British industrial policy based on concentration into "national champions" in the 1960s and 1970s, see Williams *et al.* (1983).
62 For critical discussions of the likely impact of 1992 on economies of scale, see Geroski (1989) and Thompson (1990).
63 Prais (1976) argued persuasively that the concentration of firms in the UK in the period 1909–70 had exhausted any possible production economies of scale at *plant* level and, moreover, that UK firms often concentrated by bringing under common control constellations of less than efficient plants.
64 This argument for the failure of free-market economies like the UK to create "developmental states" is powerfully articulated by Marquand (1988).
65 This argument is developed more fully in Hirst and Zeitlin (1989b) and Hirst (1990).
66 This is, of course, the current position in the UK – see Campbell *et al.* (1989).
67 This argument is developed in Hirst (1989: ch. 6).
68 See Zeitlin and Totterdill (1989) and Best (1989).

Bibliography

Adler, P. and B. Borys. 1989. "Automation and Skill: Three Generations of Research on the NC Case." *Politics and Society* 17, 3: 353–76.

Aglietta, M. 1976. *Régulation et crises du capitalisme: l'expérience des États-Unis*, Paris: Calmann-Levy.

Amin, A. 1989. "Flexible Specialisation and Small Firms in Italy: Myths and Realities." *Antipode* 21, 1: 13–34.

Amin, A. and K. Robins. 1990. "Industrial Districts and Regional Development: Limits and Possibilities." In *Industrial Districts and Inter-Firm Co-operation in Italy. see* Pyke *et al.* (1990).

Badham, R. and J. Mathews. 1989. "The New Production Systems Debate." *Labour and Industry* 2, 2: 194–246.

Bagnasco, A. 1988. *La costruzione sociale del mercato*, Bologna: Il Mulino.

Best, M. 1989. "Sector Strategies and Industrial Policy: The Furniture Industry and the Greater London Enterprise Board." In Hirst and Zeitlin (1989a).

Boyer, R. 1979. "Wage Formation in Historical Perspective: The French Experience." *Cambridge Journal of Economics* 3, 2: 99–118.

——. 1984. "Le taylorisme hier: presentation." In Montmollin and Pastré (1984).

——. 1986a. *La théorie de la régulation: une analyse critique*, Paris: La Découverte.

——. (ed.). 1986b. *Capitalismes fin de siècle*, Paris: Presses Universitaires de France.

——. 1988a. "Technical Change and the Theory of 'Régulation'." In Dosi *et al.* (1988).

——. 1988b. "Formalizing Growth Regimes." In Dosi *et al.* (1988).
——. (ed.). 1988c. *The Search for Labour Market Flexibility: The European Economies in Transition*, Oxford: Clarendon.
——. 1988d. "Alla ricerca di alternative al fordismo: gli anni ottanta." *Stato e mercato* 24: 387–423.
——. 1989. "Économie et histoire: vers de nouvelles alliances?" *Annales ESC* 6: 1397–1426.
Boyer, R. and J. Mistral. 1978. *Accumulation, inflation crises*, Paris: Presses Universitaires de France.
Braverman, H. 1974. *Labor and Monopoly Capital*, New York: Monthly Review Press.
Brusco, S. 1982. "The Emilian Model: Productive Decentralization and Social Integration." *Cambridge Journal of Economics* 6, 2: 167–84.
Brusco, S. and C. Sabel. 1981. "Artisanal Production and Economic Growth." In *The Dynamics of Labour Market Segmentation*, edited by F. Wilkinson. London: Academic Press, 99–114.
Campbell, A, W. Currie, and M. Warner. 1989. "Innovation, Skills and Training: Micro-electronics and Manpower in the United Kingdom and West Germany." In Hirst and Zeitlin (1989a).
Christopherson, S. and M. Storper. 1989. "The Effects of Flexible Specialization on Industrial Politics and the Labor Market: The Motion Picture Industry." *Industrial and Labor Relations Review* 42, 3: 331–47.
Cusumano, M. 1985. *The Japanese Automobile Industry*, Cambridge, MA: Harvard University Press.
Cutler, A., P. Hirst, B. Hindess and A. Hussain. *Marx's "Capital" and Capitalism Today*, vol. I, London: Routledge and Kegan Paul.
Dore, R. 1986. *Flexible Rigidities: Industrial Policy and Structural Adjustment in Japan*, London: Athlone Press.
——. 1987. *Taking Japan Seriously: A Confucian Perspective on Leading Economic Issues*, London: Athlone Press.
Dosi, G., C. Freeman, R. Nelson, G. Silverberg and L. Soete (eds). 1988. *Technical Change and Economic Theory*, London: Pinter.
Elam, M. 1990. "Puzzling Out the Post-Fordist Debate: Technology, Markets and Institutions." Forthcoming in *Economic and Industrial Democracy*, 11, 1.
Freeman, C. 1987. *Technology Policy and Economic Performance: Lessons from Japan*, London: Frances Pinter.
Freeman, Christopher 1988. "Japan: A New National System of Innovation." In Dosi *et al.* (1988).
Freeman, C., J. Clark and L. Soete. 1982. *Unemployment and Technical Innovation: A Study of Long Waves in Economic Development*, London: Frances Pinter.
Freeman, C. and C. Perez. 1988. "Structural Crises of Adjustment: Business Cycles and Investment Behaviour." In Dosi *et al.* (1988).
Fridenson, P. 1987. "Un tournant taylorien de la société française (1904–1918)." *Annales ESC* 5: 1031–60.
Friedman, D. 1988. *The Misunderstood Miracle: Industrial Development and Political Change in Japan*, Ithaca: Cornell University Press.
Friedman, M. 1953. "The Methodology of Positive Economics." In *idem, Essays in Positive Economics*, Chicago: Chicago University Press.
Gillespie, J. and N. Whiteside. 1989. "Deconstructing Unemployment: British Developments during the Interwar Years." unpublished paper submitted to the *Economic History Review.*

Goodman, E. and J. Bamford with P. Saynor (eds). 1989. *Small Firms and Industrial Districts in Italy*, London: Routledge.

Gorz, A. 1985. *Paths to Paradise: On the Liberation From Work*, London: Pluto Press.

Gramsci, A. 1971. "Americanism and Fordism." In *Selections from the Prison Notebooks* (ed. and trans. Quintin Hoare and Geoffrey Nowell-Smith), London: Lawrence and Wishart.

Hall, P. A. (ed.). 1989. *The Political Power of Economic Ideas: Keynesianism Across Nations*, Princeton: Princeton University Press.

Harvey, D. 1989. *The Condition of Postmodernity*, Oxford: Basil Blackwell.

Herrigel, G. 1989. "Industrial Order and the Politics of Industrial Change: Mechanical Engineering." In *Industry and Political Change in West Germany: Towards the Third Republic*, edited by Peter Katzenstein. Ithaca: Cornell University Press.

Hindess, B. 1987. *Politics and Class Analysis*, Oxford: Blackwell.

Hirst, P. 1985. "Is It Rational to Reject Relativism?" In *Reason and Morality*, edited by J. Overing. ASA Monographs no. 24, London: Tavistock.

——. 1989. *After Thatcher*, London: Collins.

——. 1990a. "Democracy: Socialism's Best Reply to the Right?" In *Responses to the Right*, edited by Barry Hindess, London: Routledge.

——. 1990b. "An Answer to Relativism." *New Formations* Spring.

Hirst, P. and Zeitlin, J. (eds). 1989a. *Reversing Industrial Decline? Industrial Structure and Policies in Britain and her Competitors*, Oxford: Berg/New York: St Martins.

——. 1989b. "Flexible Specialization and the Competitive Failure of UK Manufacturing." *Political Quarterly* 60, 3: 164–78.

Hyman, R. and W. Streeck (eds). 1988. *New Technology and Industrial Relations*, Oxford: Basil Blackwell.

Jacques, M. and S. Hall (eds). 1989. *New Times*, London: Lawrence & Wishart.

Jessop, B. 1990. "Regulation Theories in Retrospect and Prospect." Forthcoming in *Economy and Society*, 19, 2.

Johnson, C. 1982. *MITI and the Japanese Miracle: The Growth of Industrial Policy, 1925–1975*, Palo Alto: Stanford University Press.

Jones, B. 1982. "Destruction or Redistribution of Engineering Skills? The Case of Numerical Control." In *The Degradation of Work?*, edited by Stephen Wood. London: Hutchinson.

——. 1989. "Flexible Automation and Factory Politics: Britain in Comparative Perspective." In Hirst and Zeitlin (1989a).

Katz, H. and C. Sabel 1985. "Industrial Relations and Industrial Adjustment in the Car Industry." *Industrial Relations* 24, 3: 295–315.

Kenney, M. and R. Florida. 1988. "Beyond Mass Production: Production and the Labor Process in Japan." *Politics and Society* 16, 1: 121–58.

——. 1989. "Japan's Role in a Post-Fordist Age." *Futures* 21, 2: 136–51.

Kern, H. and C. Sabel. 1990. "Trade Unions and Decentralized Production: A Sketch of Strategic Problems in the West German Labor Movement." Unpublished paper, January.

Kondratiev, N. 1979. "The Major Economic Cycles." *Review* 2, 4: 519–62.

Lane, C. 1988. "Industrial Change in Europe: The Pursuit of Flexible Specialisation in Britain and West Germany." *Work, Employment and Society* 2, 2: 141–68.

Lash, S. and J. Urry 1987. *The End of Organized Capitalism*, Cambridge: Polity.

Leborgne, D. and A. Lipietz. 1988. "New Technologies, New Modes of Regulation: Some Spatial Implications." *Society and Space* 6: 263–80.

Lipietz, A. 1987. *Mirages and Miracles: The Crises of Global Fordism*, London: New Left Review.
——. 1989. *Choisir L'Audace*, Paris: La Découverte.
Lorenz, E. 1989. "The Search for Flexibility: Subcontracting Networks in French and British Engineering." In Hirst and Zeitlin (1989a).
Luria, D. 1989. "Automation, Markets and Scale: Can 'Flexible Niching' Modernize American Manufacturing?" Unpublished paper submitted to the *International Review of Applied Economics.*
Lyberaki, A. 1988. "Small Firms and Flexible Specialisation in Greek Industry." D.Phil. thesis, University of Sussex.
McClelland, K. and A. Reid. 1985. "Wood, Iron and Steel: Technology, Labour and Trade Union Organisation in the Shipbuilding Industry, 1840–1914." In *Divisions of Labour*, edited by Royden Harrison and Jonathan Zeitlin. Brighton: Harvester.
Marquand, D. 1988. *The Unprincipled Society*, London: Collins.
Mathews, J. 1989a. *Tools of Change: New Technology and the Democratization of Work*, Sydney: Pluto Press.
——. 1989b. *Age of Democracy: The Politics of Post-Fordism*, Melbourne: Oxford University Press Australia.
Maurice, M., F. Eyraud, A. d'Iribarne, and F. Rychener. 1986. *Des enterprises en mutation dans la crise: Apprentissage de technologies flexibles et emergence de nouveaux acteurs*, Aix-en-Provence: Laboratoire d'Économie et de Sociologie du Travail.
Mazier, J., M. Basle, and J.-F. Vidal. 1984. *Quand les crises durent. . .* , Paris: Économica.
Michelsons, A. 1987. "Turin Between Fordism and Flexible Specialization: Industrial Structure and Social Change, 1960–75." Ph.D. thesis, University of Cambridge.
Michelsons, A. 1989. "Local Strategies of Industrial Restructuring and the Changing Relations between Large and Small Firms in Contemporary Italy: The Case of Fiat Auto and Olivetti." In Zeitlin (1989a).
Montmollin, M. de and O. Pastré (eds). 1984. *Le taylorisme*, Paris: La Découverte.
Murray, F. 1987. "Flexible Specialisation in the 'Third Italy'." *Capital and Class*, 33: 84–95.
Murray, R. (ed.) 1987. *The Cyprus Industrial Strategy: Report of the UNDP/UNIDO Mission*, 8 vols, Institute of Development Studies, University of Sussex.
Nelson, D. 1975. *Managers and Workers: The Origins of the New Factory System in the United States, 1880–1920*, Madison: University of Wisconsin Press.
——. 1980. *Frederick W. Taylor and the Rise of Scientific Management*, Madison: University of Wisconsin Press.
New, C. and A. Myers. 1986. *Managing Manufacturing Operations in the UK, 1975–85*, Brighton: Institute of Manpower Studies.
Okimoto, D. 1989. *Betwen MITI and the Market: Japanese Industrial Policy for High Technology*, Palo Alto: Stanford University Press.
Perez, C. 1983. "Structural Change and the Assimilation of New Technologies in the Economic and Social System." *Futures* 15, 4: 357–75.
——. 1985. "Microelectronics, Long Waves and World Structural Change: New Perspectives for Developing Countries." *World Development* 13, 3: 441–63.
Piore, M. and C. Sabel (1983). "Italian Small Business Development: Lessons for US Industrial Policy." In *American Industry in International Competition*, edited by John Zysman and Laura Tyson. Ithaca: Cornell University Press.
——. 1984. *The Second Industrial Divide: Possibilities for Prosperity*, New York: Basic Books.

Pollert, A. 1988. "Dismantling Flexibility." *Capital and Class* 34: 42–75.
Prais, S. J. 1976. *The Evolution of Giant Firms in Britain*, Cambridge: Cambridge University Press.
Pyke, F., G. Becattini and W. Sengenberger (eds). 1990. *Industrial Districts and Inter-firm Co-operation in Italy*, Geneva: International Institute for Labour Studies.
Regini, M. and C. Sabel 1989. *Strategie di riaggiustimento industriale*, Bologna: Il Mulino.
Sabel, C. 1982. *Work and Politics: The Division of Labour in Industry*, Cambridge: Cambridge University Press.
——. 1984. "Industrial Reorganization and Social Democracy in Austria." *Industrial Relations* 23, 3: 344–61.
——. 1986. "Changing Models of Economic Efficiency and Their Implication for Industrialization in the Third World." In *Development, Democracy and the Art of Trespassing: Essays in Honor of Albert O. Hirschman*, edited by Alejandro Foxley, Michael McPherson and Guillermo O'Donnell. Notre Dame: Notre Dame University Press.
——. 1989a. "Flexible Specialisation and the Re-emergence of Regional Economies." In Hirst and Zeitlin (1989a).
——. 1989b. "Equity and Efficiency in the Federal Welfare State." Unpublished paper presented to the Nordic Working Group on the New Welfare State, Copenhagen, 8 August.
——. 1990. "Skills without a Place: The Reorganization of the Corporation and the Experience of Work." Unpublished paper presented to the British Sociological Association conference, Guildford, 2–4 April.
Sabel, C. and J. Zeitlin. 1985. "Historical Alternatives to Mass Production: Politics, Markets and Technology in Nineteenth-Century Industrialization." *Past and Present* 108: 133–76.
——. (eds). (forthcoming) *Worlds of Possibility: Flexibility and Mass Production in Western Industrialization.*
Sabel, C., G. Herrigel, R. Deeg and R. Kazis. 1989. "Regional Prosperities Compared: Massachusetts and Baden-Württemberg in the 1980s." In Zeitlin (1989a).
Sabel, C., H. Kern and G. Herrigel. 1989. "Collaborative Manufacturing: New Supplier Relations in the Automobile Industry and the Redefinition of the Industrial Corporation." Unpublished paper.
Salais, R. 1989. "Why Was Unemployment So Low in France during the 1930s?" In *Interwar Unemployment in International Perspective*, edited by Barry Eichengreen and T. J. Hatton. Dordrecht: Kluwer.
Salais, R. and M. Storper. 1990. "One Industry, Multiple Rationalities: Flexibility and Mass Production in the French Automobile Industry." Working Paper D901, School of Architecture and Urban Planning, University of California Los Angeles.
Salais, R., N. Baverez and B. Reynaud. 1986. *L'invention du chomage*, Paris: Presses Universitaires de la France.
Sayer, A. 1986. "New Developments in Manufacturing: The Just-in-Time System." *Capital and Class* 30: 43–72.
——. 1989. "Post-Fordism in Question." *International Journal of Urban and Regional Research* 13, 4: 666–93.
Schmitz, H. 1989. "Flexible Specialisation – A New Paradigm of Small-Scale Industrialization?" *Institute of Development Studies Discussion Paper* no. 261, University of Sussex.
Sengenberger, W. and G. Loveman 1988. *Smaller Units of Employment: A Synthesis*

Report on Industrial Reorganization in Industrialized Countries, Geneva: International Institute for Labour Studies.

Smith, K. 1984. *The British Economic Crisis*, Harmondsworth: Penguin.

Solomou, S. 1987. *Phases of Economic Growth, 1850–1973: Kondratieff Waves and Kuznets Swings*, Cambridge: Cambridge University Press.

Sorge, A., G. Hartmann, M. Warner and I. Nicholas. 1983. *Microelectronics and Manpower in Manufacturing*, Aldershot: Gower.

Sorge, A. and W. Streeck. 1988. "Industrial Relations and Technical Change: The Case for an Extended Perspective." In Hyman and Streeck (1988).

Storper, M. 1989. "The Transition to Flexible Specialization in the US Film Industry: The Division of Labour, External Economies and the Crossing of Industrial Divides." *Cambridge Journal of Economics* 13, 2: 273 305.

Storper, M. and S. Christopherson. 1987. "Flexible Specialization and Regional Industrial Agglomerations: The Case of the US Motion Picture Industry." *Annals of the Association of American Geographers* 77, 1: 104–117.

Storper, M. and B. Harrison, 1990. "Flexibility, Hierarchy and Regional Development: The Changing Structure of Industrial Production Systems and their Forms of Governance in the 1990s," Working Paper D903, School of Architecture and Urban Planning, University of California Los Angeles.

Streeck, W. 1987. "Industrial Change and Industrial Relations in the Motor Industry: An International View." *Economic and Industrial Democracy* 8, 4: 437–62.

Thompson, G. 1987. "The American Industrial Policy Debate: Any Lessons for Britain?" *Economy and Society*, 16, 1: 1–74.

Thompson, P. J. 1989. *The Nature of Work: An Introduction to Debates on the Labour Process* (2nd ed.), London: Macmillan.

Tolliday, S. and J. Zeitlin (eds). 1987. *The Automobile Industry and Its Workers: Between Fordism and Flexibility*, Cambridge: Polity Press/New York: St Martin's, 1986.

Trigilia, C. 1986. *Grandi partiti e piccole imprese*, Bologna: Il Mulino.

Tully, J. (ed.). 1988. *Meaning and Context: Quentin Skinner and his Critics*, Cambridge: Polity.

Weir, M. and T. Skocpol. 1985. "State Structures and the Possibilities for 'Keynesian' Responses to the Great Depression in Sweden, Britain and the United States." In *Bringing the State Back In*, edited by Peter Evans, Dietrich Rueschmeyer and Theda Skocpol. Cambridge: Cambridge University Press.

Williams, K., J. Williams and D. Thomas. 1983. *Why are the British Bad at Manufacturing?*, London: Routledge and Kegan Paul.

Williams, K., T. Cutler, J. Williams and C. Haslam. 1987. "The End of Mass Production?" *Economy and Society* 16, 3: 404–38.

Wood, S. 1988. "Between Fordism and Flexibility?: The Case of the US Car Industry." In Hyman and Streeck (1988).

——. (ed.). 1989. *The Transformation of Work*, London: Unwin Hyman.

Zeitlin, J. (ed.). 1989a. "Local Industrial Strategies." special issue of *Economy and Society*", 18, 4.

——. 1989b. "Italy's Success Story: Small Firms with Big-Firm Capability." *QED: Quarterly Enterprise Digest* October: 5–9.

Zeitlin, J. and P. Totterdill. 1989. "Markets, Technology and Local Intervention: The Case of Clothing." In Hirst and Zeitlin (1989a).

5 THE JAPANESE MODEL OF POST-FORDISM

Notes

1 The works of the regulation school have recently been introduced to Japanese general readers by K. Hirata et al. (1987) among others, and also by translated Japanese versions of M. Aglietta (1981), A. Lipsetz (1985), and R. Boyer (1986). I shall not explore details of various regulationist positions here but concentrate on the main conceptual framework of the school.

2 An example of this type of view is in M. Kenny and R. Florida (1988). T. Kato and R. Steven (1989) criticize it and present an international argument.

3 See E. Vogel (1979) and C. Johnson (1982) as representative works along such a tendency.

4 The annual average working days lost through labor disputes in Japanese manufacturing increased from 2·8 million in 1966–9 to 6·2 million in 1971–5.

5 See P. Armstrong, A. Glyn, J. Harrison (1984) and M. Itoh (1980, 1988, 1990) for the basic theory and more detailed analyses of the whole process of the overaccumulation crisis in the capitalist world economy.

6 Net profit share in Japanese manufacturing did not recover much from the trough 15·2 percent in 1975 to 19·1 percent in 1980 and 17·7 percent in 1985, after a big fall from 40·7 percent in 1970. Cf. P. Armstrong and A. Glyn (1986) and OECD (1988).

7 Such an effect of spreading and reducing the value of labor-power by increasing the number of wage workers in a family was noticed already by K. Marx in his *Capital*, vol. I, 1986. Harmondsworth: Penguin: 518.

8 This is an observation in W. K. Tabb (1988).

9 The public sector was traditionally stable in employment, and tended to enable stronger trade unions to grow.

10 Economic stagnation and increasing critique of a nondemocratic sociopolitical order in real-existing socialist countries ideologically played a role in strengthening such neoconservative or neoliberal belief in a free market system and reduction of social control. On the other side, the neoconservative political stance which prevailed in the capitalist world in the 1980s must have greatly influenced the direction of restructuring of Soviet and East European economies in their attempts to catch up to the advanced capitalist economies until 1990. Probably we have to think more of need and the possibility of democratic social control of the market system from both sides of the world in the 1990s.

Bibliography

(Japanese titles are romanized and translated into English.

Aglietta, M. 1981. *A Theory of Capitalist Regulation: US Experience* (translated by Fernbach) London: New Left Books.

Armstrong, P., A. Glyn, J. Harrison. 1984. *Capitalism since World War II*, London: Fontana.

Armstrong, P., A. Glyn. 1986. *Accumulation, Profits, State Spending: Data for Advanced Capitalist Countries 1952–1983*, Oxford: Institute of Economics and Statistics.

Boyer, R. 1986. *Capitalismes Fin de Siècle*, Paris: Presses Universitaires de France.

——. 1986. *La Théorie de la Régulation*, Paris: Edition de la Decouvert.

Hirata, K. (ed.). 1987. *Gendai Shimin-shakai no Senkai [A Turn in Contemporary Civil Society]*, Tokyo: Showado.

Itoh, M. 1980. *Value and Crisis*, London: Pluto Press, New York: Monthly Review Press. A 1987 French extended version, *La Crise Mondiale* (translated by C. Meillassoux), Paris: EDI.
——. 1988. *The Basic Theory of Capitalism*, London: Macmillan.
——. 1990. *The World Economic Crisis and Japanese Capitalism*, London: Macmillan.
Johnson, C. 1982. *MITI and the Japanese Miracle*, Palo Alto: Stanford University Press.
Kato, T., R. Steven. 1989. "Is Japanese Capitalism Post-Fordist?" Mimeo, translated into Japanese and now published in the Japanese journal, *Mado*, 4, 1990.
Kenney, M., R. Florida. 1988. "Beyond Mass Production: Production and the Labour Process in Japan." *Politics and Society* 16, 1.
Lipietz, A. 1985. *Mirages et Miracles*, Paris: Edition la Decouverte.
OECD 1988. *Historical Statistics 1960–86*, Paris: OECD.
Tabb, W. K. 1988. "World Cities and the Urban Problematic: The Cases of New York and Tokyo. Contrasts in Comparable Agglomeration." Mimeo, City University of New York, Department of Sociology.
Vogel, E. 1979. *Japan as Number One*, Cambridge, MA: Harvard University Press.

6 THE REVITALIZATION OF MASS PRODUCTION IN THE COMPUTER AGE

Notes

1 Among the different potential origin of the "productivity" phenomenon, specific attention was given to the rise of energy prices, the growth of service sectors with low productivity gains, or the characteristics of the new generation of capital goods, . . .

Bibliography

Bain, J. S. 1954. "Economies of Scale, Concentration and the Conditions of Entry in Twenty Manufacturing Industries," *American Economic Review*, March.
——. 1959. "*Industrial Organization*," Cambridge, MA: Harvard University Press.
Baumol, Panzar, Williwg. 1982. *Contestable Markets and the Theory of Industry Structure*, New York: Harcourt, Brace Jovanovich.
Boyer, R., B. Coriat. 1989. "De la flexibilité technique à la stabilisation macro-économique: un essai d'analyse." In *Flexibilité Information et Décision*, editors P. Cohendet, P. Llerena, Paris: Editions Economica.
Bowles, S., D. M. Gordon, T. E. Weisskopf. 1983. "*Beyond the Waste Land*", New York: Doubleday.
Cohen, S., J. Zysman. 1987. *Manufacturing Matters, The Myth of the Post-industrial Society*, New York: Basic Books.
Coriat, B. 1982. *L'Atelier et le Chronomètre* (2nd edn) Paris: C. Bourgois.
——. 1990. *L'Atelier et le Robot*, Paris: Christian Bourgois.
——. 1991. *Penser à l'Envers – Travail et Organisation dans l'Enterprise Japonaise*, Paris: Christian Bourgois.
Freeman, C. 1984. *Long Waves in the World Economy*, London: Frances Pinter.
Klein, B. 1986. *Dynamic Competition and Productivity Advances*. In *The Positive Sum Strategy: Harnessing Technology for Economic Growth*, by R. Landau and S. Rosenberg. Washington: National Academy Press.

Liebenstein, H. 1966. "Allocative Efficiency versus X-Efficiency." *American Economic Review* 56: 392–415.
Maruani, M., E. Reynaud and C. Romani (eds). 1989. *La flexibilité en Italie. Débat sur l'emploi*, Paris: Syros.
Mensch, G. O. 1978. *Stalemate in Technology*, Cambridge, MA: Ballinger.
Ohno, T. 1989. *L'esprit Toyota*, Paris: Masson.
Piore, M., S. Sabel. 1984. *The Second Industrial Divide*, New York: Basic Books.
Shingo, S. 1983. *Maîtrise de la production et Méthode Kan-Ban: le cas Toyota*, Paris: Les Editions d'Organisations.
Streeck, W. 1988. "Industrial Relations in West Germany, 1980–1987." *Labour* 2, 3, Winter.

7 TECHNOLOGICAL TRAJECTORIES AND THE CLASSICAL REVIVAL IN ECONOMICS

Notes

1 This is roughly the distinction between the American radical economists and the French Regulationists.

Bibliography

Arida, P. 1977. Unpublished paper, MIT, Cambridge, MA.
Becattini, G. 1987. "Introduzione: Il distretto industriale marshalliano: cronaca di un ritrovamento." In *Mercato E Forze Locali: Il Distretto Industriale*, edited by G. Becattini, Bologna: Il Mulino.
Bellandi, M. 1982. "Il distretto industriale in Alfred Marshall." *L'Industria*, 3.
Dei Ottati, G. 1986. "Distretto industriale, problemi delle transazioni e mercado communitario: prime considerazioni." *Economie e Politice industriale*, 51: 93–121.
Jakobson, R. 1972. *Child Language, Aphasia and Phonological Universals*, The Hague-Paris: Mouton.
Lucas, R. E. 1988. "On the Mechanics of Economic Development." *Journal of Monetary Economics* 22: 3–42.
Marglin, S. 1974. "What do bosses do? The origins and functions of hierarchy in capitalist production." *The Review of Radical Political Economy*, Part I.
Marshall, A. 1948. *Principles of Economics*, New York: MacMillan.
——. 1975. *The Pure Theory of Domestic Value. Early Economic Writings of Alfred Marshall, 1867–1890*, edited by J. Whitaker and J. King, New York: Free Press.
——. 1975a. *Industry and Trade*, 1919. *Early Economic Writings of Alfred Marshall, 1867–1890*, edited by J. Whitaker and J. King, New York: Free Press.
Marx, K. 1944. *Capital.*
Piore, M. J. 1990. 'Work, Labor, Action: Work Experience in a System of Flexible Production." In *Industrial Districts and Interfirm Cooperation in Italy*, International Institute for Labor Studies, Geneva.
——. 1991. "Corporate Reform in American Manufacturing and the Challenge to Economic Theory." In *Collected Essays on Management for the 1990s*, Oxford University Press, forthcoming.
Piore, M. J., C. Sabel. 1984. *The Second Industrial Divide*, New York: Basic Books.
Roemer, P. M. 1986. "Increasing Returns and Long Run Growth." *Journal of Political Economy* 94, 5, October: 1002–38.

Young, Allyn A. 1928. "Increasing returns and economic progress." *The Economic Journal* 38: 527–42. Reprinted in Arrow, *et al.*, *Readings in Welfare Economics*, (1969). 228–241.

8 THE STRUCTURE OF INDUSTRIAL PRODUCTION AND THE BOUNDARIES BETWEEN FIRMS AND MARKETS

Notes

1 One of us discusses these topics at greater length in Salvatore (1990).

2 Note that, putting for simplicity $\alpha = 1$ in eq. 2,

$$\mathrm{Var}[\log X_{t+n}] = \mathrm{Var}[\log X_t] + n\sigma^2:$$

the variance linearly increases over time.
Hence, for example, Kalecki (1945) modifies Gibrat's law assuming that the random component of growth decreases with firm size and Ijiri and Simon (1977) introduce appropriate probabilities of death and of entry on new firms (equal to each other). In both cases, the variance of the distribution remains constant.

3 For some empirical studies, see Hart and Prais (1956), Hymer and Pashigian (1976); Mansfield (1962), Singh and Whittington (1975), Simon and Bonini (1958).

4 True, there is here an ambiguity between *plant*- and *firm*-sizes. However, under the assumption of single-product firms it is difficult to imagine U-shaped long-term cost curves at plant level but not at firm-level.

5 Ergodic processes generated by the random multiplicative variable in eq. 2 are the simplest example.

6 Winter (1984) simulates the properties of two different "technological regimes" showing indeed the emergence of quite different industrial structures.

7 This section is largely based on Dosi, Teece, and Winter (1992), to which the reader is referred for further details.

8 Thus the conglomerate ITT was put together by Harold Geneen, only to be dismantled by his successor. Similarly, Richard Ferris put together the Allegis Group, starting with United Airlines, only to have it undone months later by his successors.

9 On these points see Hannan and Freeman (1977), (1985); Aldrich and Auster (1987).

10 Its generic form is $\beta(\alpha;\beta) = \frac{1}{B(\alpha;\beta)}\ x(\alpha-1)\ .\ (1-x)(\beta-1)$
with $B(\alpha;\beta) = \Gamma(\alpha+\beta)/\Gamma(\alpha)\Gamma(\alpha)$
Eq. 4, normalized dividing by its maximum is defined on the interval {0;1} and for positive values of the two parameters has R^+ as codomain.

11 The difference is measured as:

$$d = [f(j) - f.(j)]/f.(j)$$

with *f(.)* being the actual frequency in size class *j* and *f(.)* the estimated frequency.

Bibliography

Aldrich, H., A. Auster. 1987. "Selezione naturale e strategia d'impresa." *Sviluppo e organizzazione* 103, September–October.

Arrow, K. J. 1974. *The Limits of Organization.* New York: Norton.

Audretsch, D. B., Z. J. Acs. 1990. "Technological Regimes, Learning and Industry Turbulence," working paper.

Barca, F. 1985. "Tendenze nella struttura dimensionale dell'industria italiana: una verifica empirica del 'Modello di Specializzazione Flessibile'." *Politica Economica* 1: 71–109.

Carrol, G., R. Glenn, M. T. Hannan. 1989. *The role of density delay in the evolution of organizational populations: a model and five empirical tests*, University of California, Berkeley, Business School working paper.

Caves, R. E., M. E. Porter. 1977. "From Entry Barriers to Mobility Barriers: Conjectural Decisions and Contrived Deterrence to New Competition." *Quarterly Journal of Economic Policy* 91, 2: 241–61.

Contini, B. 1988. "Grandi e piccole imprese industriali in Italia: dinamica e performance negli anni '80 a confronto." *Moneta e Credito.*

Dosi, G. 1982. "Technological paradigms and technological trajectories. A suggested interpretation of the determinant and direction of technological change." *Research Policy* 11, 3: 147–62.

Dosi, G., C. Freeman, R. Nelson, G. Silverberg and L. Soete (eds). 1988. *Technical Change and Economic Theory*, London, Frances Pinter.

Dosi, G., R. Giannetti and P. A Toninelli. (eds). 1990a. *Technology and Enterprises in a Historical Perspective*, Oxford: Oxford University Press.

Dosi, G., D. J. Teece and S. Winter. 1992. "Towards a Theory of Corporate coherence: preliminary remarks", forthcoming in G. Dosi, R. Giannetti and P. A. Toninelli.

Dunne, T., M. J. Roberts, L. Samuelson. 1988. "Patterns of firm entry in U.S. manufacturing industries." *RAND Journal of Economics* 19, 4.

Freeman, J., M. T. Hannan. 1985. "Niche Width and the Dynamics of Organizational Population." *American Journal of Sociology* 88.

Geroski, P. A., A. Jacquemin. 1988. "The Persistence of Profits: a European Comparison", *The Economic Journal* 98, 391: 375–89.

Gibrat, R. 1931. *Les Inégalités Economiques*, Paris: Receuil Sirey.

Gort, M., S. Klepper. 1982. "Time Paths in the Diffusion of Product Innovations." *The Economic Journal* 92, 630–53.

Gort, M., A. Konakayama. 1982. "A Model of Diffusion in the Production of an Innovation." *American Economic Review* 72, 5: 1111–20.

Hannan, M. T., J. Freeman. 1977. "The Population Ecology of Organizations." *American Journal of Sociology* 82, 929–64.

Hymer, S. P. Pashingian. 1976. "Firm Size and Rate of Growth." *Journal of Political Economy* 70.

Ijiri, Y., H. A. Simon. 1977. *Skew Distribution and the Size of Business Firms*, Amsterdam: North-Holland.

Jovanovic, B. 1982. "Selection and Evolution of Industry." *Econometrica* 50: 649–70.

Kalecki, M. 1945. "On the Gibrat Distribution." *Econometrica* 13: 161–70.

Lucas, R. E. 1978. "On the Size Distribution of Business Firms." *Bell Journal of Economics*, 9.

Malerba, F., L. Orsenigo. 1989. "Technological regimes and patterns of innovation: a theoretical and empirical investigation of the Italian case." Paper presented at the Schumpeter Society Conference, "Evolution of technology and market structure in an international context," Siena.

Mansfield, E. 1962. "Entry, Gibrat's Law, Innovation and the Growth of Firms." *American Economic Review* 52: 10123–51.

Marglin, S. A. 1974. "What Do Bosses Do? The Origins and Functions of Hierarchy in Capitalist Production." *Review of Radical Political Economics* 6: 60–112.

Mueller, D. C. 1986. *Profits in the Long Run*, Cambridge: Cambridge University Press.
Nelson, R. R., S. Winter. 1982. *An Evolutionary Theory of Economic Change*, Cambridge, MA: The Belknap Press of Harvard University Press.
Orsenigo, L. 1989. *The Emergence of Biotechnology*, London: Pinter Publishers.
Pavitt, K. 1984. "Sectoral patterns of technological change: towards a taxonomy and a theory." *Research Policy*, 13, 6: 343–73.
Pavitt, K., M. Robson, J. Townsend. 1987. "The Size Distribution of Innovating Firms in the U.K.: 1945–1983". *Journal of Industrial Economics* 35, 3: 297–316.
Piore, M. J., C. F. Sabel. 1984. *The Second Industrial Divide*, New York: Basic Books.
Porter, M. 1984. *The Competitive Advantage*, New York: Free Press.
Putterman, L. 1981. "The Organization of Work: Comment." *Journal of Economic Behavior and Organization* 2.
Schumpeter, J. 1934. *The Theory of Economic Development*. Cambridge, MA: Harvard University Press.
——. J. 1942. *Capitalism, Socialism and Democracy*, New York: Harper.
Silverberg, G., G. Dosi, L. Orsenigo. 1988. "Innovation, diversity and diffusion: a self-organization model." *Economic Journal*, 98, 393: 1032–54.
Simon, H. A. 1955. "On a Class of Skewed Distribution Functions." *Biometrika* 52, 425–40.
Simon, H. A., C. P. Bonini. 1958. "The Size Distribution of Firms." *American Economic Review* 48, 607–17.
Singh, A., G. Whittington. 1975. "The Size and Growth of Firm." *Review of Economic Studies*, 42, 1: 13–26.
Teece, D. J. 1980. "Economies of Scope and the Scope of the Enterprise." *Journal of Economic Behavior and Organization*, 1, 3: 223–47.
——. D. J. 1982. "Towards an Economic Theory of the Multiproduct Firm", *Journal of Economic Behavior and Organization*, 3, 1: 39–63.
——. D. J. 1988. "Technological Change and the Nature of the Firm," in Dosi *et al.* (1988).
Tirole, J. 1989. *The Theory of Industrial Organization*, Cambridge, MA: MIT Press.
Williamson, O. E. 1975. *Markets and Hierarchies*, New York: Free Press.
——. O. E. 1985. *The Economic Institutions of Capitalism*, New York: Free Press.
Winter, S. G. 1984. "Schumpeterian competition in alternative technological regimes." *Journal of Economic Behavior and Organization* 5: 287–320.

9 TRUST, COMMUNITY, AND COOPERATION: TOWARD A THEORY OF INDUSTRIAL DISTRICTS

Notes

1 A preliminary version of this chapter was presented at the conference "Pathways to Industrialization and Regional Development in the 1990s," UCLA, March, 1990. I am indebted to Giacomo Becattini, Michael Lesser, Jean Saglio, Allen Scott, and Michael Storper for their useful comments. The research on which the chapter is based was funded in part by the Economic and Social Research Council [UK] while an ESRC Postdoctoral Fellow at the University of Cambridge, England. Its contents are the responsibility of the author and do not necessarily reflect the views of the Economic and Social Research Council.

2 On the reconsolidation of the region as a locus of economic activity, see Sabel (1989), Scott (1988), and Storper and Scott (1988).

3 Collective (or public) goods are defined as goods which in some degree are indivisible and nonexcludable.

4 Norms are defined as prescriptions or standards of behavior.

5 The use of the legalistic term "contract" does not imply the use of comprehensive written agreements. In the case of agreements between client firms and their subcontractors, for example, the order form is likely to be the only written agreement. The initial agreement is a reference point, the gaps in the planning being intentional, and it is understood that adaptations to unanticipated contingencies will have to be made in a jointly optimal way that respects an agreed distribution of benefits. For example, a common contemporary arrangement among client firms and their subcontrators in the Lyon metal-working industry is to share the risks of an unanticipated downturn in demand by agreeing that the client firm should fix the proportions of its total work it sources out and retains for in-house production. See Lorenz (1988). For comparable arrangements in the German metal-working industry, see Sabel (1989: 37–9). For the case of Nishiwaki textile industry in Japan, see Dore (1987: 171–7).

6 Due to the interdependency among firms, all potentially benefit from the innovative success of others. Correspondingly, adherence to norms of reciprocity may be addressed in terms of the collective action problem of providing public goods.

7 For some limited success in intentionally creating Silicon Valley-like industrial districts in the US, see Rogers and Larsen's (1984: 240–3) discussion of North Carolina's Research Triangle and Salt Lake City's Bionic Valley.

8 For the purposes of this chapter, *trusting behavior* consists in action that (1) increases one's vulnerability to another whose behavior is not under one's control, and (2) takes place in a situation where the future penalty suffered if the trust is abused would lead one to regret the action. In short, there is no best choice independent of trust. See Lorenz (1988). For a general discussion of how trust can be brought about, see Gambetta (1988).

9 Taylor, (1982: 44–5) defines "social order" as conformity to social norms, including security of property rights.

10 A fourth method, the use of the threat of supernatural sanctions, is not pertinent to my discussion of nineteenth and twentieth century industrial districts in Europe, the United States, and Japan. See Taylor (1982: 86–7).

11 Security of property rights, in so far as it is achieved through deterrence of *potential* violators, is a nonexcludable collective good. See Taylor (1982: 64–65).

12 The argument that adherence to social norms can always be explained rationally by the fear of being sanctioned by third parties runs up against the second order free rider problem noted above. A related point made by Elster (1989: 133) is relevant to this argument:

> Now, expressing disapproval is always costly, whatever the target behaviour . . . However, when one moves upward in the chain of actions, beginning with the original violation, the cost of receiving disapproval falls rapidly to zero. People do not frown upon others when they fail to sanction people who fail to sanction people who fail to sanction people who fail to sanction a norm violation. Consequently, some sanctions must be performed for motives rather than the fear of being sanctioned. I argued in the preceding paragraphs that sometimes there is an unmoved mover at the beginning of every chain. Here I have argued that every chain must have one.

13 This eclectic approach can legitimately be criticized on grounds of incompleteness, since I offer no explanation for why human beings hold social norms.
14 Of course, another explanation for norm violations is that some subset of individuals do not share the values held by the community at large.
15 Saglio (1989: 28–33, 1990: 59–60) similarly emphasizes the many-sided nature of relations in industrial districts.
16 My discussion focusses on the more problematic issue of assuring conformity to norms of reciprocity, rather than providing collective services. As Piore (1990: 10–11) has noted:

> The list of [collective] services, moreover, is in no way sufficient to guarantee the success of an area as a dynamic industrial district. The recent vogue of state and local economic development policy in the United States has, for example, succeeded in generating virtually every item on this, or any conceivable list, without anything like a comparable success in generating industrial districts.

17 The Sheffield "outrages" probably constitute an exception to this observation. See Pollard (1959: 152):

> In the early 1860s, when the unions were under attack for complicity in the trade outrages, their champions were claiming privileges which were elsewhere assumed to belong only to Government itself.

18 See, for example, Sabel and Zeitlin (1985: 153–4):

> Secondly, diffusion of an unspoken code of competition discouraged some forms of potentially disruptive business behavior by attaching moral sanctions to it: in a world where youngsters were often trained to study character as a condition of economic survival it was important to maintain a character that would bear scrutiny. By itself, of course, no moral code could eliminate all breaches of the code of innovative competition.

19 See Lorenz (1988) for the safeguards presently used by firms in the Lyon metalworking industry to protect themselves against the possibility of opportunistic behavior by others.
20 For a general discussion of leveling mechanisms in communities, see Taylor (1982: 104–29).
21 See Oscar Lewis' (1951) interesting study of the Mexican peasant community of Tepoztlan where inequality of income contributed to pervasive distrust and repressed hostility among families.
22 For the economically dynamic character of the Oneida community, see Carden (1969).
23 Also see Saxenian (1989: 33–4, 1990: 11–12).
24 Also see the comments of Dr. Lester Hogan, former president of Fairchild, as quoted in Rogers and Larsen (1984: 79): "The easy information-exchange here is like the scientists who feel that their technology *belongs to the world, to humanity*. It's just too important to keep this technical information to oneself or one's firm."
25 While subcontracting out to one's less successful competitors is not common practice in Silicon Valley, other practices help assure security of place in the district. Typically, unsuccessful firms are acquired by their more successful competitors, generally becoming a subsidiary or divison of the parent. Top management often stays in place, at least for a while. See Rogers and Larsen (1984: 122–4).

26 The problem of "invasion by meanies" will be more or less an issue depending on the nation under consideration. Ronald Dore's (1987: 179–83) comparative work on relations of exchange in Japan and Britain, for example, suggests that reciprocity or conditional cooperation is more likely to flourish in Japan because of a widely distributed social norm: the duty to treat others with *benevolence*. As Dore notes (1987: 95), given expectations in Japan, "a little benevolence can go a long way to evoke trust, much further than in a society where expectations are lower and suspicion of power more deeply ingrained."

27 See, for example, the comments of Liz Gallese (1990: 61) concerning the industrial district emerging in East Cambridge, Massachusetts adjacent to MIT:

> Asked about doing business in such an environment, they talk about the easy flow of ideas from campus to company, about the sharing of ideas and facilities among start-ups, even those in the same industry.
>
> But the ever-present specter of failure appears to have produced exactly the opposite effect: a nervousness, a recoiling from one another. Digate [a former Lotus Development Corp. senior vice-president], in fact, says "a notion of privacy" is really what pervades the area. "People are very loathe to talk about what they are doing."

Bibliography

Axelrod, R. 1984. *The Evolution of Cooperation*, New York: Basic Books.

Binmore, K., P. Dasgupta. 1986. "Game Theory: a survey." In *Economic Organizations as Games*, edited by K. Binmore, P. Dasgupta, Oxford: Basil Blackwell.

Brusco, S. 1986. "The Emilian Model: Productive Decentralisation and Social Integration." *Cambridge Journal of Economics* 6: 167–89.

Carden, M. L. 1969. *Oneida: Utopian Community to Modern Corporation*, Baltimore: Johns Hopkins Press.

Dore, R. 1987. *Taking Japan Seriously*. Palo Alto: Stanford University Press.

Elster, J. 1983. *Sour Grapes*, Cambridge: Cambridge University Press.

——. 1989. *The Cement of Society*, Cambridge: Cambridge University Press.

Friedman, D. 1988. *The Misunderstood Miracle: Industrial Development and Political Change in Japan*, Ithaca: Cornell University Press.

Gallese, L. 1990. "The Most Entrepreneurial Place on Earth." *Inc.* March: 60–1.

Gambetta, D. 1988. "Can We Trust Trust?" In *Trust: Making and Breaking Cooperative Relations*, edited by D. Gambetta, Oxford: Basil Blackwell.

Hawthorne, G. 1988. "Three Ironies in Trust." In *Trust: Making and Breaking Cooperative Relations*, edited by D. Gambetta, Oxford: Basil Blackwell.

Herrigel, G. 1990. "Industrial Organization and the Politics of Industry Centralized and Decentralized Production." Unpublished Massachusetts Institute of Technology Ph.D. Thesis.

Lewis, O. 1951. *Life in a Mexican Village: Tepoztlan Restudied*, Urbana, IL: University of Illinois Press.

Lorenz, E. 1988. "Neither Friends nor Strangers: Informal Networks of Subcontracting in French Engineering." In *Trust: Making and Breaking Cooperative Relations*, edited by D. Gambetta, Oxford: Basil Blackwell.

Oliver, P. 1980. "Rewards and Punishments as Selective Incentives for Collective action." *American Journal of Sociology*. 85: 1356–75.

Olson, M. 1965. *The Logic of Collective Action*, Cambridge, MA: Harvard University Press.

Piore, M. 1990. "Work, Labor and Action: Work Experience in a System of Flexible Production." Department of Economics, Massachusetts Institute of Technology.
Piore, M., C. Sabel. 1984. *The Second Industrial Divide*, New York: Basic Books.
Pollard, S. 1959, *A History of Labour in Sheffield*. Liverpool: Liverpool University Press.
Popkin, S. L. 1988. "Political Entrepreneurs and Peasant Movements in Vietnam." In *Rationality and Revolution* edited by M. Taylor, Cambridge: Cambridge University Press.
Raveyre, M., J. Saglio. 1984. "Les systèmes industriels localisés: Eléments pour une analyse sociologique des ensembles de P.M.E. industriels." *Sociologie du Travail* 2: 157–75.
Rogers, E., J. Larsen. 1984. *Silicon Valley Fever*, New York: Basic Books.
Sabel, C. 1989. "Flexible Specialization and the Re-emergence of Regional Economics." In *Reversing Industrial Decline*, edited by P. Hirst and J. Zeitlin, Oxford: Berg Publishers.
Sabel, C., J. Zeitlin. 1985. "Historical Alternatives to Mass Production: Politics, Markets and Technology in Nineteenth Century Industrialisation." *Past and Present* 108: 131–76.
Saglio, J. 1989. "Les systèmes industriels: Eléments pour une compréhension sociologique de la diversité des enterprises." Contribution au seminaire du CEPREMAP. Paris, décembre.
——. 1990. "Rationalité économique et préférences identitaires." Seminaire "Historical Alternatives to Mass Production." Université de Lyon II, janvier.
Saxenian, A. 1989. "A High Technology Industrial District: Silicon Valley in the American Context." In *Città della scienza e della tecnologia*, edited by P. Perulli, San Marco, Venezia: Arsenale Editrice.
——. 1990. "The Origins and Dynamics of Production Networks." Department of City and Regional Planning, University of California, Berkeley.
Scott, A. J. 1988. *New Industrial Spaces*, London: Pion Limited.
Sengenberger, W., G. Loveman. 1987. "Smaller Units of Employment: A Synthesis Report on Industrial Reorganisation in Industrialised Countries." International Institute for Labour Studies, Geneva. New Industrial Organisation Programme. Discussion paper #3.
Storper, M., A. J. Scott. 1988. "The Geographical Foundations and Social Regulation of Flexible Production Complexes." In *Territory and Social Reproduction*, edited by J. Wolch and M. Dear, London: Unwin Hyman.
Taylor, M. 1982. *Community, Anarchy and Liberty*, Cambridge: Cambridge University Press.
——. 1987. *The Possibility of Cooperation*, Cambridge: Cambridge University Press.

10 A REEXAMINATION OF THE ITALIAN MODEL OF FLEXIBLE PRODUCTION FROM A COMPARATIVE POINT OF VIEW

Notes

1 This is a revised version of a paper to the 1st Conference of the European Association of Labour Economists held in Turin, 8–10 September 1989. A shorter version has been published in La Lettre du Centre d'Études de l'Emploi, no. 14, Dec. 1989. The authors wish to thank gratefully Frank Pyke, Jacqueline O'Reilly and Michael Storper for their help in translating the French version.

2 See Becattini (1990).

3 See Raveyre and Saglio (1984) for the study of Oyonnax and Ganne (1983) for the study of Annonay. Those small areas have been studied extensively by French researchers from the Groupe of Labor Sociology in Lyon, who have been the first people in France, along with Houssel from the faculty of Geography of Lyon, to develop an analysis of local areas in the same way the Italian researchers have generalized it. But we have to note that Houssel was an unknown precursor because his first works were done in the early 1970s (Houssel 1972).

4 See Courault, Rérat, Weisz (1987, 1989, 1990).

5 See Bettio, Romani (1988).

6 See Brusco, Garofoli, Solinas, Villa (1987) and Solinas, Villa (1990).

7 Most of those researchers had presented their first analysis in the 1970s from empirical local studies of different regions of Italy. We have not quoted here the corresponding referee. In the 1990s, what we are discussing looks like an attempt to restrict the use of the model which has been generalized by the Anglo-Italian authors during the 1980s (Pyke *et al.* 1990). But we will not enter the internal Italian debate which arises strongly at a time where the crisis is going on and changes some aspects of the District model (see Nuti 1990).

8 See also the addenda they have written for the French translation (1989) where they have reconsidered the possibility of extending their model to the case of France.

9 Piore and Sabel have been particularly successful in establishing this idea of flexibility as a canon of the flexibility debates such that it only concerns models of internal decentralized production as seen in the Italian context.

10 Piore and Sable (1984).

11 This phenomenon illustrates the primacy of an economy of convention (Thévenot 1989) which has resulted from traditional practices. Relations of proximity and of the family are important in supporting exchange networks which constitute the social basis of production (Saglio 1989). The importance of the extended family relies on neighborhood practices; added to ancient forms of domestic labor used irregularly in temporary activities, it explains the very large flexibility of the work force in such areas.

12 This line of argument is indebted to Italian research into the footwear industry conducted by Brusco, Garofoli, Solinas and Villa (1987, 1990) as part of an international comparative survey extended to cover the United Kingdom (Rubery and Wilkinson 1987, 1989) and France (Courault, Rérat, Weisz 1987); the final report is being published (Courault 1990).

13 It is the opposite case in the knitting and general textile districts that Brusco (1982), Cappechi (1987), and many others have studied earlier (see also Lazerson 1990).

14 With the exception of certain groups of producers who have taken the initiative of integrated distribution.

15 This is the classical statement defined by Marshall in his Principles (1890). Brusco has demonstrated that economies of scale are not the most efficient to explain the growing of decentralization of production in atomistic units (see Brusco 1978). From his point, we use to speak in terms of economies of scale versus economies of scope.

16 This goes beyond the analysis of Piore and Sabel concerning the model of flexible production *sui generis*. They have limited the use of flexibility model to the cases of Italian districts, or post-Marshallian districts which are similar to them, such as the old system of the silk industry in Lyon or the system La Motte in the Northern part of France for the past, or the Silicon Valley or the Route 128 in

Boston for the recent ones, by insisting on the historical factors which have contributed to breed them.

17 This study points out that the different way localities operate is due to local rules which are not negotiated in the same manner.

18 They focus on the skills and earnings to answer the requirements of the highly skilled workers who have traditionally a large control over the labor markets in old regions of Fougères and Romans.

19 The expression of consensus and solidarity which we have encountered illustrates the importance of such an adaptation at the same time that it explains the importance that is attached to the local actor.

20 These ideas were developed by Becattini at the French-Italian Conference on Flexibility in Italy organized by the Mission Recherche Expérimentation in Paris, March 1989.

21 See Saglio (1987, 1989, 1990); Romani (1987, 1988).

22 Taking the example of the above-mentioned agreement passed in Prato (22 July 1987), it was negotiated at the level of a cluster of small and medium-sized firms in an industrial area of the Tuscany region. These firms constitute a segment of a very specific market with an industry, the woolens manufacturing.

23 These observations have been made for the industrial district of Baix Llogregat, which is bounded by the communes of L'Hospitalet, Cornellà, Sant Feliu de Llobregat, El Prat de Llobregat, Gave and San Boi de Llobregat see the "enquesta metropolitana" (1986–1987).

24 Approximately 20 percent of registered production units in the Baix Llobregat districts are units formed by autonomous workers.

Bibliography

Bagnasco, A. 1977. *Tre Italia. La problematica territoriale dello sviloppo economico italiano*, Bologna: Il Mulino.

Bagnasco, A., C. Trigilia. 1985. *Società e politica nelle area di piccola impresa: il caso della Valdesa*, Milan: Franco Angeli.

Bagnasco, A. 1985. "La construction sociale du marché: stratégie d'enterprise et expérimentation á petite échelle en Italie." *Stato e Mercato*, 13, April: 9–45.

——. 1988.*La contruzione sociale del mercato. Studi sullo sviluppo di piccola impresa in Italia*, Bologna: Il Mulino.

Barisi, G. 1988. "La flexibilité des entrepreneurs: protagoniste du renouvellement de l'industrie italienne." *Interventions Economiques* 19, 165–84.

Becattini, G. 1979. "Del settore industriale al distretto industriale." *Rivista di economia e politica industriale*, 1.

——. 1990. *Modeli locali di sviluppo*, Bologna: Il Mulino.

——. 1990. "The Marshallian industrial district as a socio-economic notion." In *Industrial districts and inter-firm co-operation in Italy*, edited by F. Pyke, G. Becattini and W. Sengenberger, Geneva: International Institute for Labour Studies – International Labour Organisation.

Becattini, G. (ed.) (1987), *Mercato e forze locali: il distretto industriale*, Bologna: Il Mulino.

Bettio, F., C. Romani. 1988. *Un dibattito atipico sulla flessibità: il caso italiano*, Rome: Report for the Mission Recherche Expérimentation, March.

Brusco, S. 1975. "Organizzazione del lavoro e decentramento produttivo nel settore metalmeccanico." In *Sindicato e Piccola Impresa*, edited by F. L. M. Bergami, Bari: De Donato.

——. 1982. "The Emilian model: productive decentralisation and social integration." *Cambridge Journal of Economics* 6: 167–84.

Brusco, S. 1990. "The idea of the industrial district: its genesis." In F. Pyke *et al.* (*see* Beccatini 1990).

Brusco, S., C. Sabel. 1981. "Artisan production and economic growth." In *The dynamics of labour market segmentation*, edited by F. Wilkinson, London: Academic Press.

Brusco, S., G. Garofoli, G. Solinas, P. Villa. 1987. *Structure productive et compétitivité dans l'industrie italienne de la chaussure*, Milan: Final Report on automation of small and medium-sized firms, international comparison, (France-Italy-Great Britain), for the Centre d'Etudes de l'Emploi, Commissariat Général au Plan, Ministère de la Recherche. Paris: Centre d'Etudes de l'Emploi.

Capecchi, V. 1987. "Formation professionnelle et petite enterprise: le developpement industriel à spécialisation flexible en Emilie Romagne." *Formation-Emploi*, 19, July: 3–18.

——. 1990. "A history of flexible specialisation and industrial districts in Emilia Romagna." In F. Pyke et al. (*see* Beccatini 1990).

Courault, B., F. Rérat. 1987. "Un modèle de production régionale en transition: le cas de la chaussure dans le choletais." *Cahiers du CEE*, no. 30. Paris: Presses Universitaires de France.

——. 1988. "La flexibilité de la production." In *Les stratégies d'entreprises face aux ressources humaines. L'après-Taylorisme*, edited by F. Stankiewicz, Paris: Economica.

——. 1989. "Polyvalence, mobilité, professionalité – Les premiers enseignements d'une étude comparative France-Italie." In *Actes du Colloque Internationale sur les formes d'emploi*, (Paris, Novembre 1988), Paris: Ed. la Documentation Française, Documents du Minstère du Travail.

Courault, B. F. Rérat, R. Weisz. 1987. *L'automatisation des PMI. Comparaison internationale France Italie Grande-Bretagne*, Final Report – France – Centre d'Etudes de l'Emploi, Commissariat Général au Plan, Ministère de la Recherche. Paris: Centre d'Etudes de l'Emploi, September.

——. 1989. "Footwear manufacturing firms in France: a typological study." *Labour and Society* 14, 2, April: 141–59.

——. 1990a. "Les districts de la chaussure, survivance du passé ou exemples de systèmes industriels locaux? Une première mise à l'épreuve du modèle du district industriel". In *L'emploi, l'entreprise et la société*, edited by F. Michon and D. Segrestin, Paris: Economica, forthcoming.

——. 1990b. *La chaussure européenne sur les voies de la flexibilité. Comparaison internationale France Grande Bretagne Italie*, Paris: Centre d'Etudes de l'Emploi, forthcoming.

Domergues, P. (ed.). 1988. *La société de partenariat, Economie, Territoire et revitalisation régionale aux U.S.A. et en France*, Paris: Afnor-Anthropos.

Enquesta Metropolitana. 1986–7. Institut d'Estudis Metropolitans de Barcelona, Mancomunitat de Municipis de l'area metropolitana de Barcelona. Barcelona.

Fontana, J. 1988. "La fi de l'antic règim i la industrializacio." In *Historia de Catalunya*, vol. V, edited by P. Vilar, Barcelona: Editorial AEBE Edicions 72.

Fuà, G. 1985. "Les voies diverses du développement en Europe." *Les Annales* 3, May–June: 579–603.

Ganne, B. 1983. *Gens du cuir, gens du papier: transformation d'Annonay depuis les années vingt*, Lyon: Editions du Centre National de la Recherche Scientifique.

——. 1989. *Industrialisation diffuse et systèmes industriels localisés. Essai de bibliographie critique du cas français*, Lyon: Glysi – Maison Rhone Alpes des Sciences de l'Homme.

Ganne, B., J. Saglio, F. Baptiste, A. Michelson, McArthur. 1988. *Milieux industriels et systèmes industriels locaux, une comparaison France-Italie*. Lyon: Groupe Lyonnais de Sociologie du Travail, Centre National de la Recherche Scientifique.

Greffe, X. 1988. *Décentraliser pour l'emploi*, Paris: Economica.

Houssel, J. P. 1972. "Essor des villes manufacturières de l'habillement et industrialisation spontanée dans l'Italie du milieu." *Revue de Géographie de Lyon*, 4: 241–69.

——. 1980. "Les industries autochtones en milieu rural." *Revue de Géographie de Lyon* 4: 305–41.

Lazerson, M. 1990. 'Subcontracting in the Modena knitwear industry." In F. Pyke et al.

Lorthiois, J., F. Rérat. 1987. "Ouvrières: savoir-faire en sursis?" *Pour*, n° 108. Toulouse: Ed. Privat.

Piore, M., C. Sabel. 1984. *The second industrial divide*, New York: Basic Books.

——. 1989. *Les chemins de la prospérité – De la production de masse à la spécialisation souple*, Paris: Hachette (French translation of the 1984 book, with a new foreword and postscript).

Pyke, F., G. Becattini and W. Sengenberger (eds). 1990. *Industrial districts, inter-firm co-operation in Italy*, Geneva: International Institute for Labour Studies – International Labour Organisation.

Raveyre, M. F., J. Saglio. 1984. "Les systèmes industriels localisés: éléments pour une analyse sociologique des ensembles de PME industriels." *Sociologie du Travail* 2: 157–75.

Regini, M. 1987. "Tendenze delle relazioni industriali di impresa in italia: alcuni punti interpretativi," Université de Milan: presentation at Conference on Mediterranean industrial relations, Cagliari.

Regini, M., C. Sabel. 1989. *Strategie di riaggiustamento industriale*, Bologna: Il Mulino.

Romani, C. 1987. *La Cassa Integrazione Guadagni, Réalités juridiques, économiques et sociales d'une institution*, Aix-en-Provence: Laboratoire d'Economie et de Sociologie du Travail, Centre National de la Recherche Scientifique.

——. 1988. *La flexibilité dans l'enterprise et le district industriel: une comparaison en France et en Italie dans le textile habillement*. Brussels: Report to the European Commision, D.G.

——. 1989. "La flexibilité en Italie: les termes du débat." In *La flexibilité en Italie*, edited by M. Maruani, C. Reynaud, C. Romani, Paris: Ed. Syros.

Rubery, J., F. Wilkinson. 1989. "Distribution, Flexibility of Production and the British Footwear Industry." *Labour and Society* 14, 2, April.

Sabel, C. 1988. "The reemergence of regional economies." Massachusetts Institute of Technology, mimeo. Communication au colloque "Logiques d'enterprise et formes de légitimité". Paris: Centre d'Etudes de l'Emploi, January.

Saglio, J. 1987. "Préliminaires pour une typologie des systèmes industriels," Communication à un séminaire du Gréco Relations Industrielles. Paris: Conservatoire National des Arts et Métiers.

——. 1989. "Les systèmes industriels, éléments pour une compréhension sociologique de la diversité des enterprises". Paris: Notes for a seminar of CEPREMAP, 4 December. Groupe Lyonnais de Sociologie du Travail.

Silvestre, J. J. 1986. "Marchés du travail et crise économique: de la mobilité à la flexibilité." *Formation-Emploi*, Paris: La Documentation Française, 14, April–June: 54–61.

Solinas, G. 1982. "Labour market segmentation and workers careers: the case of the Italian knitwear industry." *Cambridge Journal of Economics* 6: 331–52.
——. 1982. "The empirical evidence of decentralisation: the case of Italian knitwear and footwear," paper for the 3rd Conference of the Social Science Research Council, *Job Generation Workshop*, Selwyn College, 19–20 September, Cambridge.
Stankiewicz, F. (ed.) 1988. *Les stratégies d'entreprises face aux ressources Humaines – L'après taylorisme*, Paris: Economica.
Thévenot, L. 1989. Economie et politique de l'entreprise, Note INSEE n° 186/930, 18 June.
Triglia, C. 1981. *Le développement économique et les transformations sociopolitiques des systèmes territoriaux de l'économie diffuse: les cultures locales*, Milan: Fondation Feltrinelli.
——. 1986. "Small-firm development and political subcultures in Italy." *European Sociological Review*, 2, 3: 161–75.
——. 1990. "Work and politics in the third Italy's industrial districts in Emilia Romagna." In F. Pyke et al.
Wilkinson, F. 1983. "Productive systems." *Cambridge Journal of Economics*, 7, 3/4: 413–29.

11 INDUSTRIAL DEVELOPMENT AND LOCAL INDUSTRIAL SYSTEMS IN POSTWAR FRANCE

Notes

1 This part takes up the conclusions of the bibliographical work that we have done for the ILO: Ganne, Bernard, Industrialisation diffuse et systèmes industriels localisés: Essai de bibliographie critique du cas français, report for the IIES of the ILO, mimeo, GLYSI-MRASH, March 1989, p. 178.
2 Bonin, 1988: 114. For all of this part, we drew principally on Sauvy, A., *Histoire économique de la France*, 1948, Caron, F., *Histoire économique de la France XIX*° *XX*° *siècle*, 1981, and Kuisel, R. F., *Le capitalisme et l'Etat en France. Modernisation et dirigisme au XX*° *siècle*, Paris, 1984.
3 Our emphasis.
4 This kind of change was observed in the local system of Annonay, where the restructuring of the leather and paper industries broke up the old segmented local system, instituting a more unified and general mode of management of the local system overall (Ganne 1983b).
5 Cf. for example the ambiguities of the international symposium "Nouveaux regards sur l'industrialisation," Tunis, December 1987.

Bibliography

Amin, A., K. Robins. 1989. "Industrial districts and regional development: limits and possibilities." Centre for Urban and Regional Development Studies, University of Newcastle upon Tyne, mimeo. 57.
Amiot, M. 1986. *Contre l'état, les sociologues. Elements pour une histoire de la sociologie urbaine en France (1900–1980)*, Paris: EHESS.
Antonelli, C., R. Cappellin, G. Garofoli, and R. Jannaccone Pazzi. 1988. *Le politiche di sviluppo locale*, Milan: Franco Angeli.
Bagnasco, A. 1977. *Tre Italie: la problematica territoriale dello sviluppo italiano*, Bologna: Il Mulino.

——. 1988. *La costruzione soziale del mercato: studi sullo sviluppo di piccola impresa in Italia*, Bologna: Franco Angeli.

Bagnasco, A., C. Trigilia. 1984. *Società e politica nelle aree di piccola impresa: il caso di Bassano*, Milan: IRES, Fondazione Corrazin, Ed. Arsenale.

——. 1988. *Società e politica nelle aree di piccola impresa: il caso della Valdelsa*, Milan: Franco Angeli.

de Banville, E. 1984. "L'enterprise entre la PMI-sation et l'évolution des réseaux de solidarité." St Etienne, Cresal (Centre de Recherche en sciences appliquées de la Loire) report, mimeo.

Becattini, G. (ed.). 1987. *Mercato e forze locali: il distretto industriale*, Bologna: Il Mulino.

Bonin, H. 1988. *Histoire économique de la France depuis 1880*, Paris: Masson.

Browaeys, X., P., P. Chatelain. 1984. *Les France du travail*, Paris: PUF.

Brusco, S. 1982. "The Emilian model: productive decentralisation and social integration." In *Cambridge Journal of Economics* 6.

——. 1986. "Small firms and industrial districts: the experience of Italy." In *New firms and regional development in Europe*, edited by D. Keeble and Z. Weyer, London: Croom Helm.

Capecchi, V. 1987. "Les transformations du petit entreprenariat,dans une région de l'Italie: parcours masculins et féminins." In *Annales de Vaucresson* 26: 243–58.

——. 1988a. "Les facteurs de développement d'une économie régionale", duplicated copy, pp. 43.

——. 1988b. *La mobilità sociale in Emilia Romagna*, Bologna: Il Mulino.

Caron, F. 1981. *Histoire économique de la France XIX° XX° siècle*. Paris, A. Colin.

Castells, M., F. Godard. 1974. *Monopolville: l'entreprise, l'Etat. l'urbain*, Paris: Mouton.

Chanaron, J. J., J. Perrin, and B. Ruffieux. 1988. "L'efficacité des technopoles; les leçons de l'expérience de la ZIRST de Meylan." Report for the SGAR, Prefecture of the Rhône-Alpes region, Université de Sciences Sociales de Grenoble.

Cohen, E. 1989. *L'état brancardier: politiques du déclin industriel (1974–1984)*, Paris: Calmann-Lévy.

Courault, B., C. Romani. 1989. "La flexibilité locale à l'italienne." In *La Lettre d'Information de la CEE* 14.

Courlet, C., P. Judet. 1986a. "Nouveaux espaces de production en France et en Italie." *Les Annales de la Recherche Urbaine* 29: 95–103.

——. 1986b. "Industrialisation et développement: la crise des paradigmes." *Revue Tiers-Monde* 27, 107: 519–36.

Fua, G. 1985. "Les voies diverses du développement en Europe." *Annales Economies Sociétés Civilisations* 3: 579–603.

Fua, G., C. Zacchia. (eds). 1983. *Industrializzazione senza fratture*, Bologna: Il Mulino.

Ganne, B. 1983a. *Gens du cuir, gens du papier: transformations d'Annonay depuis les années 20*, Paris: CNRS.

——. 1983b. "Conflit du travail et changement urbain: transformation d'un rapport local: le cas des Tanneries d'Annonay." *Sociologie du Travail* 2: 127–46.

——. 1985. "Du notable au local: transformations d'un modèle politique." *Les Annales de la Recherche Urbaine*, 28: 23–32.

——. 1987. "Pour une sociologie des PME, ou de l'entreprise comme articulation de systèmes de relations: de quelques préalables." *Technologies. Idéologies, Pratiques*. 6, 4, and 7, 1: 129–147.

——. 1989a. "Industrialisation diffuse et systèmes industriels localisés: Essai de bibliographie critique du cas français." Report for the IIES of the ILO, mimeo: 178. Published by IIES (1990), Série bibliographique 14, Genève.

——. 1989b. "Les relations industrielles et leur contexte: à propos du politique comme 'contexte'." Paper given at the symposium "Les systèmes de relations professionnelles: trente ans après J. T. Dunlop," Paris, GRECO 41, Relations professionnelles.

——. 1989c. "Regional dynamics of innovation: a look at the Rhône-Alpes region." *Journal of Entrepreneurship and Regional Development* 1: 147–54.

——. 1989d. "PME et districts: à propos du modèle italien: de quelques réflexions critiques." *Revue Internationale PME*, 2, 2–3: 273–85.

Ganne, B., J. Saglio, F. Baptiste, A. Michelsons, and R. MacArthur. 1988. "Milieux industriels et systèmes industriels locaux: une comparison France-Italie." Lyon: Groupe Lyonnais de Sociologie Industrielle, document 1: 194.

Garofoli, G. 1981. "Lo sviluppo delle 'aree periferiche' nell'economia italiana degli anni settanta." *Industrial* 11: 391–404.

——. 1985. "Industrialisation diffuse en petite entreprise: le modèle italien des années 70." *Cahiers I.R.E.P. développement* 9: 245–56.

Grémion, P. 1976. *Le pouvoir périphérique: bureaucrates et notables dans le système politique français*, Paris: Ed. du Seuil, Sociologie collection.

Houssel, J.-P. 1980. "Les industries autochtones en milieu rural." *Revue de Géographie de Lyon* 4: 304–41.

——. 1984. "L'industrialisation spontanée face à la crise de 1973 en Europe occidentale." *Revue de Géographie de Lyon* 4: 331–46.

INSEE. 1974. *La mutation industrielle de la France*, Paris: Insee Editor.

Kuisel, R. F. 1984. *Le capitalisme et l'Etat en France, Modernisation et dirigisme au XX° siècle*, Paris.

Leborgne, D., A. Lipietz. 1988. "L'après-fordisme et son espace." *Les Temps Modernes* 501: 75–114.

Maglione, R., A. Michelsons, and S. Rossi. 1989. Autostrada numero cinque: uomini, tecnologie et imprese fra Ivrea e Torino, duplicated copy.

Maruani, M., E. Reynaud, and C. Romani. 1989. *La flexibilité en Italie*, Paris: Syros-Alternatives.

Michelsons, A., F. Baptiste. 1989. "Le modèle italien: systèmes industriels localisés et réseaux de PME." Paper given to the Développement local et ensembles de PME group, GLYSI, Lyon.

Minguet, G. 1985. *Naissance de l'Anjou industriel*, Paris: Ed. L'Harmattan, Logiques Sociales collection.

Piore, M. J., C. F. Sabel. 1984. *The Second Industrial Divide: possibilities for prosperity*, New York: Basic Books.

Ritaine, E. 1985. "Ressources sociales du développement économique: un système industriel localisé en Italie." Talence: Université de Bordeaux I, Centre d'études et de recherche sur la vie locale (C.E.R.V.L.).

——. 1987. "Prato ou l'exaspération de la diffusion industrielle." *Sociologie du Travail* 2: 139–56.

Sabel, C. F., J. Zeitlin. 1985. "Historical alternatives to mass production: politics, markets and technology in nineteenth-century industrialization." *Past and Present* 108: 133–76.

Saglio, J. 1985. "Petites et moyennes entreprises industrielles et environnement urbain: note de synthèse pour le Plan Urbain." Lyon: Groupe Lyonnais de Sociologie Industrielle, document 6: 26.

Sauvy, A. 1984. *Histoire économique de la France entre les deux guerres* (vol. 2), Paris: Economica.

Sforzi, F. 1988. "The geography of industrial districts in Italy. "Working document distributed to the working group Districts industriels et systèmes de coopération inter-entreprises at the ILO (Geneva), 11–12 July 1988.

Worms, J.-P. 1966. "Le préfet et ses notables." *Sociologie du Travail* 3: 276.

12 LOCALIZED INDUSTRIAL SYSTEMS IN FRANCE: A PARTICULAR TYPE OF INDUSTRIAL SYSTEM

Bibliography

Bagnasco, A. 1985. "La costruzione sociale del mercato: strategie di impresa e experimenti di scala in Italia." *Stato e Mercato* 13.

Bauer, M., E. Cohen. 1981. *Qui gouverne les groupes industriels?* Paris: Seuil.

Becattini. G. 1989. "Some thoughts on the Marshallian industrial districts as a socio economic notion." Paper presented at an ILO conference on industrial districts. Florence.

Bernoux, P., J. Magaud, M. F. Raveyre, J. Saglio, and G. Villegas. 1984. "Qui connait les machines? Les connaissances que les salariés ont des machines qui'ils utilisent ou pourraient utilise." Groupe Lyonnais de Sociologie Industrielle, Université Lyon II, Lyon.

Blau, P. M. 1964. *Exchange and Power in Social Life*, New York, London, Sydney: John Wiley and Sons.

Boisard, P., M. T. Letablier. 1986. *Liens locaux de production et standards industriels. Le cas du Camembert*, Paris: Centre d'Etudes de l'Emploi.

Braudel, F. 1979. *Civilisation matèrielle, économie et capitalisme XV°-XVIII° Siècle* (3 vols), Paris: A. Colin.

Bunel, J., J. Saglio. 1977. "Faiblesse de la négociation collective et pouvoir patronal." *Sociologie du travail* 4.

——. 1979. *L'Action patronale, du Cnpf au petit patron*, Paris: PUF.

Callon, M. 1986. "Eléments pour une sociologie de la traduction. La domestication des coquilles Saint Jacques et des marins-pêcheurs dans la baie de Saint Brieuc." *L'Année Sociologique 1986*: 169–208.

Castoriadis, C. 1975. *L'Institution imaginaire de la société*, Paris: Seuil.

Coase, R. H. 1937. "The Nature of the firm." *Economica* 4: 386–405.

Cohen, E. 1989. *L'Etat brancardier: Politiques du déclin industriel (1974–1984)*, Paris: Calmann-Lévy.

Commons, J. R. 1934. *Institutional Economics. Its place in political economy*, New York: Macmillan. Quoted from the 2nd edn (1959), Madison: University of Wisconsin Press.

Courault, B., F. Rerat. 1986. *Les stratégies des producteurs face à la distribution. Le cas des entreprises de chaussure du Choletais*, Paris: Centre d'Etudes de l'Emploi.

Donzelot, J. 1984. *L'invention du social*, Paris: Fayard.

Duchemin, R. 1940. *L'organisation syndicale partonale en France*, Paris: Plon.

Dunlop, J. T. 1958. *Industrial Relations Systems*, New York: Holt, Rinehart and Winston.

Eymard Duvernay, F. 1989. "Conventions de qualité et formes de coordination." *Revue économique* 40: 329–59.

Flanders, A. 1968. "Eléments pour une théorie de la négociation collective." *Sociologie du Travail*. 1.

——. 1970. *Management and Unions*. London: Faber & Faber.

Fox, A. 1975a. "Collective bargaining, Flanders and the Webbs." *British Journal of Industrial Relations* XIII: 151–74.

——. 1975b. "Travail et contrat: les cas des sociétés à confiance limitée." *Sociologie du Travail* 3: 266–72.

Ganne, B. 1983a. *Gens du cuir, gens du papier: transformations d'Annonay depuis les années 20*, Lyon: CNRS.

——. 1983b. "Conflits du travail et changement urbain: transformation d'un rapport local." *Sociologie du travail* 2.

——. 1989. *Industrialisation diffuse et systèmes industriels localisés*, Lyon: Groupe Lyonnais de Sociologie Industrielle, Université Lyon II.

Ganne, B., D. Motte. 1980. *Licenciements collectifs et reclassement: études de cas à Valence*, Lyon: Groupe Lyonnais de Sociologie Industrielle, Université Lyon II.

Hennion, A., J. P. Vignolle. 1978. *Artisans et industriels du disque*, Paris: Centre de Sociologie de I'Innovation.

Houssel, J. P. 1989. "L'industrie en milieu rural." Paper presented in Séminaire du Groupe Lyonnais de Sociologie Industrielle, Lyon.

Jacob-Ory, A. 1969. '"Formes d'organisations patronales et limitations de la concurrence." *Revue française de Sociologie* X.

Kuisel, R. F. 1981. *Capitalism and the state of Modern France, Renovation and economic management in the twentieth century*, Cambridge: Cambridge University Press.

Lequin, Y. 1977. *Les ouvriers de la région lyonnaise* (1848–1914) (2 vols), Lyon: PUL.

Lorenz, E. H. 1987. "L'offre de travail et les stratégies d'emploi dans la construction navale en France et en Grande Bretagne (1880–1970)." *Le Mouvement Social* 138: 21–44.

——. 1989. "Neither friends nor strangers. Informal networks of subcontracting in French industry." Paper presented at Diego Gambetta's Seminar on Trust, King's College, Cambridge.

Mac Arthur, J., B. Scott. 1970. *L'industrie française face aux plans*, Paris: Ed. d'organisation.

O'Brien, P., C. Keyder. 1978. *Economic Growth in Britain and France 1780–1914, Two Paths to the Twentieth Century* Boston, Sydney: George Allen and Unwin.

Oualid, W., C. Picquemard. 1928. *Salaires et Tarifs*, Paris: PUF.

Padioleau, J. R. 1981. *Quand la France s'enferre*, Paris: PUF.

Paradeise, C. 1984. "La marine marchande française: un marché du travail fermé!" *Revue française de sociologie* XXV.

Piore, M., C. Sabel. 1984. *The second Industrial divide; Possibilities for Prosperity*, New York: Basic Books.

Raveyre, M. F. 1988. *Jeux de miroirs: l'aide de Saint Gobain Développement aux PME*, Lyon: Groupe Lyonnais de Sociologie Industrielle, Université Lyon II.

Raveyre, M. F., J. Saglio. 1984. "Les Systémes Industriels Localisés: Eléments Pour une Analyse Sociologique des Ensembles de P.M.E. industriels." *Sociologie du travail*. 2: 157–75.

Reynaud, J. D. 1989. *Les règles du jeu*, Paris: A. Colin.

Sabel, C., J. Zeitlin. 1985. "Historical alternatives to Mass Production: Politics, Markets and Technology in Nineteenth Century Industrialisation." *Past and Present* 108: 133–76.

Saglio, J. 1984. "Systémes locaux de Relations Industrielles et Innovations Productives." *Cahiers S.T.S.* 3.

Saglio, Jean. 1989. *La construction sociale des marchés*, Lyon: Groupe Lyonnais de Sociologie Industrielle, Université Lyon II.

——. 1990a. "Négocier les décisions technologiques ou seulement leurs effets." In *Les systèmes de relations professionnelles. Examen critique d'une théorie*, edited by J. D. Reynaud, F. Eyraud, C. Paradeise, J. Saglio, Lyon: CNRS.

——. 1990b. "Rationalité économique et préférences identitaires." Paper presented at the workshop "Historical alternatives to mass production," Maison Rhone Alpes des Sciences de l'Homme, Lyon.

Saglio, J., P. Garrouste, M. F. Raveyre and G. Richoilley. 1984. *Les systèmes industriels localisés. Relations professionnelles, stratégies économiques et innovations technologiques dans les ensembles de PME*, Paris: Groupe Lyonnais de Sociologie Industrielle, Université Lyon II.

Saglio, J., C. Thudéroz. 1989. *Entre monopole et marché: les professions règlementées face à l'ouverture du marché européen*, Lyon: Groupe Lyonnais de Sociologie Industrielle, Université Lyon II.

Sellier, F. 1961. *Stratégie de la lutte sociale*, Paris: Ouvrières.

——. 1984a. *La Confrontation sociale en France, 1936–1981*, Paris: PUF.

——. 1984b. "La Problèmatique du niveau dans le Système français de Relations Professionnelles." Mimeo, Laboratoire d'Economie et de Sociologie du Travail, Aix en Provence.

Stoleru, L. 1969. *L'impératif industriel*, Paris: Seuil.

Szokoloczy-Syllaba, J. 1965. *Les organisations professionnelles françaises et le Marché Commun*, Paris: A. Colin.

Tolliday, S. 1988. "The Regime of Mass Production. Competition, technology and the Workplace in the US Automobile Industry since 1945." Paper presented at the workshop "Historical alternatives to mass production", Maison Rhône Alpes des Sciences de l'Homme, Lyon.

Touraine, A. 1974. *Production de la société*, Paris: Seuil.

Weber, M. 1971. *Economie et Société*, Paris: Plon. French translation of *Wirtschaft und Gesellschaft*. 1956. Tubingen: Mohr.

Williamson, O. E. 1985. *The economic institutions of capitalism: Firms, markets, relational contracting*, New York: Macmillan.

Zeitlin, J. 1987. "Surveys and speculations From Labour history to the history of industrial relations." *The Economic History Review* second series XL.

13 ALTERNATIVE ROUTES TO LABOR FLEXIBILITY

Notes

1 One is tempted to add, somewhat mischievously, "Japan," since the paternalistic and dualistic labor practices there have been depicted as a model to be emulated elsewhere.

2 For instance, a part of the growth of employment in the United Kingdom since 1985 has consisted of increasing numbers in labor market "training" schemes without employment. They are now counted in the official labor force statistics as employed, which has wondrous effects on the unemployment rate, and has helped to transform the United Kingdom into the outstandingly successful country in Europe in terms of creating employment in the past five years.

3 Is the writer alone in feeling unease at such statements as the following, "Functional flexibility requires workers to have several skills" (Brunes 1989: 15)? If a heart surgeon tells me he is a doctor of economics I hope I will have enough strength to escape the knife in time. Of course, often functional flexibility is a euphemism for a form of employer control over labor, as the text immediately following that quotation indicates: "The more skills they have, the more easily they can be switched from one activity or job to another." Exactly.

4 For instance, an Act was passed in Belgium in March 1987 permitting the

extension of working hours beyond the standard, if doing so had positive employment effects. Elsewhere, lower "recruitment" wages have been introduced, ostensibly to promote job creation.

5 On the "reemergence" of large firms, the EC Directorate of Enterprise Policy and a 1989 French study have both recently indicated that large companies have been demanding closer links to a smaller number of subcontractors, suggesting a tiering of small-scale specialists between "insiders" and "outsiders" among the outsiders!

6 It is somewhat devastating to realize that in the United Kingdom – on the edge of the political spectrum of countries pursuing the subordinated flexibility route – not only did income inequality grow in the 1980s but, it is reliably reported, earnings inequality was greater in 1989 than in 1889, the year of Charles Booth's first reliable earnings survey in the country. One could quibble, no doubt. Yet the recent growth of inequality is underestimated for various reasons, including the tendency for very low-paying firms to refuse to cooperate in the government's New Earnings Survey, giving an overestimate of the average earnings of the low-paid. Similarly, tax changes and changes in relative prices of items consumed by the rich and poor, due primarily to the supply-side "deregulation" of rents, transport, etc., have widened inequality of living standards.

7 For a very different approach, which seems to be based on pay and stability of employment and demographic characteristics, see Burchell and Rubery (1990).

8 There is little evidence to support the view that two-tier wage structures have enhanced union militancy, let alone union strength to resist changes. Hope that this resistance and strength would occur was held out by Mike Piore (1986: 51), and is an issue that deserves further analysis.

9 Esping-Anderson (1989) makes a distinction between good and bad jobs within postindustrial sectors. One is not sure how such a descriptive dichotomy can be defined in practical terms.

10 "Employee leasing" companies seems to be the most popular term in the United States. For a review of trends, see Young and Elliott (1986).

11 Conservatives, such as Murray and Mead, attribute the growth of what they call the underclass to individual behavioral traits coupled with the welfare system, which together make it hard or impossible for members of that category to enter the economic mainstream. Liberals, such as Bill Wilson (1987), tend to attribute their exclusion to the refusal of employers to hire certain groups. One question in the debate is whether such individuals would take the jobs if they were offered them. But the real issue is structural. What produces a detached social-labor stratum? A system that depends on subordinated flexibility and certain types of "skill" may well generate a far larger detached stratum than one based on strict hierarchies of jobs or one based on alternative social divisions of labor.

Bibliography

Brunes, B. 1989. "Labour Flexibility in Enterprises: A Comparison of Firms in Four European Countries." In *Labour Market Flexibility: Trends in Enterprises*, compiled by the OECD, Paris: OECD.

Burchell, B., J. Rubery. 1990. *Segmented Jobs and Segmented Workers: An Empirical Investigation*, Oxford: ESRC Social Change and Economic Life Initiative, Working Paper No. 13.

Confederation of British Industry (CBI). 1989. *Pay, Performance and Inflation*. London: Memorandum by the CBI to the National Economic Development Council, 18 Oct.

Esping-Anderson, G. 1989. "Divergent models of stratification in post-industrial society." Florence: European University Institute (EUI), mimeo.
Kelley, M., B. Harrison. 1989. "The Big Firms are Coming Out of the Corner." Cambridge, MA: Massachusetts Institute of Technology, mimeo.
Lilja, R., T. Santamäki-Vuori, and G. Standing. 1990. *Unemployment and Labour Market Flexibility: Finland*, Geneva: ILO.
de Neubourg, C. 1990. *Unemployment and Labour Market Flexibility: The Netherlands*, Geneva: ILO.
OECD. 1989. *The Future of Social Protection*, Paris: Report on a meeting of trade union experts held under the OECD Labour Management Program.
Piore, M. 1986. "A Critique of Reagan's Labor Policy." *Challenge*. 29, 1: 48–53.
Standing, G. 1983. "A Labour Status Approach to Labour Statistics," Geneva: ILO. Population and Labor Policies Program Working Paper No. 139.
——. 1989. "European Unemployment, Insecurity and Flexibility: A Social Dividend Solution," Geneva: ILO. World Employment Programme Working Paper No. 23.
Wilson, J. W. 1987. *The Truly Disadvantaged: The Inner City, The Underclass and Public Policy*, Chicago: University of Chicago Press.
Young, L. A., E. S. Elliott. 1986. *Employee Leasing in Small versus Large Business*, Washington, DC: Final Report for US Small Business Administration.

14 LABOR CONVENTIONS, ECONOMIC FLUCTUATIONS, AND FLEXIBILITY

Notes

1 For instance, refer to the methodological rule proposed by L. Thévenot (1989): "the taking into account of a complex universe, within which several forms of co-ordination are possible."

2 We translate the word "*travail*" (in "*conventions du travail*") as "labor" understood in a general sense of work activities together with their environment of institutions.

3 In research carried out by M. Storper and myself (1992) we elaborate an analytical framework for the quality of the product based upon two factors: technology and the production process on the one hand, and the market, on the other. This framework is founded upon two oppositions relative to the distinctions between standard/specialized and general/dedicated products. We have also broached the uncertainty/risk opposition on the market.
The models of labor conventions that I construct here are therefore necessary to make the connection with the study of the products and their markets.

4 F. Eymard-Duvernay (1989a) has shown that the work of J. Dunlop attempted to realize a compromise between the economic theory of the market and the observations of specialists in professional relations. For the former, the viability of a compromise founded upon the separation of areas of competence is open to question. He has developed (1989b) the analysis of conventions of quality in the goal of articulating these areas.

5 C. Littler (1982) takes up the issue of the observability of work as the object of struggles between the employers and the workers in Great Britain between 1870 and 1939. The implantation of the Bedaux system in large enterprises was an attempt to evade the "opaque" internal-contract organization of work, directed by foremen. His ambition was to arrive at a universal measure of work freed

from the barriers of visibility and to "make comparisons of the relative efficiency of workers, workshops and factories, even when the types of work were fundamentally different."

6 Flexible specialization appears then as an alternative model of production resting upon decentralization, autonomy, initiative and informal agreements between individuals.

7 A conclusion also suggested by R. Salais and M. Storper (1992). The definition of fluctuations is also largely taken from that text.

8 The word "market," here, may be a source of misunderstanding. What we intend to cope with is not at all the conventional sense of market as it is used in the neoclassical theory of the labor market. The neoclassical labor market corresponds for us to the nonquality labor conventions (see later) in which standardization suppresses any connection to qualities and identities of individuals.
In conventions of market labor, the important point is that persons have a quality which is evaluated *on the market of their products*, by exchange. Furthermore, every product is specific and dedicated to one individual demand. Thus this product market is special and resembles traditional markets (like craft products markets in towns or like what occurs in Marshallian industrial districts) where everybody knows everybody by virtue of personal relations and reputations. Another perspective is that of L. Thévenot (1989) the domestic convention Thévenot refers, not to relations within a family (which could involve us towards hierarchy and traditional authority), but to markets organized by durable interpersonal relations. It remains to build up the necessary connections between these two configurations.

9 A precise analysis necessitates an examination of the historical weight of the structuration of economic activity by branch, the diversity of national traditions in state intervention etc.

10 See G. Noiriel and R. Salais (1990) on this point.

Bibliography

Bernoux, P., D. Motte, and J. Saglio. 1973. *Trois ateliers d'O.S.*, Paris: Les Editions ouvrières.

Boltanski, L., L. Thévenot. 1987. *Les économies de la grandeur*, Paris: Cahiers du Centre d'Etudes de l'Emploi, 31, série PROTEE, Presses Universitaires de France.

Boyer, R. 1986. *La théorie de la régulation: une analyse critique*, Paris: Editions La Découverte.

Cohendet, P., and P. Llerena. 1989. "Flexibilité, risque et incertitude dans la théorie de la firme: un survey." In *Flexibilité, information et décisions*, edited by P. Cohendet and P. Llerena, Paris: Economica.

Coriat, B. 1990. *L'atelier et le robot*, Paris: Christian Bourgois.

Didry, C. 1991 "De l'Etat aux solidarités. Les itinéraires croisés de Léon Duguit et Emile Durkheim à la fin du siècle." *Genèses*, 2, January: 5–27.

Doeringer, P., M. Piore. 1971. *Internal Labor Markets and Manpower Analysis*, Lexington, MA: Heath.

Durkheim, E. 1967. *De la division du travail social*, Paris: Presses Universitaires de France, (first edition 1893).

Eymard-Duvernay, F. 1989a. "Qualifications du produit, qualifications du travail. "Communication au Colloque "Les systèmes de relations professionnelles," Paris, 2–3 March.

——. 1989b. "Conventions de qualité et formes de coordination. "Revue économique 40, 2: 329–59.

Favereau, O. 1986. "La formalisation du rôle des conventions dans l'allocation des ressources." in *Le travail. Marchés, règles, conventions*, by R. Salais and L. Thévenot, Paris: INSEE/Economica.
——. 1989. "Marchés internes, marchés externes." *Revue économique* 40, 2: 273–328.
Hayek, F. 1985. "The Use of Knowledge in Society." *American Economic Review* 35, 4: 519–30.
Liebenstein, H. 1966. "Allocative Efficiency vs. X-Efficiency." *American Economic Review*, 56, June: 392–415.
Littler, C. 1982. *The Development of Labour Process in Capitalist Societies*, London: Heinemann.
Marx, K. 1980. *Le capital*, Paris: Editions Sociales (first edition, 1867).
Mottez, B. 1966. *Systèmes de salaire et politiques patronales*, Paris: Editions du Centre National de la Recherche Scientifique.
Noiriel, G. 1984. *Longwy. Immigrés et prolétaires*, Paris: Presses Universitaires de France.
Noiriel, G., R. Salais. 1990. "Approche des questions du travail et interdisciplinarité." *Sociétés Contemporaines* 1, March: 47–69.
Piore, M. 1972. "Notes for a Theory of Labor Market Stratification." Working Paper, M.I.T., 93.
Piore, M., C. Sabel. 1984. *The Second Industrial Divide. Possibilities for Prosperity*, New York: Basic Books.
Polanyi, K. 1983. *La Grande Transformation. Aux origines politiques et économiques de notre temps*, Paris: Editions Gallimard (first edition 1944, in English).
Revue économique. 1989. Special issue on the "economie des conventions," articles by Dupuy, J. P., F. Eymard-Duvernay, O. Favereau, A. Orléan, R. Salais and L. Thévenot. 40, 2, March.
Reynaud, B. 1986a. "Diversité de la relation salariale de branche et codification des conventions collectives." In *Le travail. Marché, règles conventions*, edited by R. Salais and L. Thévenot, Paris: INSEE/Economica.
——. 1986b. "Règles et logique des relations salariales." *Economie et Statistique* 192, October: 43–63.
Sabel, C. 1982. *Work and Politics. The Division of Labor in Industry*, Cambridge: Cambridge University Press.
Salais, R. 1989. "L'analyse économique des conventions du travail." *Revue Economique*, 40, 2: 199–240.
Salais, R., N. Baverez and B. Reynaud. 1986. *L'invention du chômage. Histoire et transformations d'une catégorie en France des années 1890 aux années 1980*, Paris: Presses Universitaires de France.
Salais, R., L. Thévenot (eds). 1986. *Le travail. Marchés, règles, conventions*, Paris: INSEE/Economica.
Salais, R., M. Storper. 1989. "The four worlds of contemporary industry". Cambridge Journal of Economics 16.
Storper, M., R. Salais. 1992. "The division of labour and industrial diversity: flexibility and mass production in the French automobile industry." *International Review of Applied Economics* 6, 1: 1–37.
Thévenot, L. 1989. "Equilibre et rationalité dans un univers complexe." *Revue Economique* 40, 2: 147–97.
Verry, M. 1955. *Les laminoirs ardennais; le déclin d'une aristocratie ouvrière*, Paris: Editions du Centre National de la Recherche Scientifique.
Williamson, O., M. Wachter, and J. Harris. 1975. "Understanding the Employment Relation: the Analysis of Idiosyncratic Exchange. "Bell Journal of Economics 6, 1: 250–78.

15 LEVELS OF POLICY AND THE NATURE OF POST-FORDIST COMPETITION

Notes

1 The formal analysis of these aspects is in Bianchi (1990).
2 Detailed reviews of recent economic literature on strategic deterrence are in Jacquemin (1987), Tirole (1988) Schmalensee (1988).
3 Extensive reviews and analyses of the Product life cycle theory applied to the historic industrial growth are in Norton (1986) and De Jong (1988).
4 The emergence of these mobility barriers contributes to explain the delay of restructuring by the leading companies during the 1970s crisis. On the decline of American industry see Bluestone and Harrison (1982), Piore and Sabel (1984), Norton (1986); on Europe, see De Jong (1988).
5 For the definition of MID see Becattini (1989); more extensively on the Italian districts experiences see Goodman and Bamford (1989).
6 In June 1985, the Commission of the European Communities issued a White Paper on the completion of the European Single Market by 31 December 1992. It took more than two years to complete the process of ratifying the new integration approach; the European Single Act went into effect on 1 July 1987 and carried with it major changes in the Treaty of Rome. The various policies were reformed with specific acts and regulations. See Commission (1987) Padoa Schioppa (1987), Cecchini (1988), Commission (1989b).
7 The revision of the Community competition policy carried out several acts reforming the different aspects of this complex matter; among them, the most important are the Regulations (EEC) 4064/89 of the Council, issued on 21 December 1989 (OJ L 395, 30 Dec. 1989), introducing a guideline for concentration among European companies.
8 The Commission has also reinforced articles 92–4 controlling public aids to individual companies. The Treaty of Rome does not admit national subsidies to individual companies, because they can distort competition throughout the Community. Aids are admitted only for extraordinary intervention, to help innovation, less developed areas, small and medium-sized companies; now, all these aids have to be accepted and integrated within programs supervised by the Commission (Commission 1989a).
9 The reform of structural funds was enacted with the Single European Act, which amended the Treaty of Rome, by introducing the new article 130 A–E. This article reinforces the equity and stability goals of the Community, and reforms the structural funds, and especially the European Regional Development Fund to correct territorial imbalances of the Community; the reform of structural funds went into effect on 1 January 1989 (Commission 1989b).
10 These interventions can be financed by the European Social Funds, the European Investment Bank grants, the New Community Instruments, and with all the facilities provided by the Commission for small and medium-sized companies, such as the promotion of partnership and joint ventures among companies in different countries, an easy entry in the R & D Community Programs, the creation of Business innovation centres, etc. (Commission 1988b).

Bibliography

Becattini, G. (ed.). 1989. *Modelli locali di sviluppo*, Bologna: Il Mulino.
Bianchi, P. 1990. "Structural Change and Strategic Behaviour: the interaction

between Technology Management and Corporate Strategy moving from Mass Production to Flexibility." *International Journal of Technology Management* Special issue, August.

Bluestone, B., B., Harrison. 1982. *The Deindustrialization of America*, New York: Basic Books.

Caves, R., M. Porter. 1977. "From Entry Barriers to Mobility Barriers: Conjectural Decisions and Contrived Deterrence to New Competition." *Quarterly Journal of Economics* 97: 247–61.

Cecchini, P. 1988. *The European Challenge 1992*, Aldershot: Wildwood House.

Commission of the European Communities. 1985. *White Paper on Completing the Internal Market*. Brussels: CEC.

——. 1987. "The Single Act: A New Frontier for Europe." *Bulletin of the E.C.*, supplement.

——. 1988a. "The Economics of 1992." *European Economy* 35, March.

——. 1988b. *Operations of the European Community concerning Small and Medium-sized Enterprises*, Brussels, CEC.

——. 1988c. *Research on the Cost of Non-Europe, Basic Findings*, Brussels: CEC.

——. 1989a. *Primo Censimento degli aiuti di stato nella comunità europea*, Brussels: CEC.

——. 1989b. *Vademecum sulla riforma dei fondi strutturali*, Brussels: CEC.

De Jong, H. W. 1988. "Market Structures in the EEC." In *Structure of European Industry*, edited by H. K. De Jong, Dordrecht: Kluwer Academic Press.

Eaton, B. C., R. G. Lipsey. 1979. "The Theory of Market Preemption: The Persistence of Excess Capacity and Monopoly in Growing Spatial Markets." *Economica* 46: 149–58.

——. 1980. "Exit Barriers are Entry Barriers: The Durability as a Barrier to Entry." *Bell Journal of Economics* 10: 721–9.

Fudenberg, D., J. Tirole. 1983. "Capital as a Commitment: Strategic Investment to Deter Mobility." *Journal of Economic Theory* 31: 227–56.

Gold, B. 1981. "Changing Perspective on Size, Scale, and Returns: an Interpretative Survey." *The Journal of Economic Literature* XIX, 1 March: 5–33.

Goodman, E., J. Bamford. 1989. *Small Firms and Industrial Districts in Italy*, London: Routledge.

Jacquemin, A. 1987. *The New Industrial Organization, Market Forces and Strategic Behavior*, Cambridge, MA: MIT Press.

Leonardi, R. 1990. *The Regional Revolution in Europe: The Single European Act and the Regional Funds*, Florence: European University Institute.

Milgrom, P. R., J. Roberts. 1982. "Limit Price and Entry Under Incomplete Information." *Econometrica* 50: 443–60.

——. 1982. "Predation, Reputation, and Entry Deterrence." *Journal of Economic Theory* 27: 280–312.

Marshall, A., M. P. Marshall. 1879. *The Economics of Industry*, London: Macmillan.

Modigliani, F. 1958. "New Developments on the Oligopolistic Front." *The Journal of Political Economy* 215–32.

Norton, R. D. 1986. "Industrial Policy and American Renewal." *Journal of Economic Literature* XXIV, 1: 1–40.

Padoa Schioppa, T. 1987. *Efficiency, Stability and Equity. A Strategy for the Evolution of the Economic System of the E.C.*, Brussels: CEC.

Piore, M. J., C. F. Sabel. 1984. *The Second Industrial Divide: Possibilities for Prosperity*. New York: Basic Books.

Sabel, C. F. 1982. *Work and Politics. The Division of Labour in Industry*, Cambridge: Cambridge University Press.

Schmalensee, R. 1988. "Industrial Economics: an Overview." *Economic Journal* 98: 634–81.

Schoenberger, E. 1988. "Multinational Corporations and the New International Division of Labour: A Critical Appraisal." *International Regional Science Review* II, 105–20.

Tirole, J. 1988. *The Theory of Industrial Organization*, Cambridge, MA: MIT Press.

16 DIVERGENT PATTERNS OF BUSINESS ORGANIZATION IN SILICON VALLEY

Notes

1 There is no necessary association between size and production strategy. While mass production firms tend to be large to achieve economies of scale, there are large firms which replicate the flexibility of networks of small firms through internal decentralization and the creation of autonomous production units. Similarly, small firms are by no means all flexible specialists.

2 The region is the home of 4,800 technology enterprises in sectors ranging from computers and semiconductors to software and medical instrumentation. More than 85 percent of these firms employ fewer than 100 workers, while only 5 percent employ more than 500.

3 They rely on the region's ample supplies of venture capital, high rates of interfirm mobility, and a regional culture conducive to informal exchange, risk-taking and experimentation. See Angel (1989), Saxenian (1989b), and Florida and Kenney (1988).

4 For more detail on regional interdependencies, see Saxenian (1990a, 1990b) and Gordon, Dilts, and Kimball (1988).

5 In the US, such organizations are typically regarded as collusive efforts to fix prices and investment levels. For the literature on business associations, which focus on Europe, see Coleman and Jacek (1989), Streeck and Schmitter (1985).

6 Businessmen's interests are not determined solely by their economic position. Like other social groups, they must organize to articulate as well as to advance their interests, which are often defined and redefined through the processes of conflict and alliance building. See Saxenian (1989a).

7 SIA founders were W. J. Sanders III of Advanced Micro Devices, Charles Sporck of National Semiconductor, Robert Noyce of Intel Corporation, and Wilf Corrigan of Fairchild. Motorola joined the SIA very soon thereafter. TI did not join until the early 1980s.

8 The SIA's sought action on two claims: first that Japanese firms were engaging in unfair trade practices in the US by dumping chips below the cost of production, and second that they were denying US firms access to the domestic Japanese market.

Their policy agenda also included a call for more favorable tax treatment, intellectual property rights for chip designs, and changes in antitrust legislation to allow cooperative R&D – most of which was achieved.

9 Clyde Prestowitz, Jr., the former chief US trade negotiator with Japan, claimed that the SIA was "the most effective lobbying group that I've seen." Cited in A. Pollack "A Small Lobby's Large Voice" *The New York Times* 7 September, 1989.

10 The trade agreement, passed in 1986, established a system whereby the US Commerce Department monitors production costs and prices for Japanese

memory chips and sets fair market values – minimum floor prices – for several commodity chips. The agreement prohibits dumping in third country markets as well as the US.

The agreement also formally acknowledges the problem of market access and commits the Japanese government to assisting in opening up markets to US producers. There is some controversy over whether the agreement set an explicit target for US market share in Japan: US firms claiming that Japan agreed to a doubling of their market share to 20 percent by 1991, while Japanese firms deny any such numerical targets.

11 For more detail on the trade agreement and Sematech, see Borrus (1988).

12 Although ironically, by the time of its passage, only two US firms remained in the DRAM business.

13 WEMA was founded in Southern California in 1943 by a small group of California electronics manufacturers seeking to gain a greater share of the nation's defense contracts.

14 Interview, E. E. Ferrey, former president of the AEA, 31 July, 1986.

15 From an article on the retirement of E. E. Ferrey, AEA president from 1960 to 1985. J. Mitchell, "A valley hero readies to retire" (20 January 1985, 1985 *San Jose Mercury News*).

16 Mitchell, *ibid.* One of the most successful of early AEA activities was the 1971 Monterey Conference. The conference brought together financial investors and presidents of small "emerging" companies to interact face-to-face for a weekend – an explicit forum for deal making and networking. Investors had a chance to survey new firms while entrepreneurs courted potential investors. The annual Monterey Conference is now one of the most popular of AEA activities.

17 Granovetter (1973) discusses the role of such activities as conferences in building the weak ties that integrate communities.

18 See, for example, James S. Goldman "Pressures mount on AEA" (*The San Jose Business Journal* 25 June, 1990).

19 The firms in the semiconductor equipment and materials industry make the equipment needed to create and prepare wafers, to lay down and process microcircuitry, and to package the final chips and perform final testing.

20 Its name changed from the Semiconductor Equipment and Materials Institute to the Semiconductor Equipment and Materials International in 1988 to reflect the fact that close to 30 percent of its membership was non-US based firms.

21 Trade shows are really networks compressed in time and space: they are very efficient sources of contacts with customers and suppliers, and of information on technology and (especially) on competitors. Thanks to Marty Manley for this insight.

22 For discussion of the Homebrew Computer Club and the informal collaboration in Silicon Valley during the 1960s and early 1970s, see Freiberger and Swaine (1984), Saxenian (1989b).

23 Priscilla Rowe "SEF Breeds Software Success" (*Bay Area Computer Currents* 3–16 July, 1990).

24 The Marketing Special Interest Group, for example, has organized meetings and seminars on issues such as Dealing and Negotiating with Publishers, How to Write a Press Release, Direct Marketing, Marketing without Money, etc.

25 Cited in P. Rowe "SEF Breeds Software Success" (*Bay Area Computer Currents* 3–16 July, 1990).

26 P. Rowe, *ibid.*

27 P. Rowe, *ibid.*

28 The SIA began lobbying for trade relief in the late 1970s. The AEA made its

initial foray to Washington to fight for reduction of the capital gains tax at the same time. Prior to that, the region's entrepreneurs remained totally apolitical.

29 In reality it simply provided Japanese semiconductor producers with windfall profits which they used to finance a subsequent round of investment in new products and processes.

30 Cited in E. Richards "U.S. Computer Firms Attack Japan Trade Pact" (*San Jose Mercury News* 7 August, 1986).

31 Total entrance fees total about $2 million. Sematech requires a $1 million entry fee, and member firms must also join the Semiconductor Research Consortium for a $62,000 entry fee plus annual dues ranging from $65 million to $2.4 million (depending on firm size). In addition a minimum of five engineers must be sent to participate in Sematech activities in Austin Texas.

These fees are high even for established companies (equalling about 10 percent of their R&D budgets), but they are prohibitive for small firms; moreover, most start-ups can't afford to be without five of their best people.

32 Only 84 SEMI members initially joined SEMI/Sematech, and several small Silicon Valley equipment producers have since discontinued their participation, claiming that it favors large firms over small and is too expensive.

33 Cited in S. Moran "Sematech's hefty fees may bar small chip makers" (*San Jose Business Journal* 30 May, 1988). See also Rogers (1988, 1990) and Gilder (1989).

34 Isolation from customers and suppliers is not unique to the semiconductor industry, but rather reflects the traditional US approach to mass production. Conflicts between mass producers and their suppliers and customers are common in US industries – from textiles to autos to steel. See Dertouzos *et al.* (1989).

35 As Glasmeier (1989) notes, the mere existence of coordinating institutions alone does not guarantee successful regional adaptation in the long run. Institutions, like firms, can lose their efficacy over time. However this does not negate the need for such institutions.

Bibliography

American Electronics Association. 1989. "Strategies for Innovation, High Technology Industries and International Competitiveness". Palo Alto, CA: AEA.

Angel, D. 1989. "The Labor Market for Engineers in the U.S. Semiconductor Industry". *Economic Geography* 65, 2: 99–112.

Borrus, N. 1988. *Competing for Control: America's Stake in Microelectronics*, New York: Ballinger.

Brusco, S., E. Righi. 1989. "Local Government, Industrial Policy and Social Consensus: the Case of Modena (Italy)". *Economy and Society* 18, 4: 405–24.

Coleman, W., H. Jacek (eds) 1989. *Regionalism, Business Interests and Public Policy*, London: Sage Publications.

Congressional Budget Office. 1987. "The Benefits and Risks of Funding Sematech". Washington: Congress of the United States.

Dertouzos, M., R. Lester, and R. Solow. 1989. *Made In America: Regaining the Productive Edge* Cambridge: MA: MIT Press.

Frieberger, P., M. Swaine. 1984. *Fire in the Valley*, Berkeley, CA: Osborne, McGraw-Hill.

Gilder, G. 1989. *Microcosm*, New York: Simon & Schuster.

Glasmeier, A. 1990. "Technological Discontinuities and Flexible Production Networks: The Case of Switzerland and the World Watch Industry". Unpublished paper, University of Texas at Austin.

Gordon, R., A. Dilts, and L. Kimball. 1988. "High Technology Innovation and the Global Milieu: Small and Medium-sized Enterprises in Silicon Valley". Silicon Valley Research Group, University of California, Santa Cruz.

Granovetter, M. 1973. "The Strength of Weak Ties". *The American Journal of Sociology* 78, 6: 1360–80.

——. 1985. "Economic Action and Social Structure: The Problem of Embeddedness". *The American Journal of Sociology* 91, 3: 481–510.

Herrigel, G. 1988. "Industrial Order and the Politics of Industrial Change". In *Toward A Third Republic?*, edited by P. Katzenstein, Ithaca, NY: Cornell University Press.

Hirst, P., J. Zeitlin. 1990. "Flexible Specialization vs. Post-Fordism: Theory, Evidence and Policy Implications". See Chapter 4 of this volume.

Hollingsworth, J., L. Lindberg. 1985. "The governance of the American economy: the role of markets, clans, hierarchies, and associative behaviour". In *Private Interest Government: Beyond Market and State*, edited by W. Streek and P. Schmitter, London: Sage Publications.

Maillat, D., B. Lecoq. 1990. "New Technologies and the Transformation of Regional Structures in Europe: the Role of the Milieu" Unpublished paper, Institut de Recherches Economiques et Regionales (IRER), University of Neuchatel.

Mowery, D., N. Rosenberg. 1989. "New Developments in U.S. Technology Policy: Implications for Competitiveness and International Trade Policy." *California Management Review* 32, 6: 107–24.

Piore, M., C. Sabel. 1984. *The Second Industrial Divide*, New York: Basic Books.

Rogers, T. J. 1988. "Return to the Microcosm". Letter to the Editor, *The Harvard Business Review* July–August: 139–40.

——. 1990. "Landmark Messages from the Microcosm". *The Harvard Business Review* January/February: 24–30.

Sabel, C. 1988. "Flexible specialisation and the reemergence of regional economies". In *Reversing Industrial Decline*, edited by P. Hirst and J. Zeitlin, Oxford: Berg.

Salais, R., M. Storper. 1990. "One Industry, Multiple Rationalities: Flexibility and Mass Production in the French Automobile Industry". Working Paper D901, School of Architecture and Urban Planning, University of California, Los Angeles.

Saxenian, A. 1989a. "In Search of Power: The Organization of Business Interests in Silicon Valley and Route 128". *Economy and Society* 18, 1: 25–70.

——. 1989b. "A High Technology Industrial District: Silicon Valley in the American Context". In *Città della Scienza e della Tecnologia*, edited by P. Perulli, Venice: Arsenale Editrice.

——. 1990a. "Regional Networks and the Resurgence of Silicon Valley". *The California Management Review*, forthcoming.

——. 1990b. "The Origins and Dynamics of Production Networks in Silicon Valley". Working Paper No. 516, Institute of Urban and Regional Development, University of California at Berkeley.

Schmitter, P., W. Streeck. 1981. "A Research Design to Study the Associative Action of Business in the Advanced Industrial Societies of Western Europe".

Scott, A., A. Paul. 1990. "Collective order and economic coordination in industrial agglomerations: the technopoles of Southern California". *Environment and Planning C: Government and Policy* 8, 2: 179–93.

Scranton, P. 1989. *Figured Tapestry: Production, Markets and Power in Philadelphia Textiles, 1885–1941*, Cambridge: Cambridge University Press.

Semiconductor Equipment and Materials International. 1990. "SEMI Membership Profile", Mountain View, CA: SEMI.

Semiconductor Industry Association. 1989. "About the SIA: Semiconductor Industry Association", Cupertino, CA: SIA.

Storper, M., A. J. Scott. 1988. "The Geographical Foundations and Social Regulation of Flexible Production Complexes". In *Territory and Social Reproduction*, edited by J. Wolcher and M. Dear, London: Allen Unwin.

Storper, M., R. Walker. 1989. *The Capitalist Imperative*, New York: Basil Blackwell.

Streeck, W., P. Schmitter (eds). 1985) *Private Interest Government: Beyond Market and State*, London: Sage.

Walker, R. 1988. "The Geographical Organization of Production Systems". *Environment and Planning D* 6, 4: 377–408.

Yoffie, D. 1984. "Interest Groups vs. Individual Action: An Analysis of Corporate Political Strategies". Working Paper, Harvard Business School, Division of Research.

——. 1988. "How an Industry Builds Political Advantage". *The Harvard Business Review* May–June: 82–9.

Vogel, D. 1978. "Why Businessmen Distrust Their State: The Political Consciousness of American Corporate Executives". *The British Journal of Political Science* 8, 1: 45–78.

17 CONCEPTUAL FALLACIES AND OPEN QUESTIONS ON POST-FORDISM

Notes

1 No scholar would recognize himself/herself in such a diagnosis of doom. Yet this caricature reminds us of such books as Bluestone and Harrison (1982), Bowles, Gordon, and Weiskopf (1983), and many others.

2 These ideas came out of the debate on Third Italy and were popularized by the book of Piore and Sabel (1984). This presentation is necessarily schematic.

3 See Storper and Scott (1989), Moulaert and Swyngedouw (1989).

4 See Bluestone and Harrison (1989), Amin and Robins (1989), Martinelli and Schoenberger (1989). A good example with the problems of these critiques is that of Pollert (1988), who tries to prove that: a) "Post-Fordism theorization" does not work . . . because it does not work in Great Britain; b) Moreover, it must be anti-women and anti-workers, because Thatcher's Britain is so. But what if Great Britain is neo-Fordist, and if post-Fordism develops elsewhere? The point is precisely that Great Britain seems to be a marginalized, subcontracting, deindustrializing country. As Hudson (1989) correctly describes its old industrial regions: "Maybe flexibility, but not Flexible Accumulation."

5 A clear distinction between "neo-Fordism" and "post-Fordism" is made in Badham and Mathews (1989). For the "French regulationist positions," see Boyer (1989), Leborgne and Lipietz (1987), Lipietz (1989a). The Radical American "Social Structure of Accumulation" school shares the same problematics (see Bowles, Gordon and Weisskopf 1986).

6 "Technological paradigm" should be understood according to its Greek etymology: a pattern of wisdom (*logos*) about craft (*techne*). The current paradigmatic shift in technology is less about "rigid versus flexible" or "mass versus diversified" than about the new forms of socialization of productive knowledge. Yet it would also be an error to restrict it to its "cognitive" aspects. The capitalist labor process consists of both a cognitive structure and a coordination/subordination structure. See Marglin (1990).

7 See Aoki (1987, 1988). Long ago, Andrew Friedman (1977) contrasted "responsible autonomy" to "direct control" as two *permanent* conflicting tendencies in capitalist organization of labor.

8 See Piore and Sabel (1984: 252).

9 This diagram should be well understood. It is *not* a diagram of relative positions of different countries. It is a representation of the scope of *directions* out of Fordism, represented as a single origin. These directions are illustrated by advanced capitalist countries (from Lipietz 1990). Since the *real* Fordist starting point is different for these various countries, a diagram of relative *positions* would look different.

10 See Doeringer and Piore (1971). The term "market" may be confusing: even the (external) labor market is not a real market, and the "inner market" is certainly not a market at all. Yet the opposition "inner/external" (or "Hierarchy vs. Market" *à la* Williamson (1985) is rooted in a long tradition established by Marx (*Das Kapital*, Chap. XIV) as "manufacturing vs. social division of labor." As shown in Lipietz (1979), Fordism tends to project towards the market the "organized" character of intrafirm relations, but the reverse tendency also exists: and this is what flexibility is about.

11 In the standard wage contract of principal Japanese firms, the rigidity of the wage contract used to encompass the wage level (seniority wage). Nowadays the individual "meritocratic" component of the wage is developing, though it does not yet interfere with internal flexibility (Lecler 1989).

12 This may go so far (as in Sweden in January 1990) that a social-democratic government, with the support of the National Trade Union (L.O.), may try to impose a wage-freeze *against* the bargaining power of some fractions of the waged population in a context of full employment.

13 This is *not* an empirical statement but a logical outcome of the necessity for *some* compensation for workers' involvement. No union would negotiate workers' involvement in the battle for productivity if the gains in productivity would entail their own redundancy, except if some social compensation appears, for instance "national interest" in a very serious crisis (like in Poland). That is the reason why this site may be empirically occupied for a transitory period.

14 A systematic world analysis of the evolution of capital–labor relations is carried on within a project coordinated by Stephen Marglin and Juliet Schor at the World Institute for Development Economics Research (United Nations University). See Lipietz (1990).

15 The coexistence of negotiated involvement and neo-Taylorism within the same society should be distinguished from the combination of (external) flexibility and involvement in the *same* segment of the labor force *à la* Moulaert and Swyngedouw (1989). The social possibility and macroeconomic outcomes of this coexistence are discussed later. It should be noted here that this coexistence is the basis for some polemics (echoed by the Japanese review *MADO* about Japan's after-Fordism. On the one hand, Kenney and Florida (1988) insist rightly upon the "progressive" character of the new capital–labor truce in Japanese major firms ("fujitsuism"), and on its roots in the bargaining power of the working-class in post-WWII Japan. But they forget the "other" side of Japan: the conditions of labor and living in the secondary segment of the labor market, the competition between workers for the access to the first market, the absence of a generalized Welfare State, and so on.

On the other hand, Kato and Steven (1989) incorrectly seize these gloomy aspects to argue that there are no enlightened aspects in Japan, that Toyotism is nothing more than Thatcherism, and that Japan represents "the cruellest and the most oppressive system of capitalist domination over labour." An appreciation which is nearly an insult to the major part of the world working class! Moreover, they argue that Florida and Kenney's position could be the result of "the ignorance by regulation theory of the openendedness of class struggles," and of its "unconscious assumption that the new mode of regulation must have represented a higher stage of development than the old one." We hope that the present chapter (see also Lipietz 1985, Chap. 1) will clear up this undocumented assertion. Kato and Steven oppose the "cruel" situation of Japanese workers to the "nationwide solidarity" of labor relations in New Zealand. We have no basis on which to judge whether New Zealand is shifting toward Kalmarism, but we are ready to admit that Kalmarism is socially superior to Toyotism!

16 In a previous attempt to construct a taxonomy, Leborgne and Lipietz (1987), following Messine (1987), had labeled *Californian*: flexibility + individually negotiated involvement. In fact the "Californian model" appears to be a form of incitation within neo-Taylorian context each time the involvement of the wage-earner is required, such as high skills or front office jobs. Messine (1987) had proposed "Saturnian" for the combination of rigid labor contract + collective involvement, but the difficulties of the General Motors Saturn project has proved the difficulty of its implementation in a *single* firm. Thus, according to a suggestion by Rianne Mahon, we moved to the "Kalmarian" label. In fact, involvement of workers has been successfully negotiated by Toyota with the union UAW at NUMMI (Fremont, California), and unsuccessfully by GM at Van Nuys (Brown and Reich 1987). In other situations in the USA, Japanese managers prefer to avoid negotiations with unions (Mair, Florida, Kenney 1988).

17 These *two* aspects (costs and investments) are made clear in the "Make or Buy" procedure of Renault (quoted by Laigle 1989). Two conditions are required to choose automatically one of the two solutions.

a) The difference of prices between the two solutions should be at least 10 percent.
b) The cost-mimizing solution should not imply an increase of investment more than 20 percent or 1 million francs by comparison to the other solution.

18 In economic literature, there are two meanings for "horizontal integration":

a) Integration between two firms producing the same good,
b) Integration between firms producing goods in different departments of the social division of labor, one being *not* a subset of the other.

In this text we use this second meaning, being understood that "social division of labor" may be historically moving. The difference with "vertical" division of labor is that "horizontal" division of labor defines different specialities which are not usually merged into a single labor process. For instance: cars and tires, airplanes, and motors have always been different professions. They are different *production*, though they participate to the same *commodity*, and may be combined independently by the customer (e.g.: Air France may order an Airbus with a SNECMA-G.E. motor *or* with a Rolls-Royce motor: these are independent productions).

19 The higher efficiency of oblique or horizontal quasi-integration, by comparison to less sophisticated or industrial organization, induces a "quasi-rent" (or, in Marxist terms an "extra surplus-value"), that is a higher total return to total capital. But the subordinated agent (or the partner) is in a position to negotiate a share of this revenue accruing from its participation to the improvement of the whole process, and from its specific skills. In turn, this share of the quasi-rent

allows it more investments and R & D. In turn the upgrading of the partner increases the efficiency of the whole network, hence of the quasi-rent. On the contrary, in the case of vertical subcontracting at the lowest price, the subcontractor cannot invest and improve its own production process, which in turn weakens the efficiency of the whole network. This is the well-known "peripherizing process" of the French car industry in the 1970s.

20 See the differences between USA and Italy in the sector of industrial equipment goods in Leborgne (1987). In Japan, the "firm-level negotiated involvement" allows for a hierarchy between the principal firms, the first layer of subcontractors, and lower layers. Whereas the first layers adapt to difficulties through an "enlarged inner flexibility" (the redundant workers of the principal firms being hired to subcontractors), the lowest layers have to adapt through external flexibility (Lecler and Mercier, 1989). This mobility of highly skilled workers between principal and subcontractors is a major aspect of "partnership" in Japan. By contrast, just-in-time partnership is hard to implement in the USA (Mair, Florida, Kenney 1988).

21 This struggle will not be focused only upon labor relations, but more widely on "societal paradigms" (Lipietz 1989a).

22 Obviously, wages are not the only component of effective demand. Warfare expenditures may pull the demand (as in USA in the 1980s), but with constraint on the external balance.

23 See Leclerc (1989), Lecler and Mercier (1989).

24 This macroeconomics of the Toyotist model is outlined by Itoh (1990): the quasi-rents accruing to Japanese firms from their more efficient technological paradigm are transformed into external surpluses and are diffused with great difficulties through Japanese society.

25 "Sustainable" is the term adopted for "long-term ecologically consistent" in the report of the United Nations Commission on Environment coordinated by Ms. Gro Brundtland (1987).

26 When strong unions take into account the "outsiders," they include reduction of work-time in their objectives in order to fight unemployment *and* to improve the quality of life. See the German I.G. Metall strategy, the textile agreement (1983) about employment sharing.

27 These models had been labeled "primitive Taylorization" and "peripheral Fordism" in Lipietz (1985). The analogy between the macroeconomics of Toyotism (see above) and "peripheral Fordism" proclaimed by Kato and Steven (1989) is completely irrelevent. Japan is certainly not "peripheral" (but central), its regime is inward-oriented, and its industrial core is no more Fordist. "Peripheral Fordisms" (as Korea and Brazil in the 1980s) are regimes based on the Fordist technological paradigm, importing production goods and exporting the products towards external mass markets.

Bibliography

Amin, A., K. Robins. 1989. "Industrial Districts and Regional Development." *Society & Space* 8, 1: 7–34.

Armstrong, P., A. Glyn and J. Harrison. 1984. *Capitalism since World War II*, London: Fontana.

Aoki, M. 1986. "Horizontal vs Vertical Structures of the Firm." *American Economic Review* 76, 5: 971–83.

—— 1988. "A New Paradigm of Work Organization and Coordination: Lessons from Japanese Experiences." In *The Golden Age of Capitalism*, edited by S. Marglin and Schor, Oxford: Clarendon Press.

Badham, R., J. Mathews. 1989. "The New Production Systems Debate." *Labour & Industry* 2, 2: 194–246.

Bianchi, P. 1985. "Nuova concurrenza dinamica e potere di mercato," presented to the conference *Restrutturazioni industriali degli anni '80*, Trieste, September. Published in *L'Industria*, special issue n°1, 1986.

Bluestone, B., B. Harrison. 1982. *The Deindustrialization of America*, New York: Basic Books.

——. 1989. *The great U-Turn: corporate restructuring and the polarizing of America*, New York: Basic Books.

Bowles, S., D. Gordon and T. Weisskopf. 1983. *Beyond the Waste Land: A Democratic Alternative to Economic Decline*, Garden City, NY: Doubleday.

——. 1986. "Power and Profits: The Social Structure of Accumulation and the Profitability of the Postwar US Economy." *Review of Radical Political Economics* 18: 132–67.

Boyer, R. 1989. *New Directions in Management Practices and Work Organization.* Report to OECD, Helsinki Conference, December – Forthcoming OECD.

Braverman, H. 1974. *Labor and Monopoly Capital. The Degradation of Work in the XXth Century*, New York: Monthly Review Press.

Brown, C., M. Reich. 1987. "When Does Union–Management Cooperation Work? A look at NUMMI and GM-Van Nuys." *California Management Review* 31, 4: 24–42.

Brundland, G. 1987. *Our Common Future*, Oxford: Oxford University Press.

Davis, M. 1986. *Prisoners of the American Dream: politics and economy in the history of the U.S. Working class*, London: Verso.

Doeringer, P. B., M. J. Piore. 1971. *International Labor Markets and Manpower Analysis*, New York: Sharpe (revised 1985).

Friedman, A. 1977. *Industry and Labour*, London: MacMillan.

Glyn, A., A. Hughes, A. Lipietz and A. Singh. 1988. "The Rise and Fall of the Golden Age". In: *The Golden Age of Capitalism*, edited by S. Marglin and Schor, Oxford: Clarendon: 36–125.

Hudson, R. 1989. "Labour-market changes and new forms of work in old industrial regions: maybe flexibility but not flexible accumulation." *Society and Space* 17, 1: 5–30.

Itoh, M. 1990. "The Japanese Model of post-Fordism." See Chapter 5 of this volume.

Jenson, J. 1989. "The Talents of Women, the Skills of Men: Flexible Specialization and Women." In *The Transformation of Work*, edited by Wood, London: Hutchinson.

Kato, T., R. Steven. 1989. "Is Japanese Capitalism Post-Fordism?," presented to the *8th New Zealand Asian Studies Conference*, Christchurch, 17–19 August.

Kenney, M., R. Florida. 1988. "Beyond Mass Production: Production and the Labour Process in Japan." *Politics and Society*, 16, 1: 121–58.

Laigle, L. 1989. *La réorganisation du réseau des équipementiers de l'industrie automobile: de la sous-traitance au partenariat.* Diplôme d'Etudes Approfondies Univ. Paris VII, mimeo.

Leborgne, D. 1987. *Equipements flexibles et organisations productives: les relations industrielles au coeur de la modernisation. Eléments de comparaison internationale.* Mimeo, Centre d'Etudes Prospectives d'Economie Mathématique Appliquée à la Planification (CEPREMAP).

Leborgne, D., A. Lipietz 1987. "New Technologies, New Modes of Regulation: Some Spatial Implications." *Society and Space*, 6, 3: 263–80.

——. 1989. "Flexibilité défensive, flexibilité offensive." Forthcoming in *Industrialisation and Regional Development in the 1990s*, edited by Benko and Dunford, London: Belhaven Press.

Lecler, Y. 1989. "Le meritocratisme pragmatique. Vers une nouvelle politique salariale au Japon?," *Annales des Mines – Gérer et Comprendre*, Septembre: 69–77.

Lecler, Y., C. Mercier. 1989. "Vers une gestion globale? Le partenariat dans l'industrie japonaise." *Annales des Mines – Gérer et comprendre* 17: 53–61.

Lipietz, A. 1985a. *Mirages et miracles. Problèmes de l'industrialisation dans le Tiers-Monde*. Paris: La Découverte. (English Translation,: London: Verso (1987)).

——. 1989a. *Choisir l'Audace. Une alternative pour le XXIème siècle*. Paris: La Découverte. English translation, Oxford: Polity Press. (forthcoming).

——. 1989b. "The Debt Problem, European Integration, and the New Phase of the World Crisis." *New Left Review* 176: 37–50.

——. 1990a. "The French Economy: Waiting for New Methods of Regulation." In *In Search of the New France*, edited by Hollifield and Ross, New York: Routledge (1992).

——. 1990b. "Capital–Labour relations at the Dawn of 21st Century." Report to World Institute for Development Economic Research (WIDER). In French: Paris: *Couverture Orange CEPREMAP* 9016.

Mahon, R. 1987. "From Fordism to ? New Technologies, Labor Market and Unions." *Economic and Industrial Democracy* 8: 5·60.

Mair, A., R. Florida and M. Kenney 1988. "The New Geography of Automobile Production: Japanese Transplants in North America". *Economic Geography* 64, 4, October.

Marglin, S. (ed.) 1990. *Dominating Knowledge*, Oxford: Clarendon Press.

Martinelli, F., E. Schoenberger. 1989. "Oligopoly Alive and Well. Notes for a Broader Discussion on Flexible Accumulation." Colloque *Les nouveaux espaces industriels*, Paris, 21–22 March.

Mathews, J. 1989. "New Production Concepts." *Prometheus* 7, 1: 129–48.

——. 1989. *Age of Democracy. The Politics of Post-Fordism*, Melbourne: Oxford University Press.

Messine, P. 1987. *Les Saturniens*, Paris: La Découverte.

Moulaert, F., E. Swyngedouw. 1989. "A regulation approach to the geography of flexible production systems." *Society and Space* 7, 3: 327–45.

Piore, M. J., C. F. Sabel. 1984. *The Second Industrial Divide: Possibilities for Prosperity*, New York: Basic Books.

——. 1989. *The Second Industrial Divide. Possibilities for Prosperity*, New York: Basic Books.

Pollert, A. 1988. "Dismantling Flexibility." *Capital and Class* 34: 42–75.

Sabel, C. 1989. "Flexible Specialisation and Re-emergence of Regional Economics." In *Reversing industrial decline?*, edited by D. Hirst and J. Zeitlin, Oxford: Berg.

Storper, M., A. J. Scott. 1989. "Work organization and local Labour Markets in an Era of Flexible Production." Geneva: ILO, *World Employment Program Research Working Paper 30*.

Williamson, O. E. 1985. *The Economic Institutions of Capitalism: Firms, Markets, Relational Contracting*, New York: The Free Press – MacMillan.

INDEX